D0084653

Oakland on
Quality Management

For Susan

Oakland on Quality Management

John S. Oakland
PhD, CChem, MRSC, FIQA, FSS, FInstD, MASQ, MCMI

Professor of Business Excellence and Quality Management
Leeds University Business School

Executive Chairman, Oakland Consulting plc

ELSEVIER
BUTTERWORTH
HEINEMANN

AMSTERDAM   BOSTON   HEIDELBERG   LONDON   NEW YORK   OXFORD
PARIS   SAN DIEGO   SAN FRANCISCO   SINGAPORE   SYDNEY   TOKYO

Elsevier Butterworth-Heinemann
Linacre House, Jordan Hill, Oxford OX2 8DP
200 Wheeler Road, Burlington, MA 01803

First published 2004

British Library Cataloguing in Publication Data
A catalogue record for this book is available from the British Library

Library of Congress Cataloguing in Publication Data
A catalogue record for this book is available from the Library of Congress

ISBN 0 7506 5741 3

For information on all Elsevier Butterworth-Heinemann publications
visit our website at http://books.elsevier.com

Printed and bound in Great Britain by
Biddles Ltd, Kings Lynn, Norfolk

Contents

Preface

When I wrote the first edition of *Total Quality Management* in 1988, there were very few books on the subject. Since its publication, the interest in quality and business performance improvement has exploded. There are now many texts on quality management and its various aspects, including business excellence and process-people management.

So much has been learned during the last 20 years of managing quality that it has been necessary to rewrite the book and revise it again. The content and illustrative case studies in this edition have changed substantially to reflect the developments, current understanding, and experience gained over this period.

Increasing the satisfaction of customers and other stakeholders through effective goal deployment, cost reduction, people development and process improvement has proved essential for organizations to stay in operation. We cannot avoid seeing how quality has developed into the most important competitive weapon, and many organizations have realized that 'quality' is *the* way of managing for the future. Quality management is far wider in its application than assuring product or service quality – it is a way of managing organizations to improve every aspect of performance, both internally and externally.

This book is about how to manage in a total quality way. It is structured in the main around four parts of a new framework – improving Performance through better Planning and management of People and the Processes in which they work. The core of the model will always be performance in the eyes of the customer, but this must be extended to include performance measures for all stakeholders. This new core still needs to be surrounded by *commitment* to quality and meeting the customer requirements, *communication* of the quality message, and recognition of the need to change the *culture* of most organizations to create total quality. These are the 'soft foundations' which must encase the 'hard management necessities' of planning, people and processes.

Under these headings the essential steps for successfully implementing quality are set out in what I hope is still a meaningful and practical way. The book should guide the reader through the language of quality, all the recent developments, and set down a clear way to proceed for most organizations.

Many of the new approaches related to quality and improving performance appear to present different theories. In reality they are talking the same 'language' but they use different dialects; the basic principles of defining quality and taking it into account throughout all the activities of the 'business' are common. Quality has to be managed – it does not just happen. Understanding and commitment by senior management, effective leadership, teamwork, and good process management, are fundamental parts of the recipe for success. I have tried to use my extensive research and consultancy experience to take what is to many a jigsaw puzzle and assemble a comprehensive, practical, working model – the rewards of which are greater efficiencies, lower costs, improved reputation and customer loyalty. Moreover, I have tried to show how holistic quality management now is embracing the most recent models of 'excellence', six sigma, and a host of other management methods and teachings. To support the text, there are many short illustrative case studies.

The book is aimed at directors and managers in all types of organization. It should also meet the requirements of the increasing number of students who need to understand the part quality may play in their courses on science, engineering, or management. I hope that those engaged in the pursuit of professional qualifications in the management of quality, such as memberships of the Institute of Quality Assurance, the American Society for Quality, or the Australian Organization for Quality, will make this book an essential part of their library. With its companion book, *Statistical Process Control* (now in its fifth edition), *Oakland on Quality Management* documents a comprehensive approach, one that has been used successfully in many organizations throughout the world.

I would like to thank my colleagues in the European Centre *for* Business Excellence and Oakland Consulting plc for the sharing of ideas and help in their development. The book is the result of many years' collaboration in assisting organizations to introduce good methods of management and embrace the concepts of total quality.

John S. Oakland

Part 1

The Foundations of Quality Management

Good order is the foundation of
all good things.

Edmund Burke, 1729–1797, from
'Reflections on the Revolution in France'

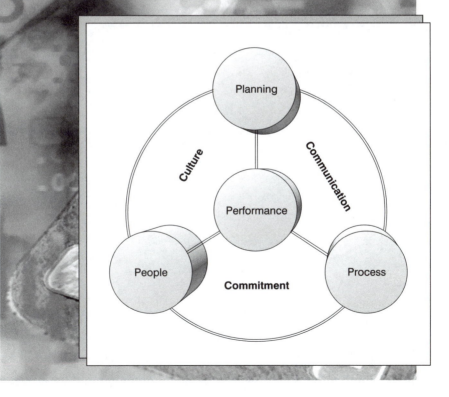

Chapter 1

Understanding quality

■ Quality, competitiveness, reputation and customers

Quality – does it matter any more? _____

If you need an answer to this question, just ask the customer of any product or service in any organization, in any sector. Yet to read or listen to some people on the subject you could begin to believe that we have 'moved on from quality'. Apparently, approaches such as Total Quality Management (TQM) and Statistical Process Control (SPC) have been replaced by other concepts, including (alphabetically) ABC costing, ABCD checklists, benchmarking, BPR (business process re-engineering), business excellence, FMEA (failure mode and effect analysis), Hoshim Kanri, JIT (just-in-time), Kaizen, Lean (manufacturing), process management, QFD (quality function deployment) and Six Sigma (including lean sigma!).

There has been no shortage of newcomers in the management fad procession but, as successful organizations know, the fundamentals of quality and its management are key to competing successfully.

Whatever type of organization you work in – a hospital, a university, a bank, an insurance company, local government, an airline, a factory – competition is rife: competition for customers, for students, for patients, for resources, for funds. Any organization basically competes on its *reputation* – for quality, reliability and price – and most people recognize that quality is still the most important of these competitive weapons. If you doubt that, just look at the way some organizations, even whole industries in certain countries, have used quality to take the heads off their competitors. American, British, French, German, Italian, Japanese, Spanish, Swiss, Swedish organizations, and organizations from other countries have used quality strategically to win customers, steal business resources or funding, and be competitive. Moreover, this sort of attention to quality improves performance in reliability and price.

For any organization, there are several aspects of reputation which are important:

1 It is built upon the competitive elements of quality, reliability and price, of which quality has become strategically the most important.
2 Once an organization acquires a poor reputation for quality, it takes a very long time to change it.
3 Reputations, good or bad, can quickly become national reputations.
4 The management of the competitive weapons, such as quality, can be learned like any other skill, and used to turn round a poor reputation, in time.

Before anyone will buy the idea that quality is still a vital consideration in business, they would have to know what was meant by it.

What is quality? _____

'Is this a quality watch?' Pointing to my wrist, I ask this question of a class of students – undergraduates, postgraduates, experienced managers – it matters not who. The answers vary:

- ■ 'No, it's made in Japan.'
- ■ 'No, it's cheap.'
- ■ 'No, the face is scratched'
- ■ 'How reliable is it?'
- ■ 'I wouldn't wear it.'

My watch has been insulted all over the world – London, New York, Paris, Sydney, Brussels, Amsterdam, Leeds! Clearly, the quality of a watch depends on what the wearer requires from it – perhaps a piece of jewelry to give an impression of wealth; a timepiece that gives the required data, including the date, in digital form; or one with the ability to perform at 50 meters under the sea? These requirements determine the quality.

Quality is often used to signify 'excellence' of a product or service – people talk about 'Rolls-Royce quality' and 'top quality'. In some manufacturing companies the word may be used to indicate that a piece of material or equipment conforms to certain physical dimensional characteristics often set down in the form of a particularly 'tight' specification. In a hospital it might be used to indicate some sort of 'professionalism'. If we are to define quality in a way that is useful in its *management*, then we must recognize the need to include in the assessment of quality the true requirements of the 'customer' – the needs and expectations.

Quality then is simply *meeting the customer requirements*, and this has been expressed in many ways by other authors:

- 'Fitness for purpose or use' – Juran, an early doyen of quality management.
- 'The totality of features and characteristics of a product or service that bear on its ability to satisfy stated or implied needs' – BS 4778: 1987 (ISO 8402, 1986) *Quality Vocabulary*: Part 1, *International Terms*.
- 'Quality should be aimed at the needs of the consumer, present and future' – Deming, another early doyen of quality management.
- 'The total composite product and service characteristics of marketing, engineering, manufacture and maintenance through which the product and service in use will meet the expectation by the customer' – Feigenbaum, author of 'Total Quality Control.'
- 'Conformance to requirements' – Crosby, an American quality management consultant famous in the 1980s.
- 'Degree to which a set of inherent characteristics fulfils requirements' – ISO (EN) 9000:2000 *Quality Management Systems – fundamentals and vocabulary*.

Another word that we should define properly is *reliability*. 'Why do you buy a BMW car?' 'Quality and reliability' comes back the answer. The two are used synonymously, often in a totally confused way. Clearly, part of the acceptability of a product or service will depend on its ability to function satisfactorily *over a period of time*, and it is this aspect of performance that is given the name *reliability*. It is the ability of the product or service to *continue* to meet the customer requirements. Reliability ranks with quality in importance, since it is a key factor in many purchasing decisions where alternatives are being considered. Many of the general management issues related to achieving product or service quality are also applicable to reliability.

It is important to realize that the 'meeting the customer requirements' definition of quality is not restrictive to the functional characteristics of products or services. Anyone with children knows that the quality of some of the products they purchase is more associated with *satisfaction in ownership* than some functional property. This is also true of many items, from antiques to certain items of clothing. The requirements for status symbols account for the sale of some executive cars, certain bank accounts and charge cards, and even hospital beds! The requirements are of paramount importance in the assessment of the quality of any product or service.

By *consistently* meeting customer requirements, we can move to a different plane of satisfaction – *delighting the customer*. There is no doubt that many organizations have so well ordered their capability to meet their customers' requirements, time and time again, that this has created a reputation for 'excellence'. A development of this thinking regarding customers and their satisfaction is *customer loyalty*, an important variable in an organization's success. Research shows that focus on customer loyalty can provide several commercial advantages:

■ Customers cost less to retain than acquire.
■ The longer the relationship with the customer, the higher the profitability.
■ A loyal customer will commit more spend to its chosen supplier.
■ About half of new customers come through referrals from existing clients (indirectly reducing acquisition costs).

Companies like 3M use measures of customer loyalty to identify customers which are 'completely satisfied', would 'definitely recommend' and would 'definitely repurchase'.

■ Understanding and building the quality chains

The ability to meet the customer requirements is vital, not only between two separate organizations, but within the same organization.

When the air stewardess pulled back the curtain across the aisle and set off with a trolley full of breakfasts to feed the early morning travelers on the short domestic flight into an international airport, she was not thinking of quality problems. Having stopped at the row of seats marked 1ABC, she passed the first tray onto the lap of the man sitting by the window. By the time the second tray had reached the lady beside him, the first tray was on its way back to the hostess with a complaint that the bread roll and jam were missing. She calmly replaced it in her trolley and reached for another – which also had no roll and jam.

The calm exterior of the girl began to evaporate as she discovered two more trays without a complete breakfast. Then she found a good one and, thankfully, passed it over. This search for complete breakfast trays continued down the aeroplane, causing inevitable delays, so much so that several passengers did not receive their breakfasts until the plane had begun its descent. At the rear of the plane could be heard the mutterings of discontent. 'Aren't they slow with breakfast this morning?' 'What is she doing with those trays?' 'We will have to eat so quickly before landing we will get indigestion.'

The problem was perceived by many on the aeroplane to be one of delivery or service. They could smell food but they weren't getting any of it, and they were getting really wound up! The air hostess, who had suffered the embarrassment of being the purveyor of defective product and service, was quite wound up and flushed herself, as she returned to the curtain and almost ripped it from the hooks in her haste to hide. She was heard to say through clenched teeth, 'What a bloody mess!'

A problem of quality? Yes, of course, requirements not being met, but where? The passengers or customers suffered from it on the aircraft, but in another part of the organization there was a man whose job it was to assemble the breakfast trays. On this day the system had broken down – perhaps he ran out of bread rolls, perhaps he was called away to refuel the aircraft (it was a small airport!), perhaps he didn't know or understand, perhaps he didn't care.

Three hundred miles away in a chemical factory ... 'What the hell is Quality Control doing? We've just sent 15 000 liters of lawn weed killer to CIC and there it's back at our gate – they've returned it as out of spec.' This was followed by an avalanche of verbal abuse, which will not be repeated here, but poured all over the shrinking quality control manager as he backed through his office door, followed by a red faced technical director advancing menacingly from behind the bottles of sulphuric acid racked across the adjoining laboratory.

'Yes, what is QC doing?' thought the production manager, who was behind a door two offices along the corridor, but could hear the torrent of language now being used to beat the QC man into an admission of guilt. He knew the poor devil couldn't possibly do anything about the rubbish that had been produced except test it, but why should he volunteer for the unpleasant and embarrassing ritual now being experienced by his colleague – for the second time this month. No wonder the QC manager had been studying the middle pages of the *Telegraph* on Thursday – what a job!

Do you recognize these two situations? Do they not happen every day of the week – possibly every minute somewhere in manufacturing or the service industries? Is it any different in banking, insurance, the health service? The inquisition of checkers and testers is the last bastion of desperate systems trying in vain to catch mistakes, stop defectives, hold lousy materials, before they reach the external customer – and woe betide the idiot who lets them pass through!

Two everyday incidents, but why are events like these so common? The answer is the acceptance of one thing – *failure*. Not doing it right the first time at every stage of the process.

Why do we accept failure in the production of artefacts, the provision of a service, or even the transfer of information? In many walks of life we do not accept it. We do not say, 'Well, the nurse is bound to drop the odd baby in a thousand – it's just going to happen.' We do not accept that!

In each department, each office, even each household, there are a series of suppliers and customers. The PA is a supplier to the boss. Are the requirements being met? Does the boss receive error-free information set out as it is wanted, when it is wanted? If so, then we have a quality PA service. Does the air steward receive from the supplier in the airline the correct food trays in the right quantity, at the right time?

Throughout and beyond all organizations, whether they be manufacturing concerns, banks, retail stores, universities, hospitals or hotels, there is a series of *quality chains* of customers and suppliers (Figure 1.1) that may be broken at any point by one person or one piece of equipment not meeting the requirements of the customer, internal or external. The interesting point is that this failure usually finds its way to the interface between the organization and its outside customers, and the people who operate at that interface – like the air hostess – usually experience the ramifications. The concept of internal and external customer–supplier chains form the *core* of quality management.

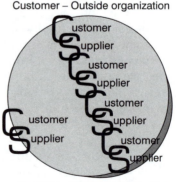

Customer – Outside organization

■ **Figure 1.1**
The quality chains

Supplier – Outside organization

A great deal is written and spoken about employee motivations as a separate issue. In fact the key to motivation *and* quality is for everyone in the organization to have well-defined customers – an extension of the word beyond the outsider that actually purchases or uses the ultimate product or service to anyone to whom an individual gives a part, a service, information – in other words the results of his or her work.

Quality has to be managed – it will not just happen. Clearly it must involve everyone in the process and be applied throughout the organization. Many people in the support functions of organizations never see, experience, or touch the products or services that their organizations buy or provide, but they do handle or produce things like purchase orders or invoices. If every fourth invoice carries at least one error, what image of quality is transmitted?

Failure to meet the requirements in any part of a quality chain has a way of multiplying and a failure in one part of the system creates problems elsewhere, leading to yet more failure, more problems and so on. The price of quality is the continual examination of the requirements and our ability to meet them. This will lead to a 'continuing improvement' philosophy. The benefits of making sure the requirements are met at every stage, every time, are truly enormous in terms of increased competitiveness and market share, reduced costs, improved productivity and delivery performance, and the elimination of waste.

Case study ■■■ ▬▬▬▬▬▬▬▬▬▬▬▬

■ The relaunch of quality in BT Retail

BT was created in 1981 when the telecommunications arm of the British Post Office was reformed as a separate entity in preparation for privatization in 1984. Since then BT has operated in one of the most open telecommunications markets in the world. BT faces competition within the UK for local services from cable TV companies, while other network operators vie for its long haul national and international traffic. BTs day-to-day operations are subject to regulation by OFTEL, a government appointed regulatory body which has major impact on key aspects of BTs business. For example, in a number of key markets BT is required to keep price increases significantly below the level of retail price inflation. BTs very survival has depended on successful performance in this highly competitive yet tightly regulated environment. Following privatization BT faced the imperative of transforming itself from bureaucratic monopoly to customer-centric service provider, while growing income, reducing costs and minimizing loss of market share.

In 2000 after a decade of international expansion, BT decided to refocus on the UK and Europe and carried out a major corporate reorganization which resulted in the formation of the BT

Group and the demerger of mobile (MO2) and the directory publishing (Yell) businesses. The BT Group consists of BT Wholesale, responsible for BT's telecommunications network, BT Ignite, delivering sophisticated IT solutions for large businesses across Europe, and BT Openworld, specializing in the internet mass market, and BT Retail.

Formed in October 2000, BT Retail is the largest unit in the BT Group with almost 60 000 employees. Its role is to provide communications solutions to 21 million customers in the UK – from residential consumers to the largest businesses and its vision is 'Connecting your World, Completely'. BT Retail's first CEO, Pierre Danon, possessed a strong personal commitment to quality improvement that stemmed from his experiences at a previous European Quality Award winner, Xerox Europe. Pierre and a new leadership team were building a 'new' customer centric distribution business with a remit to *deliver a superb experience to a huge customer base*. They recognized the benefits and necessity of taking a quality approach to support achievement of some very challenging goals. It was also acknowledged that the major business and organizational changes that took place in 1999 and the early part of 2000 had inevitably meant that many people in BT Retail had not been focusing on quality quite as much as in previous years. Within a few months of BT Retail's inception a Revitalizing Quality program was launched to drive an unremitting focus on improvement. The ongoing drive and commitment of the CEO and the leadership team has been pivotal in driving the success of this quality program.

The approach to 'Revitalizing Quality' is based on seven steps to 'real' quality:

- put customers at the heart of what we do;
- reduce the cost of failure;
- develop and deploy strategy;
- get the basics right – quality for everyone;
- quality approach to major change;
- get the workforce involved;
- innovation.

Put customers at the heart of what we do _____

All quality programs have to have, at their center, a very clear focus on customers. Delivering customer satisfaction is the primary goal for BT Retail and the approach is inherently simple – listen to customers and respond to what they say. BT Retail has a wide range of methods for listening to their customers, ranging from market research to asking thousands of customers detailed questions about how they felt about a specific transaction with BT. From this data BT Retail has built quantitative models of the drivers of customer satisfaction (Figure 1.2) which enable them to ensure that internal measures are aligned with what customers really want.

One major shift in approach made early in the life of BT Retail was to change which senior managers were targeted (and bonused) against a customer satisfaction measure. Traditionally customer satisfaction had been the responsibility of the customer service manager with revenue being the responsibility of the channel managers. Now, everyone who deals with customers has a customer satisfaction target, normally with the same importance as financial

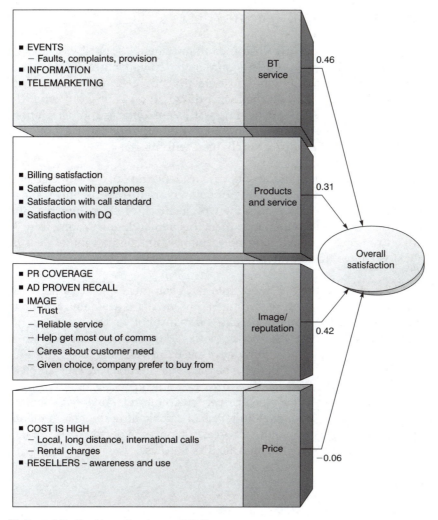

EVENTS
 — Faults, complaints, provision
INFORMATION
TELEMARKETING

BT service   0.46

Billing satisfaction
Satisfaction with payphones
Satisfaction with call standard
Satisfaction with DQ

Products and service   0.31

Overall satisfaction

PR COVERAGE
AD PROVEN RECALL
IMAGE
 — Trust
 — Reliable service
 — Help get most out of comms
 — Cares about customer need
 — Given choice, company prefer to buy from

Image/ reputation   0.42

COST IS HIGH
 — Local, long distance, international calls
 — Rental charges
RESELLERS – awareness and use

Price   −0.06

Figure 1.2   The drivers of customer satisfaction

targets. BT Retail also changed their primary customer satisfaction measure from 'overall satisfaction' to 'satisfaction compared with competitors' so that benchmarking is built in to this key measure.

As well as this fundamental shift in measurement methodology a number of strategic change programs were introduced to enable process and system improvement. In addition there was a massive increase in focus on the behaviors that support customers having a great experience with BT. BT Retail have introduced '10 Golden Rules for Customer Satisfaction' and developed new approaches to recruiting, training and coaching people to ensure that everything that delivers customer satisfaction is aligned and focused.

Meeting the requirements _____

If quality is meeting the customer requirements, then this has wide implications. The requirements may include availability, delivery, reliability, maintainability and cost-effectiveness, among many other features. The first item on the list of things to do is find out what the requirements are. If we are dealing with a customer/supplier relationship crossing two organizations, then the supplier must establish a 'marketing' activity or process charged with this task.

The marketing process must of course understand not only the needs of the customer but also the ability of its own organization to meet them. If my customer places a requirement on me to run 1500 meters in 4 minutes, then I know I am unable to meet this demand, unless something is done to improve my running performance. Of course I may never be able to achieve this requirement.

Within organizations, between internal customers and suppliers, the transfer of information regarding requirements is frequently poor to totally absent. How many executives really bother to find out what their customers' – their PAs' or secretaries' – requirements are? Can their handwriting be read, do they leave clear instructions, does the PA/secretary always know where the boss is? Equally, does the PA/secretary establish what the boss needs – error-free word processing, clear messages, a tidy office? Internal supplier/customer relationships are often the most difficult to manage in terms of establishing the requirements. To achieve quality throughout an organization, each person in the quality chain must interrogate the interfaces as follows:

Customers

- Who are my immediate customers?
- What are their true requirements?
- How do or can I find out what the requirements are?
- How can I measure my ability to meet the requirements?
- Do I have the necessary capability to meet the requirements?
  (If not, then what must change to improve the capability?)
- Do I continually meet the requirements?
  (If not, then what prevents this from happening, when the capability exists?)
- How do I monitor changes in the requirements?

Suppliers

- Who are my immediate suppliers?
- What are my true requirements?

- How do I communicate my requirements?
- How do I, or they, measure their ability to meet the requirements?
- Do my suppliers have the capability to meet the requirements?
- Do my suppliers continually meet the requirements?
- How do I inform them of changes in the requirements?

The measurement of capability is extremely important if the quality chains are to be formed within and without an organization. Each person in the organization must also realize that the supplier's needs and expectations must be respected if the requirements are to be fully satisfied.

To understand how quality may be built into a product or service, at any stage, it is necessary to examine the two distinct but interrelated aspects of quality:

- Quality of design.
- Quality of conformance to design.

Quality of design

We are all familiar with the old story of the tree swing (Figure 1.3), but in how many places in how many organizations is this chain of activities taking place? To discuss the quality of, say, a chair it is necessary to describe its purpose. What it is to be used for? If it is to be used for watching TV for three hours at a stretch, then the typical office chair will not meet this requirement. The difference between the quality of the TV chair and the office chair is not a function of how it was manufactured, but its *design*.

Quality of design is a measure of how well the product or service is designed to achieve the agreed requirements. The beautifully presented gourmet meal will not necessarily please the recipient if she is travelling on the motorway and has stopped for a quick bite to eat. The most important feature of the design, with regard to achieving quality, is the specification. Specifications must also exist at the internal supplier–customer interfaces if one is to achieve a quality performance. For example, the company lawyer asked to draw up a contract by the sales manager requires a specification as to its content:

1 Is it a sales, processing or consulting type of contract?
2 Who are the contracting parties?
3 In which countries are the parties located?
4 What are the products involved (if any)?
5 What is the volume?
6 What are the financial aspects, e.g. price escalation?

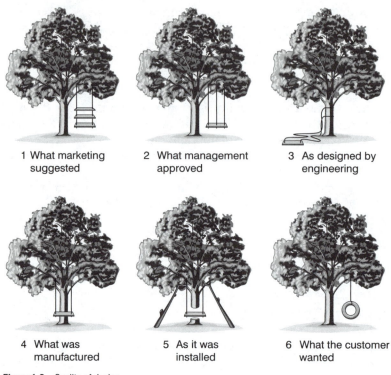

1 What marketing
  suggested

2 What management
  approved

3 As designed by
  engineering

4 What was
  manufactured

5 As it was
  installed

6 What the customer
  wanted

■ **Figure 1.3**   Quality of design

The financial controller must issue a specification of the information he needs, and when, to ensure that foreign exchange fluctuations do not cripple the company's finances. The business of sitting down and agreeing a specification at every interface will clarify the true requirements and capabilities. It is the vital first stage for a successful total quality effort.

There must be a corporate understanding of the organization's quality position in the market place. It is not sufficient that marketing specifies the product or service 'because that is what the customer wants'. There must be an agreement that the operating departments can achieve that requirement. Should they be incapable of doing so, then one of two things must happen: either the organization finds a different position in the market place or substantially changes the operational facilities.

Quality of conformance to design

This is the extent to which the product or service achieves the quality of design. What the customer actually receives should conform to the design, and operating costs are tied firmly to the level of conformance

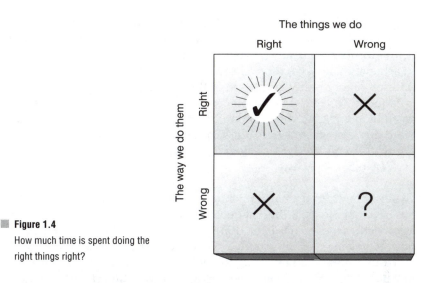

Figure 1.4

How much time is spent doing the right things right?

achieved. Quality cannot be inspected into products or services; the customer satisfaction must be designed into the whole system. The conformance check then makes sure that things go according to plan.

A high level of inspection or checking at the end is often indicative of attempts to inspect in quality. This may well result in spiraling costs and decreasing viability. The area of conformance to design is concerned largely with the quality performance of the actual operations. It may be salutary for organizations to use the simple matrix of Figure 1.4 to assess how much time they spend doing the right things right. A lot of people, often through no fault of their own, spend a good proportion of the available time doing the right things wrong. There are people (and organizations) who spend time doing the wrong things very well, and even those who occupy themselves doing the wrong things wrong, which can be very confusing!

Managing quality

Every day two men who work in a certain factory scrutinize the results of the examination of the previous day's production, and begin the ritual battle over whether the material is suitable for despatch to the customer. One is called production manager, the other the quality control manager. They argue and debate the evidence before them, the rights and wrongs of the specification, and each tries to convince the other of the validity of his argument. Sometimes they nearly start fighting.

This ritual is associated with trying to answer the question, '*Have we done the job correctly?*', correctly being a flexible word, depending on the interpretation given to the specification on that particular day. This is not quality *control*, it is *detection* – wasteful detection of bad product before it hits the customer. There is still a belief in some quarters that to achieve quality we must check, test, inspect or measure – the ritual pouring on of quality at the end of the process. This is nonsense, but it is frequently practiced. In the office one finds staff checking other people's work before it goes out, validating computer data, checking invoices, word processing, etc. There is also quite a lot of looking for things, chasing why things are late, apologizing to customers for lateness, and so on. Waste, waste, waste!

To get away from the natural tendency to rush into the detection mode, it is necessary to ask different questions in the first place. We should not ask whether the job has been done correctly, we should ask first '*Are we capable of doing the job correctly?*' This question has wide implications, and this book is devoted largely to the various activities necessary to ensure that the answer is yes. However, we should realize straight away that such an answer will only be obtained by means of satisfactory methods, materials, equipment, skills and instruction, and a satisfactory 'process'.

Quality and processes

As we have seen, quality chains can be traced right through the business or service processes used by any organization. A process is the transformation of a set of inputs into outputs that satisfy customer needs and expectations, in the form of products, information or services. Everything we do is a process, so in each area or function of an organization there will be many processes taking place. For example, a finance department may be engaged in budgeting processes, accounting processes, salary and wage processes, costing processes, etc. Each process in each department or area can be analyzed by an examination of the inputs and outputs. This will determine some of the actions necessary to improve quality. There are also cross-functional processes.

The output from a process is that which is transferred to somewhere or to someone – the *customer*. Clearly to produce an output that meets the requirements of the customer, it is necessary to define, monitor and control the inputs to the process, which in turn may be supplied as output from an earlier process. At every supplier/customer interface then there resides a transformation process (Figure 1.5), and every single task throughout an organization must be viewed as a process in this way.

Once we have established that our process is capable of meeting the requirements, we can address the next question, '*Do we continue to do*

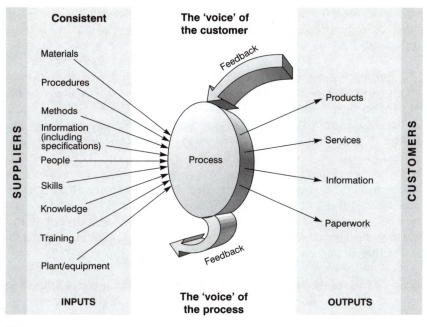

Consistent

The 'voice' of
the customer

Feedback

SUPPLIERS

Materials

Procedures

Methods

Information
(including
specifications)

People

Skills

Knowledge

Training

Plant/equipment

Process

Products

Services

Information

Paperwork

CUSTOMERS

Feedback

INPUTS

The 'voice' of
the process

OUTPUTS

Figure 1.5   A process

the job correctly?', which brings a requirement to monitor the process and the controls on it. If we now re-examine the first question, 'Have we done the job correctly?', we can see that, if we have been able to answer the other two questions with a yes, we *must* have done the job correctly. Any other outcome would be illogical. By asking the questions in the right order, we have moved the need to ask the 'inspection' question and replaced a strategy of *detection* with one of *prevention*. This concentrates all the attention on the front end of any process – the inputs – and changes the emphasis to making sure the inputs are capable of meeting the requirements of the process. This is a managerial responsibility and is discharged by efficiently organizing the inputs, the resources and controlling the processes.

These ideas apply to every transformation process; they all must be subject to the same scrutiny of the methods, the people, skills, equipment and so on to make sure they are correct for the job. A person giving a lecture whose audio/visual equipment will not focus correctly, or whose teaching materials are not appropriate, will soon discover how difficult it is to provide a lecture that meets the requirements of the audience.

In every organization there are some very large processes – groups of smaller processes often called *core business processes*. These are activities the organization must carry out especially well if its mission and objectives are to be achieved. This area will be dealt with in some detail later on in the book. It is crucial if the management of quality is to be integrated into the strategy for the organization.

The *control* of quality can only take place at the point of operation or production – where the letter is word processed, the sales call made, the patient admitted, or the chemical manufactured. The act of *inspection is not quality control*. When the answer to 'Have we done the job correctly?' is given indirectly by answering the questions of capability and control, then we have *assured* quality, and the activity of checking becomes one of *quality assurance* – making sure that the product or service represents the output from an effective *system* to ensure capability and control. It is frequently found that organizational barriers between functional or departmental empires encourage the development of testing and checking of services or products in a vacuum, without interaction with other departments.

Quality control then is essentially the activities and techniques employed to achieve and maintain the quality of a product, process, or service. It includes a monitoring activity, but is also concerned with finding and eliminating causes of quality problems so that the requirements of the customer are continually met.

Quality assurance is broadly the prevention of quality problems through planned and systematic activities around processes (including documentation). These will include the establishment of a good quality management system and the assessment of its adequacy, the audit of the operation of the system, and the review of the system itself.

■ Quality starts with understanding the needs

The marketing processes of an organization must take the lead in establishing the true requirements for the product or service. Having determined the need, the organization should define the market sector and demand, to determine such product or service features as grade, price, quality, timing, etc. For example, a major hotel chain thinking of opening a new hotel or refurbishing an old one will need to consider its location and accessibility before deciding whether it will be predominantly a budget, first-class, business or family/holiday hotel.

The organization will also need to establish customer requirements by reviewing the market needs, particularly in terms of unclear or unstated expectations or preconceived ideas held by customers. It is central to identify the key characteristics that determine the suitability of the product or service in the eyes of the customer. This may, of course, call for the use of market research techniques, data gathering, and analysis of customer complaints.

If necessary, quasi-quantitative methods may be employed, giving proxy variables that can be used to grade the characteristics in importance, and decide in which areas superiority over competitors exists. It is often useful to compare these findings with internal perceptions.

Excellent communication between customers and suppliers is the key to quality performance; it will eradicate the 'demanding nuisance/ idiot' view of customers, which still pervades some organizations. Poor communications often occur in the supply chain between organizations, when neither party realizes how poor they are. Feedback from both customers and suppliers needs to be improved where dissatisfied customers and suppliers do not communicate their problems. In such cases, non-conformance of purchased products or services is often due to customers' inability to communicate their requirements clearly. If these ideas are also used within an organization, then the internal supplier/ customer interfaces will operate much more smoothly.

All the efforts devoted to finding the nature and timing of the demand will be pointless if there are failures in communicating the requirements throughout the organization promptly, clearly, and accurately. The marketing processes should be capable of producing a formal statement or outline of the requirements for each product or service. This constitutes a preliminary set of *specifications*, which can be used as the basis for service or product design. The information requirements include:

1 Characteristics of performance and reliability – these must make reference to the conditions of use and any environmental factors that may be important.
2 Aesthetic characteristics, such as style, color, smell, taste, feel, etc.
3 Any obligatory regulations or standards governing the nature of the product or service.

The organization must also establish systems for feedback of customer information and reaction, and these systems should be designed on a continuous monitoring basis. Any information pertinent to the product or service should be collected and collated, interpreted, analyzed, and communicated, to improve the response to customer experience and expectations. These same principles must also be applied to other processes in the organization if continuous improvement at every interface is to be achieved. If one function or department in a company has problems recruiting the correct sort of staff, for example, and HR has not established a good process for gathering, analyzing, and responding to information on new employees, then frustration and conflict will replace communication and co-operation.

One aspect of the analysis of market demand that extends back into the organization is the review of market readiness of a new product or service. Items that require some attention include assessment of:

1 The suitability of the distribution and customer-service processes.
2 Training of personnel in the 'field'.
3 Availability of 'spare parts' or support staff.
4 Evidence that the organization is capable of meeting customer requirements.

All organizations receive a wide range of information from customers through invoices, payments, requests for information, letters of complaint, responses to advertisements and promotion, etc. An essential component of a system for the analysis of demand is that this data is channeled quickly into the appropriate areas for action and, if necessary, response.

There are various techniques of research, which are outside the scope of this book, but have been well documented elsewhere. It is worth listing some of the most common and useful general methods that should be considered for use, both externally and internally:

- Surveys – questionnaires, etc.
- Panel or focus group techniques.
- In-depth interviews.
- Brainstorming and discussions.
- Role rehearsal and reversal.
- Interrogation of trade associations.

The number of methods and techniques for researching the market is limited only by imagination and funds. The important point to stress is that the supplier, whether the internal or external individual or organization, keeps very close to the customer. Good research, coupled with analysis of complaints data, is an essential part of finding out what the requirements are, and breaking out from the obsession with inward scrutiny that bedevils quality.

■ Quality in all functions

For an organization to be truly effective, each of its components must work properly together. Each part, each activity, each person in the organization affects and is in turn affected by others. Errors have a way of multiplying, and failure to meet the requirements in one part or area creates problems elsewhere, leading to yet more errors, yet more problems, and so on. The benefits of getting it right first time everywhere are enormous.

Everyone experiences – almost accepts – problems in working life. This causes people to spend a large part of their time on useless activities – correcting errors, looking for things, finding out why things are late, checking suspect information, rectifying and reworking, apologizing to customers for mistakes, poor quality and lateness. The list is endless, and it is estimated that about one-third of our efforts are still wasted in this way. In the service sector it can be much higher.

Quality, the way we have defined it as meeting the customer requirements, gives people in different functions of an organization a common language for improvement. It enables all the people, with different abilities and priorities, to communicate readily with one another, in pursuit of a common goal. When business and industry were local, the craftsman could manage more or less on his own. Business is now so complex and employs so many different specialist skills that everyone has to rely on the activities of others in doing their jobs.

Some of the most exciting applications of quality management have materialized from groups of people that could see little relevance when first introduced to its concepts. Following training, many different parts of organizations can show the usefulness of the techniques. Sales staff can monitor and increase successful sales calls, office staff have used quality improvement methods to prevent errors in word processing and inputting to computers, customer-service people have monitored and reduced complaints, distribution has controlled lateness and disruption in deliveries.

It is worthy of mention that the first points of contact for some outside customers are the telephone operator, the security people at the gate, or the person in reception. Equally the e-business, paperwork and support services associated with the product, such as websites, invoices and sales literature and their handlers, must match the needs of the customer. Clearly, 'quality' cannot be restricted to the 'production' or 'operations' areas without losing great opportunities to gain maximum benefit.

Managements that rely heavily on exhortation of the workforce to 'do the right job right the first time', or 'accept that quality is your responsibility', will not only fail to achieve quality but may create division and conflict. These calls for improvement infer that faults are caused only by the workforce and that problems are departmental or functional when, in fact, the opposite is true – most problems are interdepartmental. The commitment of all members of an organization is a requirement of 'organization-wide quality improvement'. Everyone must work together at every interface to achieve improved performance and that can only happen if the top management is really committed.

Chapter highlights

■■■

Quality, competitiveness, reputation and customers

- The reputation enjoyed by an organization is built by quality, reliability and price. Quality is the most important of these competitive weapons.
- Reputations for poor quality last for a long time, and good or bad reputations can become national or international. The management of quality can be learned and used to improve reputation.
- Quality is meeting the customer requirements, and this is not restricted to the functional characteristics of the product or service.
- Reliability is the ability of the product or service to continue to meet the customer requirements over time.
- Organizations 'delight' the customer by consistently meeting customer requirements, and then achieve a reputation of 'excellence' and customer loyalty.

Understanding and building the quality chains

- Throughout all organizations there are a series of internal suppliers and customers. These form the so-called 'quality chains', the core of 'company-wide quality improvement'.
- The internal customer/supplier relationships must be managed by interrogation, i.e. using a set of questions at every interface. Measurement of capability is vital.
- There are two distinct but interrelated aspects of quality, design and conformance to design. *Quality of design* is a measure of how well the product or service is designed to achieve the agreed requirements. *Quality of conformance to design* is the extent to which the product or service achieves the design. Organizations should assess how much time they spend doing the right things right.

Managing quality

- Asking the question 'Have we done the job correctly?' should be replaced by asking 'Are we capable of doing the job correctly?' and 'Do we continue to do the job correctly?'.

- Asking the questions in the right order replaces a strategy of *detection* with one of *prevention*.
- Everything we do is a process, which is the transformation of a set of inputs into the desired outputs.
- In every organization there are some core business processes that must be performed especially well if the mission and objectives are to be achieved.
- Inspection is not *quality control*. The latter is the employment of activities and techniques to achieve and maintain the quality of a product, process or service.
- *Quality assurance* is the prevention of quality problems through planned and systematic activities around processes.

Quality starts with understanding the needs

- Marketing processes establish the true requirements for the product or service. These must be communicated properly throughout the organization in the form of specifications.
- Excellent communications between customers and suppliers is the key to a quality performance – the organization must establish feedback systems to gather customer information.
- Appropriate research techniques should be used to understand the 'market' and keep close to customers and maintain the external perspective.

Quality in all functions

- All members of an organization need to work together on organization-wide quality improvement. The co-operation of everyone at every interface is necessary to achieve improvements in performance, which can only happen if the top management is really committed.

Chapter 2

Models and frameworks for quality management

■ Early quality management frameworks

In the early 1980s when organizations in the West started to be seriously interested in quality and its management there were many attempts to construct lists and frameworks to help this process.

In the West the famous American 'gurus' of quality management, such as W. Edwards Deming, Joseph M. Juran and Philip B. Crosby, started to try to make sense of the labyrinth of issues involved, including the tremendous competitive performance of Japan's manufacturing industry. Deming and Juran had contributed to building Japan's success in the 1950s and 1960s and it was appropriate that they should set down their ideas for how organizations could achieve success.

Deming had 14 points to help management as follows:

1 Create constancy of purpose towards improvement of product and service.
2 Adopt the new philosophy. We can no longer live with commonly accepted levels of delays, mistakes, defective workmanship.
3 Cease dependence on mass inspection. Require instead statistical evidence that quality is built in.
4 End the practice of awarding business on the basis of price tag.
5 Find problems. It is management's job to work continually on the system.

6 Institute modern methods of training on the job.
7 Institute modern methods of supervision of production workers. The responsibility of foremen must be changed from numbers to quality.
8 Drive out fear, so that everyone may work effectively for the company.
9 Break down barriers between departments.
10 Eliminate numerical goals, posters, and slogans for the workforce asking for new levels of productivity without providing methods.
11 Eliminate work standards that prescribe numerical quotas.
12 Remove barriers that stand between the hourly worker and his right to pride of workmanship.
13 Institute a vigorous program of education and retraining.
14 Create a structure in top management that will push every day on the above 13 points.

Juran's ten steps to quality improvement were:

1 Build awareness of the need and opportunity for improvement.
2 Set goals for improvement.
3 Organize to reach the goals (establish a quality council, identify problems, select projects, appoint teams, designate facilitators).
4 Provide training.
5 Carry out projects to solve problems.
6 Report progress.
7 Give recognition.
8 Communicate results.
9 Keep score.
10 Maintain momentum by making annual improvement part of the regular systems and processes of the company.

Phil Crosby, who spent time as Quality Director of ITT, had four absolutes:

- Definition – conformance to requirements.
- System – prevention.
- Performance standard – zero defects.
- Measurement – price of non-conformance.

He also offered management 14 steps to improvement:

1 Make it clear that management is committed to quality.
2 Form quality improvement teams with representatives from each department.
3 Determine where current and potential quality problems lie.
4 Evaluate the cost of quality and explain its use as a management tool.
5 Raise the quality awareness and personal concern of all employees.
6 Take actions to correct problems identified through previous steps.

7 Establish a committee for the zero defects program.
8 Train supervisors to actively carry out their part of the quality improvement program.
9 Hold a 'zero defects day' to let all employees realize that there has been a change.
10 Encourage individuals to establish improvement goals for themselves and their groups.
11 Encourage employees to communicate to management the obstacles they face in attaining their improvement goals.
12 Recognize and appreciate those who participate.
13 Establish quality councils to communicate on a regular basis.
14 Do it all over again to emphasize that the quality improvement program never ends.

A comparison

One way to compare directly the various approaches of the three American gurus is in Table 2.1 which shows the differences and similarities clarified under 12 different factors.

Our understanding of quality management developed through the 1980s, and in earlier editions of this author's books on Total Quality Management (TQM), a broad perspective was given, linking the TQM approaches to the direction, policies and strategies of the business or organization. These ideas were captured in a basic framework – the TQM model (Figure 2.1) which was widely promoted in the UK through the activities of the Department of Trade and Industry (DTI) 'Quality Campaign' and 'Managing into the 90s' programs. These approaches brought together a number of components of the quality approach, including quality circles (teams), problem solving and statistical process control (tools) and quality systems, such as BS 5750 and later ISO 9000 (systems). It was recognized that *culture* played an enormous role in whether organizations were successful or not with their TQM approaches. Good *communications*, of course, were seen to be vital to success but the most important of all was *commitment*, not only from the senior management but from everyone in the organization, particularly those operating directly at the customer interface. The customer/supplier or 'quality chains' and the processes that lived within them were the core of this TQM model.

Many companies and organizations in the public sector found this simple framework useful and it helped groups of senior managers throughout the world get started with managing quality. The key was to integrate the managing quality activities based on the framework,

Table 2.1 The American quality gurus compared

	Crosby	Deming	Juran
Definition of quality	Conformance to requirements	A predictable degree of uniformity and dependability at low cost and suited to the market	Fitness for use
Degree of senior management responsibility	Responsible for quality	Responsible for 94% of quality problems	Less than 20% of quality problems are due to workers
Performance standard/motivation	Zero defects	Quality has many scales. Use statistics to measure performance in all areas. Critical of zero defects	Avoid campaigns to do perfect work
General approach	Prevention, not inspection	Reduce variability by continuous improvement. Cease mass inspection	General management approach to quality – especially 'human' elements
Structure	Fourteen steps to quality improvement	Fourteen points for management	Ten steps to quality improvement
Statistical process control (SPC)	Rejects statistically acceptable levels of quality	Statistical methods of quality control must be used	Recommends SPC but warns that it can lead to too-driven approach
Improvement basis	A 'process', not a program. Improvement goals	Continuous to reduce variation. Eliminate goals without methods	Project-by-project team approach. Set goals
Teamwork	Quality improvement teams. Quality councils	Employee participation in decision making. Break down barriers between departments	Team and quality circle approach
Costs of quality	Cost of non-conformance. Quality is free	No optimum – continuous improvement	Quality is not free – there is an optimum
Purchasing and goods received	State requirements. Supplier is extension of business. Most faults due to purchasers themselves	Inspection too late – allows defects to enter through AQLs. Statistical evidence and control charts required	Problems are complex. Carry out formal surveys
Vendor rating	Yes *and* buyers. audits useless	No – critical of most systems	Yes, but help supplier improve
Single sources of supply		Yes	No – can neglect to sharpen competitive edge

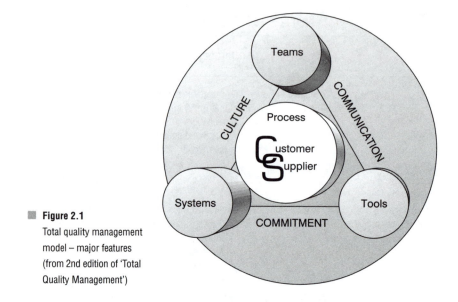

into the business or organization strategy, and this has always been a key component of the author's approach.

■ Quality award models

Starting in Japan with the Deming Prize, companies started to get interested in quality frameworks that could be used essentially in three ways:

- as the basis for awards;
- as the basis for a form of 'self-assessment';
- as a descriptive 'what-needs-to-be-in-place' model.

The earliest approach to a total quality audit process is that established in the Japanese 'Deming Prize', which is based on a highly demanding and intrusive process. The categories of this award were established in 1950 when the Union of Japanese Scientists and Engineers (JUSE) instituted the prize for 'contributions to quality and dependability of product'.

The emphasis of the Deming Prize now is on finding out how effectively the applicant is implementing TQM by focusing on the quality of its products and services. The examiners are looking to see if 'TQM has been implemented properly to achieve business objectives and strategies', and that outstanding results have been obtained.

The Deming Prize 'examination viewpoints' now include:

1 Top management leadership and organizational vision and strategies.
2 TQM frameworks:
   ■ organizational structure and its operations;
   ■ daily management;
   ■ policy management;
   ■ relationship to ISO 9000 and ISO 14000;
   ■ relationship to other management improvement programs;
   ■ TQM promotion and operation.
3 Quality Assurance System:
   ■ QA system;
   ■ new product and new technology development;
   ■ process control;
   ■ test, quality evaluation and quality audits;
   ■ activities covering the whole life cycle;
   ■ purchasing, subcontracting, and distribution management.
4 Management systems for business elements:
   ■ cross-functional management and its operations;
   ■ quantity/delivery management;
   ■ cost management;
   ■ environmental management;
   ■ safety, hygiene and work environment management.
5 Human resources development:
   ■ positioning of people in management;
   ■ education and training;
   ■ respect for people's dignity.
6 Effective utilization of information:
   ■ positioning of information in management;
   ■ information systems;
   ■ support for analysis and decision making;
   ■ standardization and configuration management.
7 TQM concepts and values:
   ■ quality;
   ■ maintenance and improvement;
   ■ respect for humanity.
8 Scientific methods:
   ■ understanding and utilization of methods;
   ■ understanding and utilization of problem-solving methods.
9 Organizational powers:
   ■ core technology;
   ■ speed;
   ■ vitality.
10 Contribution to realization of corporate objectives:
   ■ customer relations;
   ■ employee relations;

- social relations;
- supplier relations;
- shareholder relations;
- realization of corporate mission;
- continuously securing profits.

There is a general 'TQM Features (Shining Examples)' piece at the end which looks for the company promoting TQM activities that are unique and suitable to its own conditions, and contributing to the development of new TQM concepts, methodologies and technologies in anticipation of its future needs.

The recognition that quality management is a broad culture change vehicle with internal and external focus embracing behavioral and service issues, as well as quality assurance and process control, prompted the United States in the late 1980s to develop one of the most famous and now widely used frameworks, the Malcolm Baldrige National Quality Award (MBNQA). The award itself, which is composed of two solid crystal prisms 14 inches high, is presented annually to recognize companies in the USA that have excelled in quality management and quality achievement. But it is not the award itself, or even the fact that it is presented each year by the President of the USA which has attracted the attention of most organizations, it is the excellent framework for quality and organizational self-assessments.

The Baldrige National Quality Program Criteria for Performance Excellence, as it is now known, aims to:

- help improve organizational performance practices, capabilities and results;
- facilitate communication and sharing of best practices information;
- serve as a working tool for understanding and managing performance and for guiding, planning and opportunities for learning.

The award criteria are built upon a set of interrelated core values and concepts:

- visionary leadership;
- customer-driven excellence;
- organizational and personal learning;
- valuing employees and partners;
- agility;
- focus on the future;
- managing for innovation;
- management by fact;
- public responsibility and citizenship;

- focus on results and creating value;
- systems developments.

These are embodied in a framework of seven categories which are used to assess organizations:

1 Leadership:
  - organizational leadership;
  - public responsibility and citizenship.
2 Strategic planning:
  - strategy development;
  - strategy deployment.
3 Customer and market focus:
  - customer and market knowledge;
  - customer relationships and satisfaction.
4 Information and analysis:
  - measurement and analysis of organizational performance;
  - information management.
5 Human resource focus:
  - work systems;
  - employee education training and development;
  - employee well-being and satisfaction.
6 Process management:
  - product and service processes;
  - business processes;
  - support processes.
7 Business results:
  - customer-focused results;
  - financial and market results;
  - human resource results;
  - organizational effectiveness results.

Figure 2.2 shows how the framework's system connects and integrates the categories. This has three basic elements: organizational profile, system, and information and analysis. The main driver is the senior executive leadership which creates the values, goals and systems, and guides the sustained pursuit of quality and performance objectives. The system includes a set of well-defined and well-designed processes for meeting the organization's direction and performance requirements. Measures of progress provide a results-oriented basis for channeling actions to deliver ever-improving customer values and organization performance. The overall goal is the delivery of customer satisfaction and market success leading, in turn, to excellent business results. The seven criteria categories are further divided into items and areas to address. These are described in some detail in the 'Criteria for Performance

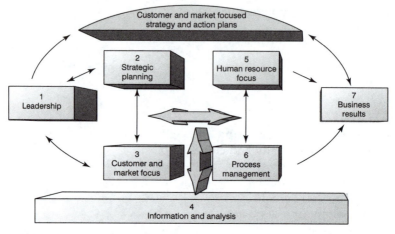

■ **Figure 2.2**   Baldrige criteria for performance excellence framework – a systems perspective
(Source: Malcolm Baldrige National Quality Award, 'Criteria for Performance Excellence', US National
Institute of Standards and Technology, Gaithesburg, USA)

Excellence' available from the US National Institute of Standards and Technology (NIST), in Gaithesburg, USA.

The Baldrige Award led to a huge interest around the world in quality award frameworks that could be used to carry out self-assessment and to build an organization-wide approach to quality, which was truly integrated into the business strategy. It was followed in Europe in the early 1990s by the launch of the European Quality Award by the European Foundation for Quality Management (EFQM). This framework was the first one to include 'Business Results' and to really represent the whole business model.

Like the Baldrige, the EFQM model recognized that processes are the means by which an organization harnesses and releases the talents of its people to produce results/performance. Moreover, improvement in performance can be achieved only by improving the processes by involving the people. This simple model is shown in Figure 2.3.

Figure 2.4 displays graphically the 'non-prescriptive' principles of the full Excellence Model. Essentially customer results, employee results and favorable society results are achieved through leadership driving policy and strategy, people partnerships, resources and processes, which lead ultimately to excellence in key performance results – the enablers deliver the results which in turn drive innovation and learning. The EFQM have provided a weighting for each of the criteria which may be used in scoring self-assessments and making awards (see Chapter 8 for more detail on the Excellence Model).

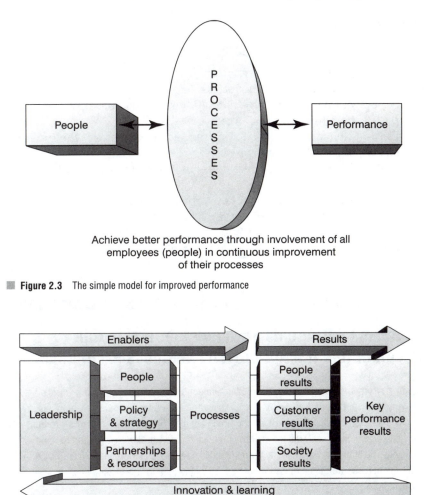

Achieve better performance through involvement of all
employees (people) in continuous improvement
of their processes

Figure 2.3   The simple model for improved performance

Figure 2.4   The EFQM Excellence Model

Through usage and research, the Baldrige and EFQM Excellence models continued to grow in stature throughout the 1990s. They were recognized as descriptive holistic business models, rather than just quality models and mutated into frameworks for (Business) Excellence.

The NIST and EFQM have worked together well over recent years to learn from each other's experience in administering awards and supporting programs, and from organizations which have used their frameworks 'in anger'.

The EFQM publication for the new millennium of the so-called 'Excellence Model' captures much of this learning and provides a framework which organizations can use to follow ten new steps:

1 Set direction through leadership.
2 Establish the results they want to achieve.
3 Establish and drive policy and strategy.
4 Set up and manage appropriately their approach to processes, people, partnerships and resources.
5 Deploy the approaches to ensure achievement of the policies, strategies and thereby the results.
6 Assess the 'business' performance, in terms of customers, their own people and society results.
7 Assess the achievements of key performance results.
8 Review performance for strengths and areas for improvement.
9 Innovate to deliver performance improvements.
10 Learn more about the effects of the enablers on the results.

Case study ■ ■ ■

■ Texas Instruments Europe – leadership and ■ commitment to quality and business excellence

Texas Instruments Incorporated is a global semiconductor company and the world's leading designer and supplier of digital signal processing and analog technologies, the engines driving the digitalization of electronics. Headquartered in Dallas, Texas, the company's products and services also include material and controls, education and productivity solutions, and digital imaging. The company has manufacturing or sales operations in more than 25 countries. Texas Instruments Incorporated employs more than 35 000 people worldwide and has net revenues in excess of $8bn.

The 'chip' has revolutionized our everyday lives. It has increased what we are able to do, the speed at which we can do it, and has created profound benefits for society. The integrated circuit was invented at Texas Instruments (TI) in 1958, one of many significant inventions contributing both to the growth of TI and the electronics industry worldwide – an industry destined to grow to $2 trillion by the year 2000. TI's technological innovations, in addition to the integrated circuit, include the first hand-held calculator, the single chip microcomputer, forward-looking infrared vision systems and the first quantum-effect transistor. These innovations have been the catalyst for the different businesses of TI, their growth, contributions to society and the way we all live, learn, work and play.

TI began in Europe in Bedford, England, back in 1956, the first US-based company to manufacture semiconductors in Europe. Today TI Europe, a wholly owned subsidiary of TI Incorporated, has responsibility to manage all operations in the European region in 15 different countries and employs more than 2300 people. The semiconductor business accounts for over 90 percent of TI Europe's revenue and over 90 percent of its people in Europe.

Total quality culture – a cornerstone of TI's philosophy

During the 1990s it became clear to TI that, while technological innovation was vital to future success, it was insufficient on its own. The company had to find a way to enable its customers to gain access to the innovations and be supported and satisfied in that process. The adoption of total quality was TI's chosen route to becoming more customer oriented, while retaining technological excellence.

The journey began in the 1980s with the first concepts and has developed over time into the way TI people do business with customers and each other. Total quality has permeated all TI companies, thousands of people having received continuous training, and it has become the TI way of life. The TQ journey took a major step forward in 1993 when the EFQM model was adopted for TI Europe.

The approach was continuously refined by adopting a common program of never-ending improvement against the EFQM criteria, under the banner of 'Total Customer Satisfaction Through Business Excellence'. As part of this program, all of TI Europe's business and support organizations' self-assessment were to fundamentally transform the company and shape the organization for the future.

A key element of the new TI Europe was its management structure, entirely based on the EFQM criteria so as to ensure maximum synergy between its component teams, a clear, common focus on TQ/business excellence and a common purpose and direction with a clear, shared vision.

The four Ps and three Cs – a new model for quality management

We have seen in Chapter 1 how *processes* are the key to delivering quality of products and services to customers. It is clear from Figure 2.4 that *processes* are a key linkage between the enablers of *planning* (leadership driving policy and strategy, partnerships and resources), through *people* into the *performance* (measured by people, society, customers, and key outcomes).

These 'four Ps' form the basis of a simple model for quality management and provide the 'hard management necessities' to take organizations successfully into the twenty-first century. These form the structure of the remainder of this book.

From the early quality management frameworks, however, we must not underestimate the importance of the three Cs – Culture, Communication and Commitment. The new model is complete when these 'soft

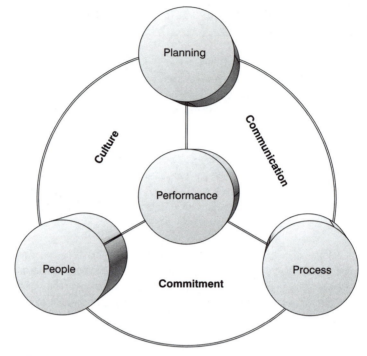

■ Figure 2.5   The new framework for quality management

outcomes' are integrated into the four P's framework to move organizations successfully forward (Figure 2.5).

This new model for quality management, based on all the excellent work done during the last century, provides a simple framework for excellent performance, covering all angles and aspects of an organization and its operation.

Performance is achieved, using a 'business excellence' approach, and by planning the involvement of people in the improvement of processes. This has to include:

- *Planning* – the development and deployment of policies and strategies; setting up appropriate partnerships and resources; and designing in quality.
- *Performance* – establishing a performance measure framework – a 'balanced scorecard' for the organization; carrying out self-assessment, audits, reviews and benchmarking.
- *Processes* – understanding, management, design and redesign; quality management systems; continuous improvement.

- *People* – managing the human resources; culture change; teamwork; communications; innovation and learning.

Wrapping around all this to ensure successful implementation is, of course, effective leadership and commitment, the subject of the next chapter.

Contact details
For Deming Prize: www.deming.org
For Baldrige Award: www.quality.nist.gov
For EFQM Excellence: www. efqm.org or www.quality-foundation.co.uk

Case study ■ ■ ■ ▬▬▬▬▬▬▬▬▬▬▬▬▬▬▬▬▬

■ Sustainable business improvement in a global
■ corporation – Shell Services

Setting up a new global organization is a challenge in itself. To do this by harmonizing existing but different business operations across the world into a single, global organization adds another level of complexity. Shell Services enabled such a transformation by developing and putting in place a set of tools, processes and systems that became known as the Shell Services Quality Framework, or SQF. To put the organization into context, Shell Services comprised several companies across the globe employing some 6500 staff with a turnover in excess of $1 bn.

With a clear focus on becoming a customer-centric organization, there was a need to look at the core processes required to sustain improved business performance as perceived by customers. At the same time, it was recognized that without helping the people in the organization to embrace the values, behaviors and competencies necessary to become customer-centric, the vision could not be achieved. Finally, both people and process improvements had to be underpinned by a quality framework that could be used to define standards, targets and metrics as well as tracking performance improvements over time (Figure 2.6).

With such a diverse and complex organization, no one existing quality model was seen as offering a suitable basis for harmonization and inclusivity. Although some proprietary models were favored locally, there was seen to be benefit in seeking to bring together the best of these into a Shell specific product. Criteria such as simplicity with completeness, inclusion of best practice, availability of supporting tools and suitability for self-assessment were chosen and several well-known quality improvement approaches were researched to arrive at the SQF (Figure 2.7). Each model contributed attributes and strengths, but no single model offered the power, simplicity and completeness of the SQF.

At the top level, the SQF is a simple but powerful construct consisting of five key chevrons. Four of these are enablers – namely Purpose, People, Resources and Process. The fifth is the

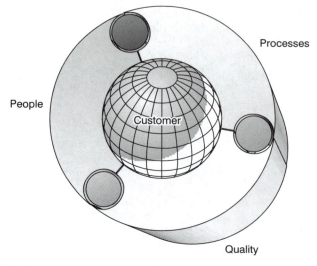

Processes

People

Customer

Quality

Figure 2.6   Components of a customer-centric strategy

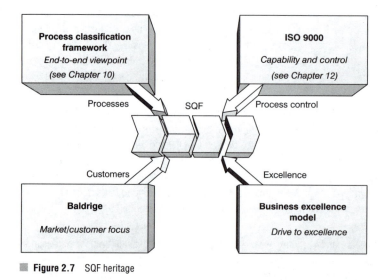

Process classification framework
End-to-end viewpoint
(see Chapter 10)

ISO 9000
Capability and control
(see Chapter 12)

Processes   SQF   Process control

Customers   Excellence

Baldrige
Market/customer focus

Business excellence model
Drive to excellence

Figure 2.7   SQF heritage

Results chevron, which focuses on tracking performance improvement as a result of implementing the framework (Figure 2.8).

Although this may seem a simple construct it has proved tremendously valuable, even at the top level, to ask a simple question about each of the five chevrons. A satisfactory answer is somewhat more difficult to provide than a business leader might expect.

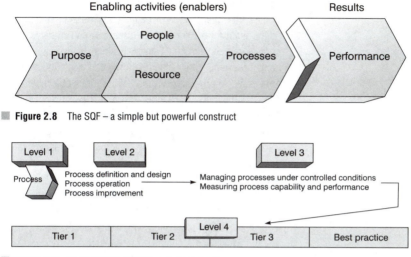

Figure 2.8 The SQF – a simple but powerful construct

Figure 2.9 Cascading the SQF down to best practice

Each of the chevrons is broken down into level 2 and level 3 components in order to define key descriptors, for which tiers of practice, including best practice, can be defined at level four. This is best illustrated in Figure 2.9, where the Process chevron is taken down to level 4 of the framework.

Chapter highlights

Early quality management frameworks

- There have been many attempts to construct lists and frameworks to help organizations understand how to implement good quality management.
- The 'quality gurus' in America, Deming, Juran and Crosby, offered senior management 14 points, ten steps and four absolutes (plus 14 steps), respectively. These similar but different approaches may be compared using a number of factors, including definition of quality, degree of senior management responsibility and general approach.
- The understanding of quality developed and, in Europe and other parts of the world, the author's early TQM model, based on a customer/supplier chain and process core surrounded by systems, tools and teams, linked through culture, communications and commitment, gained wide usage.

Quality award models

- Quality frameworks may be used as the basis for awards, for a form of 'self-assessment', or as a description of what should be in place.

- The Deming Prize in Japan was the first formal quality award framework established by JUSE in 1950. The examination viewpoints include: top management leadership and strategies; TQM frameworks, concepts and values; QA and management systems; human resources; utilization of information; scientific methods; organizational powers; realization of corporate objectives.
- The USA Baldrige Award aims to promote performance excellence and improvement in competitiveness through a framework of seven categories which are used to assess organizations: leadership; strategic planning; customer and market focus; information and analysis; human resource focus; process management; business results.
- The European (EFQM) Excellence Model operates through a simple framework of performance improvement through involvement of people in improving processes.
- The full Excellence Model is a non-prescriptive framework for achieving good results – customers, people, society, key performance – through the enablers – leadership, policy and strategy, people, processes, partnerships and resources. The framework includes proposed weightings for assessment.

The four Ps and three Cs – a new model for quality management

- Planning, People and Processes are the keys to delivering quality products and services to customers and generally improving overall Performance. These four Ps form a structure of 'hard management necessities' for a new simple quality management model which forms the structure of this book.
- The three Cs of Culture, Communication, and Commitment provide the glue or 'soft outcomes' of the model which will take organizations successfully into the twenty-first century.

Leadership and commitment

The quality management approach

'What is quality management?' Something that is best left to the experts is often the answer to this question. But this is avoiding the issue, because it allows executives and managers to opt out. Quality is too important to leave to the so-called 'quality professionals'; it cannot be achieved on a company-wide basis if it is left to the experts. Equally dangerous, however, are the uninformed who try to follow their natural instincts because they 'know what quality is when they see it'. This type of intuitive approach can lead to serious attitude problems, which do no more than reflect the understanding and knowledge of quality that are present in an organization.

The organization which believes that the traditional quality control techniques and the way they have always been used will resolve their quality problems may be misguided. Employing more inspectors, tightening up standards, developing correction, repair and rework teams do not improve quality. Traditionally, quality has been regarded as the responsibility of the QA or QC department, and still it has not yet been recognized in some organizations that many quality problems originate in the commercial, service or administrative areas.

Good quality management is far more than shifting the responsibility of *detection* of problems from the customer to the producer. It requires a comprehensive approach that must first be recognized and then implemented if the rewards are to be realized. Today's business environment is such that managers must plan strategically to maintain a hold on market share, let alone increase it. We have known for years that consumers place a higher value on quality than on loyalty to suppliers. Price is often not the major determining factor in consumer choice and this is true in industrial, service, hospitality, and many other markets. Yet continuously improving efficiencies and reducing costs is key to competitiveness and this must accompany a quality performance.

Good quality management improves the competitiveness, effectiveness and flexibility of a whole organization. It is essentially a way of planning, organizing and understanding each activity, and depends on each individual at each level. For an organization to be truly effective, each part of it must work properly together towards the same goals, recognizing that each person and each activity affect and in turn are affected by others. This rids people's lives of wasted effort by bringing everyone into the processes of improvement, so that results are achieved in less time. The methods and techniques of quality management can be applied throughout any organization. They are equally useful in the manufacturing, public service, health care, education and hospitality industries.

The impact on an organization of a focus on quality is, first, to ensure that the management adopts a strategic overview of quality. The approaches must then focus on developing a *problem-prevention* mentality, but it is easy to underestimate the effort that is required to change attitudes and behaviors. Many people will need to undergo a complete change of 'mindset' to unscramble their intuition, which rushes into the detection/inspection mode to solve quality problems – 'We have a quality problem, we had better check every letter – take two samples out of each sack – check every widget twice', etc.

A different mindset may be achieved by looking at the sort of barriers that exist in key areas. Staff may need to be trained and shown how to reallocate their time and energy to studying their processes in teams, searching for causes of problems, and correcting the causes, not the symptoms, once and for all. This often requires of management a positive, thrusting initiative to promote the right-first-time approach to work situations. Through *quality or process performance improvement teams*, these actions will reduce the inspection/rejection syndrome in due course. If things are done correctly first time round, the usual problems that create the need for inspection for failure should disappear.

The managements of many firms may think that their scale of operation is not sufficiently large, that their resources are too slim, or that the need for action is not important enough to justify managing quality 'formally'. Before arriving at such a conclusion, however, they should examine their existing performance by asking the following questions:

1 Is any attempt made to assess the costs arising from errors, defects, waste, customer complaints, lost sales, etc.? If so, are these costs minimal or insignificant?
2 Is the standard of management adequate and are attempts being made to ensure that quality is given proper consideration at the design stage?
3 Are the organization's quality management systems – documentation, processes, operations, etc. – in good order?
4 Have people been trained in how to prevent errors and problems? Do they anticipate and correct potential causes of problems, or do they find and reject?
5 Do job instructions contain the necessary quality elements, are they kept up to date, and are employers doing their work in accordance with them?
6 What is being done to motivate and train employees to do work right first time?
7 How many errors and defects, and how much wastage occurred last year? Is this more or less than the previous year?

If satisfactory answers can be given to most of these questions, an organization can be reassured that it is already well on the way to using adequate quality management. Even so, it may find that a management review of quality causes it to reappraise activities throughout.

If answers to the above questions indicate problem areas, it will be beneficial to review the top management's attitude to quality. Time and money spent on quality-related activities are *not* limitations of profitability; they make significant contributions towards greater efficiency and enhanced profits.

Commitment and policy

To be successful in promoting business efficiency and effectiveness, the management of quality must be truly organization-wide, and it must start at the top with the chief executive or equivalent. The most senior directors and management must all demonstrate that they are serious about quality. The middle management have a particularly important role to play, since they must not only grasp the principles, they must go on to explain them to the people for whom they are responsible, and ensure that their own commitment is communicated and spreads effectively throughout the organization. This level of management also needs to ensure that the efforts and achievements of their subordinates obtain the recognition, attention and reward that they deserve.

The chief executive of an organization should accept the responsibility for and commitment to a quality policy in which he/she must really believe. This commitment is part of a broad approach extending well beyond the accepted formalities of the quality assurance function. It creates responsibilities for a chain of quality interactions between the marketing, design, production/operations, purchasing, distribution and service functions. Within each and every department of the organization at all levels, starting at the top, basic changes of attitude may be required. If the owners or directors of the organization do not recognize and accept their responsibilities then these changes will not happen. Controls, systems and techniques are very important but they are not the primary requirement. It is more an attitude of mind, based on pride in the job and teamwork, and it requires from the management total commitment to quality, which must then be extended to all employees at all levels and in all departments.

Senior management commitment should be obsessional, not lip service. It is possible to detect real commitment; it shows on the shop floor, in the offices, in the hospital ward – at the point of operation. Going into organizations sporting poster campaigning for quality instead of belief, one is quickly able to detect the falseness. The people are told not to worry if problems arise, 'just do the best you can', 'the customer will never notice'. The opposite is an organization where quality means something, can be seen, heard, felt. Things happen at this operating interface as a result of *real* commitment. Material problems are corrected with suppliers, equipment difficulties are put right by improved maintenance programs or replacement, people are trained, change takes place, partnerships are built, continuous improvement is achieved.

The quality policy

A sound quality policy, together with the organization and facilities to put it into effect, is a fundamental requirement. Every organization should develop and state its policy on quality, together with arrangements for its implementation. The content of the policy should be made known to all employees. The preparation and implementation of a properly thought out quality policy, together with continuous monitoring, make for smoother production or service operation, minimize errors and reduce waste.

Management should be dedicated to the regular improvement of quality, not simply a one-step improvement to an acceptable plateau.

These ideas can be set out in a *quality policy* that requires top management to:

1 Identify the customer's needs (including perception).
2 Assess the ability of the organization to meet these needs economically.
3 Ensure that bought-in materials and services reliably meet the required standards of performance and efficiency.
4 Concentrate on the prevention rather than detection philosophy.
5 Educate and train for quality improvement.
6 Measure customer satisfaction.
7 Review the quality management systems to maintain progress.

The quality policy should be the concern of all employees, and the principles and objectives communicated as widely as possible so that it is understood at all levels of the organization. Practical assistance and training should be given, where necessary, to ensure the relevant knowledge and experience are acquired for successful implementation of the policy.

Case study

TQM implementation at ST Microelectronics

ST Microelectronics (formerly SGS-Thomson Microelectronics) is a global, independent semiconductor company which designs, develops, manufactures and markets a broad range of integrated circuits and discrete devices for a wide variety of microelectronic applications including telecommunications and computer systems, consumer equipment, automotive products, industrial automation and control systems.

In 1997 the company won the European Quality Award. This marked the progress made in developing as a world class organization and also coincided with the tenth anniversary of the formation of the company created by the merger of Thomson Semiconductor and SGS Microellecttronica.

In 1987 the two founder companies of ST saw themselves in a difficult position since neither were large enough to become truly global world class players and yet both had a reasonably broad product and technology base. Therefore, the decision was taken to merge the two bodies into one creating a company.

While this achieved a critical mass the financial results were not encouraging and much work was clearly needed to transform the company into the organization which was the vision of the senior management team. The first years of the program were devoted to rationalization and consolidation. At the same time, however, advantage was taken of the complementarity of the product and technology portfolios, customers, market strengths and production capacities. Attention was focused on eliminating the weakness and exploiting the strengths. Two of the early key goals were defined as being a rapid increase in sales and market share together with a slimming down of production sites and the number of employees.

Unfortunately as the program developed the market suddenly hit one of the down cycles which the industry experiences and, in 1990, the improvements in financial results halted and, in fact, worsened. Immediately the 'traditional' management action program was brought into play. There was a rapid 'downsizing' program which hit people, product portfolio and, ultimately, market share. By examining this process in action, both within ST and other companies, the relationship rapidly dawned of the danger of it developing into a 'vicious spiral'. This brought about a review of the focus of the company and the determination to find a new way of proceeding which would give rise to the term 'a virtuous spiral'.

In 1991 ST launched a TQM initiative based on the European Foundation for Quality Management (EFQM) model. In launching this program there was total commitment from the CEO and all his executive staff. In fact in December 1991 Pasquale Pistorio, CEO, stated that: 'TQM is a mandatory way of life in the corporation. SGS-Thomson will become a champion of this culture in the Western world.' These words needed to be backed by action and resource – both financial and people. Very quickly there was a framework put in place, based on an analysis, which determined that the key components of successful implementation of TQM should be:

- Organization
- Common framework
- Local initiatives
- Culture change
- Mechanisms
- Policy deployment

Also the program needed to be driven from the top down, not by dictate, but by example.

There was already in existence a corporate mission statement but it was not closely linked in the minds of the staff with their day-to-day activities. Furthermore it had been written shortly after the merger and did not totally reflect the needs of the company, the shareholders, the employees or the customers. It was, therefore, revised and became the key launching point for all the decisions which affect the future of the corporation.

The mission statement is both short and clear reading:

> To offer strategic independence to our partners worldwide, as a profitable and viable broad range semiconductor supplier.

This statement had implications regarding the size and dynamics of the corporation, resulting directly from the structure and investment needs of the semiconductor industry.

Following the revitalization of the mission statement there quickly followed publication of the corporation's:

■ Objectives
■ Strategic guidelines
■ Guiding principles
■ TQM principles
■ Statement of the future

All of these were published in a leaflet titled 'Shared Values' which was circulated to all employees worldwide.

■ Creating or changing the culture

The culture within an organization is formed by a number of components:

1 Behaviors based on people interactions.
2 Norms resulting from working groups.
3 Dominant values adopted by the organization.
4 Rules of the game for 'getting on'.
5 The climate.

Culture in any 'business' may be defined then as the beliefs that pervade the organization about how business should be conducted, and how employees should behave and should be treated. Any organization needs a *vision framework* that includes its *guiding philosophy, core values and beliefs* and a *purpose*. These should be combined into a *mission*, which provides a vivid description of what things will be like when it has been achieved (Figure 3.1).

The *guiding philosophy* drives the organization and is shaped by the leaders through their thoughts and actions. It should reflect the vision of an organization rather than the vision of a single leader, and should evolve with time, although organizations must hold on to the *core* elements.

The *core values and beliefs* represent the organization's basic principles about what is important in business, its conduct, its social responsibility and its response to changes in the environment. They should act as a guiding force, with clear and authentic values, which are focused on employees, suppliers, customers, society at large, safety, shareholders, and generally stakeholders.

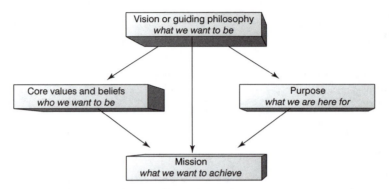

Figure 3.1   Vision framework for an organization

The *purpose* of the organization should be a development from the vision and core values and beliefs and should quickly and clearly convey how the organization is to fulfill its role.

The *mission* will translate the abstractness of philosophy into tangible goals that will move the organization forward and make it perform to its optimum. It should not be limited by the constraints of strategic analysis, and should be proactive not reactive. Strategy is subservient to mission, the strategic analysis being done after, not during, the mission setting process.

Two examples of how leaders of organizations – one in the private sector and one in the public sector – develop their vision, mission and values and are role models of a culture of quality/excellence are given in the inset below:

Private sector

To enable the company to set direction and achieve its vision, the senior management team address priorities for improvement. These are driven by a business improvement process, which consists of: articulate a vision, determine the actions to realize the vision, define measures and set targets, then implement a rigorous review mechanism.

Each member of team takes responsibility for one of the Excellence Model criteria. They develop improvement plans and personally ensure that these are properly resourced and implemented, and that progress is monitored. Improvements identified at local level are prioritized and resourced by local management against the organization's annual business plan.

Public sector

The purpose and direction of the organization – the mission – is developed by a task team. Senior, middle and junior managers review and update the mission, vision and values annually to ensure they support policy and strategy.

Leaders invite input from stakeholders via the employee involvement initiative, monthly update meetings and customer service seminars. The values have been placed on help cards for every employee and are continually re-emphasized at monthly update meetings.

Leaders act as role models and have a list of role model standards to follow, which they are measured against in their performance management system. All managers include TQM objectives in their performance agreements and personal development plans, which are reviewed through the review.

Control

The effectiveness of an organization and its people depends on the extent to which each person and department perform their role and move towards the common goals and objectives. Control is the process by which information or feedback is provided so as to keep all functions on track. It is the sum total of the activities that increase the probability of the planned results being achieved. Control mechanisms fall into three categories, depending upon their position in the managerial process:

Before the fact	*Operational*	*After the fact*
Strategic plan	Observation	Annual reports
Action plans	Inspection and correction	Variance reports
Budgets	Progress review	Audits
Job descriptions	Staff meetings	Surveys
Individual performance objectives	Internal information and data systems	Performance review
Training and development	Training programs	Evaluation of training

Many organizations use after-the-fact controls, causing managers to take a reactive rather than a proactive position. Such 'crisis orientation' needs to be replaced by a more anticipative one in which the focus is on preventive or before-the-fact controls.

Attempting to control performance through systems, procedures, or techniques *external* to the individual is not an effective approach, since it relies on 'controlling' others; individuals should be responsible for their own actions. An externally based control system can result in a high degree of concentrated effort in a specific area if the system is overly structured, but it can also cause negative consequences to surface:

1 Since all rewards are based on external measures, which are imposed, the 'team members' often focus all their effort on the measure itself, e.g. to have it set lower (or higher) if possible, to manipulate the

information which serves to monitor it, or to dismiss it as someone else's goal not theirs. In the budgeting process, for example, distorted figures are often submitted by those who have learned that their 'honest projections' will be automatically altered anyway.

2 When the rewards are dependent on only one or two limited targets, all efforts are directed at those, even at the expense of others. If short-term profitability is the sole criterion for bonus distribution or promotion, it is likely that investment for longer-term growth areas will be substantially reduced. Similarly, strong emphasis and reward for output or production may result in lowered quality.

3 The fear of not being rewarded, or even being criticized, for performance that is less than desirable may cause some to withhold information that is unfavorable but nevertheless should be flowing into the system to improve quality.

4 When reward and punishment are used to motivate performance, the degree of risk-taking may lessen and be replaced by a more cautious and conservative approach. In essence, the fear of failure replaces the desire to achieve.

The following problem situations have been observed by the author and his colleagues, within companies that have taken part in research and consultancy projects:

- The goals imposed are seen or known to be unrealistic. If the goals perceived by the subordinate are in fact accomplished, then the subordinate has proved himself wrong. This clearly has a negative effect on the effort expended, since few people are motivated to prove themselves wrong!

- Where individuals are stimulated to commit themselves to a goal, and where their personal pride and self-esteem are at stake, then the level of motivation is at a peak. For most people the toughest critic and the hardest taskmaster they confront is not their immediate boss but themselves.

- Directors and managers are often afraid of allowing subordinates to set the goals for fear of them being set too low, or loss of control over subordinate behavior. It is also true that many do not wish to set their own targets, but prefer to be told what is to be accomplished.

Quality management is concerned with moving the focus of control from outside the individual to within, the objective being to make everyone accountable for their own performance, and to get them committed to attaining quality in a highly motivated fashion. The assumptions a director or manager must make in order to move in this direction are simply that people do not need to be coerced to perform well, and that people want to achieve, accomplish, influence activity, and challenge their abilities. If there is belief in this, then only the techniques remain to be discussed.

Quality management is user-driven – it cannot be imposed from outside the organization, as perhaps can a quality system X standard or statistical process control. This means that the ideas for improvement must come from those with knowledge and experience of the processes, activities and tasks; this has massive implications for training and follow-up. Although the effects of successful quality management will reduce costs and improve productivity, it is concerned chiefly with changing attitudes and skills so that culture of the organization becomes one of preventing failure – doing the right things, right first time, every time.

■ Effective leadership

Some management teams have broken away from the traditional style of management; they have made a 'managerial breakthrough'. Their approach puts their organization head and shoulders above others in the fight for sales, profits, resources, funding and jobs. Many public service organizations are beginning to move in the same way, and the successful quality-based strategy they are adopting depends very much on effective leadership.

Effective leadership starts with the chief executive's and his top team's vision, capitalizing on market or service opportunities, continues through a strategy that will give the organization competitive or other advantage, and leads to business or service success. It goes on to embrace all the beliefs and values held, the decisions taken and the plans made by anyone anywhere in the organization, and the focusing of them into effective, value-adding action.

Together, effective leadership and good quality management result in the company or organization doing the right things, right first time.

The five requirements for effective leadership are the following:

1 Develop and publish clear documented vision, corporate values/beliefs, purpose and a mission statement

Executives should express values and beliefs through a clear vision of what they want their company to be and its purpose – what they specifically want to achieve in line with the basic beliefs. Together, they define what the organization is all about. The senior management team will need to spend some time away from the 'coal face' to do this and develop their plans for implementation.

Clearly defined and properly communicated beliefs and objectives, which can be summarized in the form of vision and mission statements, are essential if the directors, managers and other employees are to work together as a winning team. The beliefs and objectives should address:

- The definition of the business, e.g. the needs that are satisfied or the benefits provided.
- A commitment to effective leadership and quality.
- Target sectors and relationships with customers, and market or service position.
- The role or contribution of the company, organization, or unit, e.g. profit-generator, service department, opportunity-seeker.
- The distinctive competence – a brief statement which applies only to that organization, company or unit.
- Indications for future direction – a brief statement of the principal plans which would be considered.
- Commitment to monitoring performance against customers' needs and expectations, and continuous improvement.

The vision, mission statement and the broad beliefs and objectives may then be used to communicate an inspiring vision of the organization's future. The top management must then show *TOTAL COMMITMENT* to it.

Case study ■ ■ ■ ▬▬▬▬▬▬▬▬▬▬▬▬▬▬

As BT emerged from the public sector it was realized that to be successful, a significant cultural change would have to be stimulated and managed within the organization. Accordingly in 1986 BT embraced enthusiastically the philosophy of total quality management (TQM) to drive continuous improvement through a focus on customer requirements, team working and problem solving. Led personally by the chairman, TQM was implemented through a series of workshops involving all managers and their teams.

At the same time BT launched the BT Values to define the desired culture of the organization. Despite many organizational changes the five BT Values remain unaltered and continue to guide behaviors within the company. The BT Values are:

- We put our customers first.
- We are professional.
- We respect each other.
- We work as one team.
- We are committed to continuous improvement.

BT is imbued with a strong management by objectives climate and this was later refined with the adoption of a balanced corporate scorecard approach to translate BT's strategy into action through a set of key objectives, measures and targets.

2 Develop clear and effective strategies and supporting plans for achieving the mission _____

The achievement of the company or service vision and mission requires the development of business or service strategies, including the strategic positioning in the 'market place'. Plans can then be developed for implementing the strategies. Such strategies and plans can be developed by senior managers alone, but there is likely to be more commitment to them if employee participation in their development and implementation is encouraged.

3 Identify the critical success factors and critical processes _____

The next step is the identification of the *critical success factors* (CSFs), a term used to mean the most important subgoals of a business or organization. CSFs are what must be accomplished for the mission to be achieved. The CSFs are followed by the key, *core business processes* for the organization – the activities that must be done particularly well for the CSFs to be achieved. This process is shown in overview in Figure 3.2 and described in more detail in later chapters.

4 Review the management structure _____

Defining the corporate vision, values mission, strategies, CSFs and core processes might make it necessary to review the organizational structure.

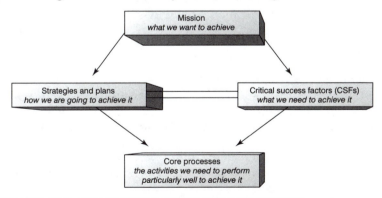

Figure 3.2   Mission into action through strategies, CSF's and core processes

This may include both the definition of responsibilities for the organization's management and the operational procedures they will use. These should be the agreed best ways of carrying out the core processes.

The review of the management structure may also include the establishment of a process improvement team structure throughout the organization.

5 Empowerment – encouraging effective employee participation

For effective leadership it is necessary for management to get very close to the employees. They need to develop effective communications – up, down and across the organization – and take action on what is communicated; and they should encourage good communications between all suppliers and customers.

Particular attention should be paid in particular to attitudes, abilities and participation:

Attitudes

The key attitude for managing any winning company or organization may be expressed as follows: 'I will personally understand who my customers are and what are their needs and expectations and I will take whatever action is necessary to satisfy them fully. I will also understand and communicate my requirements to my suppliers, inform them of changes and provide feedback on their performance.' This attitude should start at the top – with the chairman or chief executive. It must then percolate down, to be adopted by each and every employee. That will happen only if managers lead by example. Words are cheap and will be meaningless if employees see from managers' actions that they do not actually believe or intend what they say.

Abilities

Every employee should be able to do what is needed and expected of him or her, but it is first necessary to decide what is really needed and expected. If it is not clear what the employees are required to do and what standards of performance are expected, how can managers expect them to do it?

Train, train, train and train again. Training is very important, but it can be expensive if the money is not spent wisely. The training should be related to needs, expectations, and process improvement. It must be planned and its effectiveness *always* reviewed.

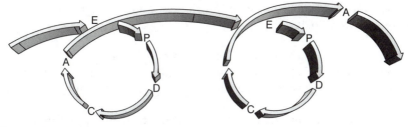

Figure 3.3 The helix of never-ending improvement

Participation

If all employees are to participate in making the company or organization successful (directors and managers included), then they should also be trained in the basics of disciplined management.

They need training to:

E Evaluate – the situation and define their objectives.
P Plan – to achieve those objectives fully.
D Do – implement the plans.
C Check – that the objectives are being achieved.
A Amend – take corrective action if they are not.

The word 'disciplined' applied to people at all levels means that they will do what they say they will do. It also means that in whatever they do they will go through the full process of Evaluate, Plan, Do, Check and Amend, rather than the more traditional and easier option of starting by doing rather than evaluating. This will lead to a never-ending improvement helix (Figure 3.3).

This basic approach needs to be backed up with good project management, planning techniques and problem-solving methods, which can be taught to anyone in a relatively short period of time. The project management enables changes to be made successfully and the people to remove the obstacles in their way. Directors and managers need this training as much as other employees.

Excellence in leadership

A vehicle for achieving excellence in leadership is quality management. We have seen that it should cover the entire organization, all the people and all the functions, including external organizations and suppliers.

In the first three chapters, several facets of quality management have been reviewed, including:

■ Recognizing customers and discovering their needs.
■ Setting standards that are consistent with customer requirements.
■ Controlling processes, including systems, and improving their capability.
■ Management's responsibility for setting the guiding philosophy, quality policy, etc., and providing motivation through leadership and equipping people to achieve quality.
■ Empowerment of people at all levels in the organization to act for quality improvement.

Implementing quality management properly into an organization can be daunting and the chief executive and directors faced with it may become confused and irritated by the proliferation of theories and packages. A simplification is required. The *core* must be the customer/supplier interfaces, both internally and externally, and the fact that at each interface there are processes to convert inputs to outputs. Clearly, there must be commitment to building in quality through management of the inputs and processes.

How can senior managers and directors be helped further in their understanding of what needs to be done to become committed to quality and implement the vision? The American and Japanese quality 'gurus' each set down a number of points or absolutes – words of wisdom in management and leadership – and many organizations have used these to establish a policy based on quality.

Similarly, the EFQM have defined the criterion of leadership and its subcriteria as part of their model of Excellence. A fundamental principle behind all these approaches is that the behaviors of the leaders in an organization need to create clarity and constancy of purpose. This may be achieved through development of the vision, values, purpose and mission needed for longer-term performance success.

Using as a construct the new 'Oakland model' for quality management, the four Ps and the three Cs plus a fourth C – Customers (which resides in 'performance'), the main items for attention to deliver excellence in leadership are given below:

Planning

■ Develop the vision and mission needed for constancy of purpose and for long-term success.

- Develop, deploy and update policy and strategy.
- Align organizational structure to support delivery of policy and strategy.

Performance

- Ensure key performance results are measured, reviewed and improved.
- Help people to know how well they are doing against the customer and performance goals.

Processes

- Ensure a system for identifying and managing processes is developed.
- Ensure through personal involvement that the management system is implemented and continuously improved.
- Prioritize improvement activities and ensure they are planned on an organization-wide basis.

People

- Help and support people to achieve plans, goals, objectives and targets.
- Stimulate empowerment ('experts') and teamwork to encourage creativity and innovation.
- Encourage and support training, education and learning activities.
- Motivate, support and recognize the organization's people – both individually and in teams.
- Respond to people's ideas and encourage them to participate in improvement activities.

Customers

- Be involved with customers and other stakeholders.
- Ensure customer (external and internal) needs are understood and responded to.
- Establish and participate in partnerships – as a customer demand continuous improvement in everything.

Commitment

- Be personally and actively involved in quality and improvement activities.
- Review and improve effectiveness of own leadership.

Culture _____

- ■ Develop the values and ethics to support the creation of a total quality culture.
- ■ Implement the values and ethics through actions and behaviors.
- ■ Ensure creativity, innovation and learning activities are developed and implemented.

Communications _____

- ■ Stimulate and encourage communication and collaboration.
- ■ Personally communicate the vision, values, mission, policies and strategies.
- ■ Be accessible and actively listen.

Quality management requires strong leadership with clear direction and a carefully planned and fully integrated strategy derived from the vision. One of the greatest tangible benefits of excellence in leadership is the improved overall performance of the organization. The evidence for this can be seen in some of the major consumer and industrial markets of the world. Moreover, effective leadership leads to improvements and superior quality which can be converted into premium prices. Research now shows that leadership and quality clearly correlate with profit but the less tangible benefit of greater employee participation is equally, if not more, important in the longer term. The pursuit of continual improvement must become a way of life for everyone in an organization if it is to succeed in today's competitive environment.

■
■ Reference

British Quality Foundation, *The Model in Practice*, Vols 1 and 2, BQF, London, 2002.

Chapter highlights
■ ■ ■

The quality management approach

- ■ Quality management provides a comprehensive approach to improving competitiveness, effectiveness and flexibility through planning, organizing and understanding each activity, and involving each individual at each level. It is useful in all types of organization.
- ■ It ensures that management adopts a strategic overview of quality and focuses on prevention, not detection, of problems.

■ It often requires a mindset change to break down existing barriers. Managements that doubt its applicability should ask questions about the operation's costs, errors, wastes, standards, systems, training and job instructions.

Commitment and policy

■ Quality starts at the top, where serious obsessional commitment and leadership must be demonstrated. Middle management also has a key role to play in communicating the right messages.

■ Every chief executive should accept the responsibility for commitment to a quality policy that deals with the organization for quality, the customer needs, the ability of the organization, supplied materials and services, education and training, and review of the management systems for never-ending improvement.

Creating or changing the culture

■ The culture of an organization is formed by its beliefs, behaviors, norms, dominant values, rules and climate.

■ Any organization needs a vision framework, comprising its guiding philosophy, core values and beliefs, purpose, and mission.

■ The effectiveness of an organization depends on the extent to which people perform their roles and move towards the common goals and objectives.

■ Quality management is concerned with moving the focus of control from the outside to the inside of individuals, so that everyone is accountable for his/her own performance.

Effective leadership

■ Effective leadership starts with the chief executive's vision and develops into a strategy for implementation.

■ Top management should develop the following for effective leadership: clear beliefs and objectives in the form of a mission statement; clear and effective strategies and supporting plans; the critical success factors and core processes; the appropriate management structure; employee participation through empowerment, and the EPDCA helix.

Excellence in leadership

■ Quality management provides a vehicle for achieving excellence in leadership. Using the construct of the new Oakland model, the four Ps and four Cs provide a framework for this: Planning, Performance, Processes, People, Customers, Commitment, Culture, Communications.

Part 2

Planning

A mighty maze!
but not without a plan.

Alexander Pope, 1733,
from 'An Essay on Man'

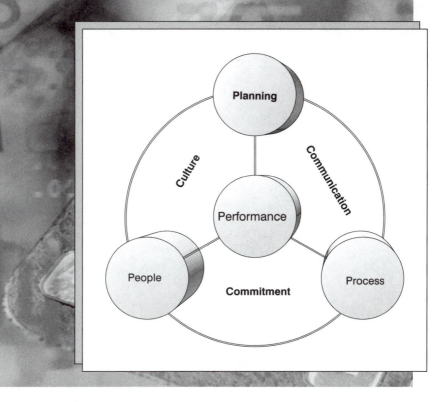

Chapter 4

Policy, strategy and goal deployment

■ Integrating quality into the policy and strategy

In the previous chapter on leadership the main message was that leaders should have a clear sense of direction and purpose, which they communicate effectively throughout the organization. This involves the development of the vision, values and mission which are clearly aspects of policy and strategy. Included in the EFQM Excellence Model, the criterion policy and strategy is concerned with:

> How the organization implements its mission and vision via a clear stakeholder-focused strategy, supported by relevant policies, plans, objectives, targets and processes.

For this to happen the vision and mission and their deployment must be based on the needs and expectations of the organization's stakeholders – present and future. This in turn requires information from research and learning activities and, even more importantly, performance measurement, on which to base the policies and strategies. Of course, time and the world around us do not stand still, so the policy and strategies must be reviewed, updated and generally developed to meet the changing needs of the organization.

There are six basic steps for achieving this and providing a good foundation for the good quality management:

Step 1 Develop a shared vision and mission for the business/organization _____

Once the top team is reasonably clear about the direction the organization should be taking it can develop vision and mission statements that will help to define process-alignment, roles and responsibilities. This will lead to a co-ordinated flow of analysis of processes that crosses the traditional functional areas at all levels of the organization, without changing formal structures, titles, and systems which can create resistance. The vision framework was introduced in Chapter 3 (Figure 4.1).

The mission statement gives a purpose to the organization or unit. It should answer the questions 'what are we here for?' or 'what is our basic purpose?' and 'what have we got to achieve?' It therefore defines the boundaries of the business in which the organization operates. This will help to focus on the 'distinctive competence' of the organization, and to orient everyone in the same direction of what has to be done. The mission must be documented, agreed by the top management team, sufficiently explicit to enable its eventual accomplishment to be verified, and ideally be no more than four sentences. The statement must be understandable, communicable, believable, and usable.

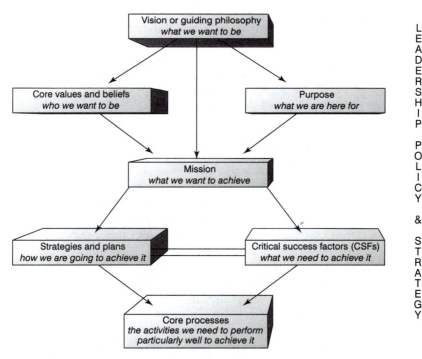

Figure 4.1   Vision framework for an organization

The mission statement is:

- an expression of the aspiration of the organization;
- the touchstone against which all actions or proposed actions can be judged;
- usually long term;
- short term if the mission is survival.

Typical content includes a statement of:

- the role or contribution of the business or unit – for example, profit generator, service department, opportunity seeker;
- the definition of the business – for example, the needs you satisfy or the benefits you provide. Do not be too specific or too general;
- your distinctive competence – this should be a brief statement that applies only to your specific unit. A statement, which could apply equally to any organization, is unsatisfactory;
- indications for future direction – a brief statement of the principal things you would give serious consideration to.

Some questions that may be asked of a mission statement are, does it:

- define the organization's role?
- contain the need to be fulfilled:
  - is it worthwhile/admirable?
  - will employees identify with it?
  - how will it be viewed externally?
- take a long-term view, leading to, for example, commitment to a new product or service development, or training of people?
- take into account all the 'stakeholders' of the organization?
- ensure the purpose remains constant despite changes in top management?

It is important to establish in some organizations whether or not the mission is survival. This does not preclude a longer-term mission, but the short-term survival mission must be expressed, if it is relevant. The management team can then decide whether they wish to continue long-term strategic thinking. If survival is a real issue it is inadvisable to concentrate on long-term planning initially.

There must be open and spontaneous discussion during generation of the mission, but there must in the end be convergence on one statement. If the mission statement is wrong, everything that follows will be wrong too, so a clear understanding is vital.

Step 2 Develop the 'mission' into its critical success factors (CSFs) to coerce and move it forward

The development of the mission is clearly not enough to ensure its implementation. This is a 'danger gap' which many organizations fall into because they do not foster the skills needed to translate the mission through its CSFs into the core processes. Hence, they have 'goals without methods' and change is not integrated properly into the business.

Once the top managers begin to list the CSFs they will gain some understanding of what the mission requires. The first step in going from mission to CSFs is to brainstorm all the possible impacts on the mission. In this way 30 to 50 items ranging from politics to costs, from national cultures to regional market peculiarities may be derived.

The CSFs may now be defined – *what* the organization must accomplish to achieve the mission, by examination and categorization of the impacts. This should lead to a balanced set of deliverables for the organization in terms of:

- financial and non-financial performance;
- customer/market satisfaction;
- people/internal organization satisfaction;
- environmental/societal satisfaction.

There should be no more than eight CSFs, and no more than four if the mission is survival. They are the building blocks of the mission – minimum key factors or subgoals that the organization **must have** or **needs** and which together will achieve the mission. They are the **whats** not the **hows**, and are not directly manageable – they may be in some cases statements of hope or fear. But they provide direction and the success criteria, and are the end product of applying the processes. In CSF determination, a management team should follow the rule that each CSF is **necessary** and together they are **sufficient** for the mission to be achieved.

Some examples of CSFs may clarify their understanding:

- We must have right-first-time suppliers.
- We must have motivated, skilled workers.
- We need new products that satisfy market needs.
- We need new business opportunities.
- We must have best-in-the-field product quality.

The list of CSFs should be an agreed balance of strategic and tactical issues, each of which deals with a 'pure' factor, the use of 'and' being forbidden. It will be important to know when the CSFs have been

Divisional or functional CSFs	CSF No.	1	2	3	4	5	6	7	8	KPIs
	1									
	2									
	3									
	4									
	5									
	6									
	7									
	8									

(Corporate CSFs / KPOs shown across the top angled columns)

Figure 4.2   Interaction of corporate and divisional CSFs

achieved, but an equally important step is to use the CSFs to enable the identification of the processes.

Senior managers in large complex organizations may find it necessary or useful to show the interaction of divisional CSFs with the corporate CSFs in an impact matrix (see Figure 4.2 and discussion under Step 6).

Step 3 Define the key performance outcomes as being the quantifiable indicators of success in terms of the mission and CSFs

The mission and CSFs provide the **what** of the organization, but they must be supported by measurable key performance outcomes (KPOs) that are tightly and inarguably linked. These will help to translate the directional and sometimes 'loose' statements of the mission into clear **targets**, and in turn to simplify management's thinking. The KPOs will be used to monitor progress and as evidence of success for the organization, in every direction, internally and externally.

Each CSF should have an 'owner' who is a member of the management team that agreed the mission and CSFs. The task of an owner is to:

- define and agree the KPOs and associated targets;
- ensure that appropriate data is collected and recorded;
- monitor and report progress towards achieving the CSF (KPOs and targets) on a regular basis;
- review and modify the KPOs and targets where appropriate.

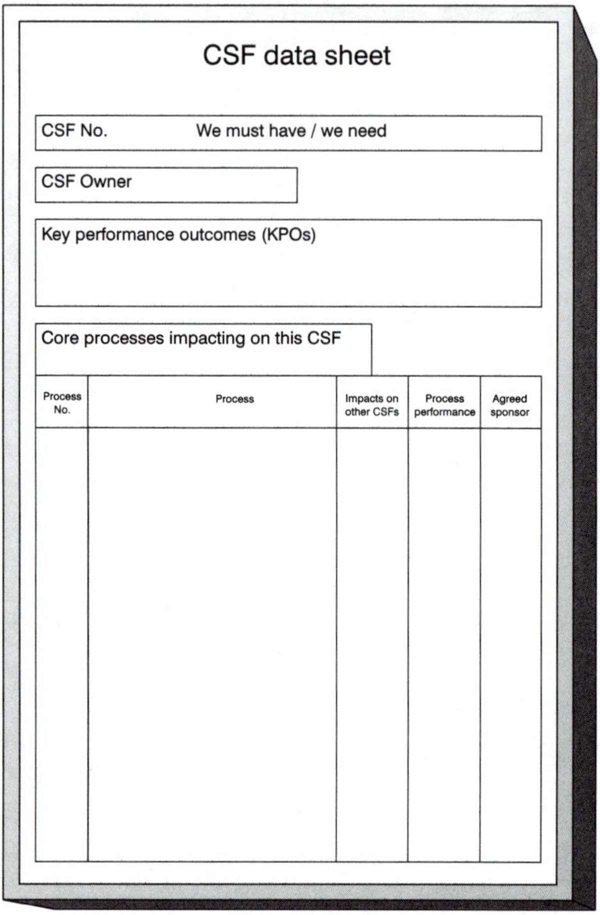

CSF data sheet

CSF No.	We must have / we need

CSF Owner

Key performance outcomes (KPOs)

Core processes impacting on this CSF

Process No.	Process	Impacts on other CSFs	Process performance	Agreed sponsor

■ **Figure 4.3**
CSF data sheet

A typical CSF data sheet for completion by owners is shown in Figure 4.3.

The derivation of KPOs may follow the 'balanced scorecard' model, proposed by Kaplan, which divides measures into financial, customer, internal business and innovation and learning perspectives (see Chapter 7).

Step 4 Understand the core processes and gain process sponsorship

This is the point when the top management team have to consider how to institutionalize the mission in the form of processes that will continue to be in place, until major changes are required.

The core business processes describe what actually is or needs to be done so that the organization meets its CSFs. As with the CSFs and the mission, each process which is **necessary** for a given CSF must be identified, and together the processes listed must be **sufficient** for all the CSFs to be accomplished. To ensure that **processes** are listed, they should be in the form of verb plus object, such as research the market, recruit competent staff, or manage supplier performance. The core processes identified frequently run across 'departments' or functions, yet they must be measurable.

Each core process should have a sponsor, preferably a member of the management team that agreed the CSFs.

The task of a sponsor is to:

- ensure that appropriate resources are made available to map, investigate and improve the process;
- assist in selecting the process improvement team leader and members;
- remove blocks to the team's progress;
- report progress to the senior management team.

The first stage in understanding the core processes is to produce a set of processes of a common order of magnitude. Some smaller processes identified may combine into core processes, others may be already at the appropriate level. This will ensure that the change becomes entrenched, the core processes are identified and that the right people are in place to sponsor or take responsibility for them. This will be the start of getting the process team organization up and running.

The questions will now come thick and fast; is the process currently carried out? By whom? When? How frequently? With what performance and how well compared with competitors? The answering of these will force process ownership into the business. The process sponsor may form a process team which takes quality improvement into the next steps. Some form of prioritization using process performance measures is necessary at this stage to enable effort to be focused on the key areas for improvement. This may be carried out by a form of impact matrix analysis (see Figure 4.4). The outcome should be a set of 'most critical processes' (MCPs) which receive priority attention for improvement, based on the number of CSFs impacted by each process and its performance on a scale A to E.

Step 5 Break down the core processes into subprocesses, activities and tasks and form improvement teams around these

Once an organization has defined and mapped out the core processes, people need to develop the skills to understand how the new process

No.	Process	CSF No.								Number of CSF impacts	A-E ranking
		A-E process ranking: A-Excellent; B-Good; C-Average; D-Poor; E-Embryonic									

Figure 4.4   Process/CSF matrix

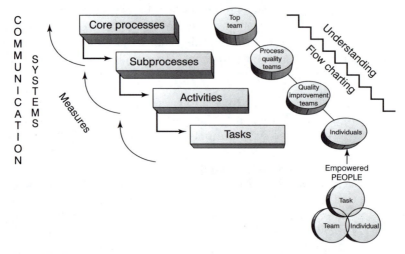

Figure 4.5   Breakdown of core processes into subprocesses, activities and tasks

structure will be analyzed and made to work. The very existence of new process teams with new goals and responsibilities will force the organization into a learning phase. The changes should foster new attitudes and behaviors.

An illustration of the breakdown from mission through CSFs and core processes, to individual tasks may assist in understanding the process required (Figure 4.5) (*continued on p. 75*).

Case study ■■■

■ DRIVER for change in BBC Resources

Background _____

London Operations, part of BBC Resources Ltd, provides studio, outside broadcast and post-production facilities to customers both within and outside the BBC. It was hemorrhaging money at the rate of over £7 m (c. $10 m) a year. It was overstaffed and locked into inefficient, outmoded work practices. Under 'Producer Choice', it was being increasingly ignored by BBC program-makers who were going outside the corporation to obtain better terms for production facilities. Under political pressure, the Corporation was so concerned that it was considering selling off all or part of BBC Resources. The company's management required insight and plans to determine whether Resources Ltd could become competitive, and how it could rapidly implement the changes needed to transform the business from its current loss-making situation.

Resources management carried out a program of improvement that began with a review of the London Operations to assess current performance, recommend the necessary steps to achieve profitability and to plan and implement the changes.

As improvements and changes were being implemented and as the senior management became more aware of commercial pressures it was recognized that, for these changes to have any durability and long-lasting impact on the business, it was vitally important that everyone in the organization understood the part that they had to play in helping turn the business around. Furthermore the management team needed to have a clear understanding of what they were doing, why they were doing it and how they needed to do it. To this end the senior management team identified the need to understand and further develop the mission and vision for the business. Then to be able to cascade these down through the organization, focused around a small number of factors that were deemed critical to the achievement of the mission and vision:

- Vision – turning ideas into reality.
- Mission – we will enrich the BBC creatively and financially by helping customers create the sounds and vision of the future. Relied upon for innovation, efficiency and service, working with us will be inspirational and fun.

Defining measurable objectives _____

From the mission statement the key words were identified to form the basis for the development of a strategic framework:

- Cash
- Creativity
- Innovation

■ Service
■ Efficiency

Using these key words eight factors critical to the achievement of the mission were identified

1 Skilled, motivated and flexible people.
2 Key talent that is industry recognized.
3 Focused investment in products and services.
4 Profitable revenue growth.
5 Efficient and effective processes.
6 Effective customer relationships.
7 Strong leadership, clarity of direction and co-operation.
8 Industry recognized customer base.

To help the business to remain focused on the achievement of the eight critical factors a set of guiding principles were defined (Figure 4.6).

Through a series of senior management workshops the eight factors were further developed to identify their key activities and performance measures. These performance indicators were then arranged into a balanced set of measures and appropriate targets for the coming year defined for each (Figure 4.7).

To assist in the development of these key activities the senior management team used a CSF planning document (Figure 4.8). One planning sheet is detailed for every measure for each CSF.

The CSF itself defines **what** must be achieved. In the example Resources 'must have skilled, motivated and flexible people'. This is linked to one of the performance measures (KPI) and an appropriate description of what that KPI represents is provided. In addition the current performance is given, where applicable, together with its target.

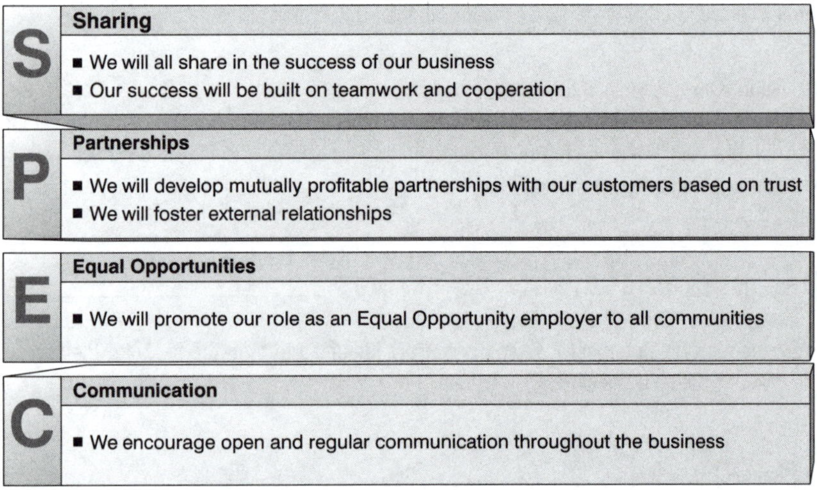

S **Sharing**
  ■ We will all share in the success of our business
  ■ Our success will be built on teamwork and cooperation

P **Partnerships**
  ■ We will develop mutually profitable partnerships with our customers based on trust
  ■ We will foster external relationships

E **Equal Opportunities**
  ■ We will promote our role as an Equal Opportunity employer to all communities

C **Communication**
  ■ We encourage open and regular communication throughout the business

■ **Figure 4.6** Guiding principles

The bottom section of the document identifies **how** the KPI will be achieved. By doing this the business identifies a lower level of specific actions that should help to achieve the specific success factor. Each of these actions is allocated an owner and a date for completion.

As the '**whats**' are cascaded down to the '**hows**', responsibility is likely to be cascaded down to the most appropriate level within the organization. For example, the KPI is owned by a member of the senior management team, as are the four identified actions. However, these four actions, if cascaded to the next level of detail, would become the '**whats**' that would require their own series of '**hows**' to be defined and probably be owned functionally by a department or business unit.

Implementation of this process allowed for a link to be created from the highest level of critical success factor right down to individual or team objectives and goals. Furthermore, it provides a means of feedback through the chain to the CSFs and enables performance to be monitored and aligned to corporate objectives.

Achievements

The project helped London Operations to dramatically enhance its understanding of the business and its performance and identified opportunities to reduce costs by nearly 20 percent, while maintaining levels of customer satisfaction and market share. Furthermore, the approach has led to these changes being locked into the future working of the business. Many attitudes have changed and barriers broken down to secure the future of Resources Ltd. The schedule

Customer	**Finance**
Customer Satisfaction Survey Results	Return on Sales
No. Of Customer Complaints Resolved	Return on Capital Employed
Post Contract Review Results	Market Share – external and internal
Commendations Resulting in Awards	Variance From Cash Flow Budget
Strength of Customer Relationship	Performance to Investment Budget
Market Profile	
Staff Utilisation	Staff Satisfaction Survey Results
Facilities Utilisation	% of Staff Within Appraisal Process
Quotation Turn-round Time	Turnover of Key Talent
Invoices Issued Within 5 Days	Number of Identified Leaders Within a Leadership Development Programme
	Staff Turnover
	Absence Rate
Processes	**People**

Figure 4.7 Balanced Scorecard of Measures

CSF 1 – We must have skilled, motivated and flexible people.

Owner – A.N. Other.

No.	KPI	Definition	Current Measure	Target	Due Date
1	Staff satisfaction survey results.	Overall staff rating against satisfaction/motivation index.	N/A.	50%	June 2001.

The activities that will take place to address the identified performance gaps

No.	KPI No.	Activity	Resp	Due	Driver Measure
1	1	Compile London Operations specific staff survey which asks staff to identify 3 priority issues that motivate/demotivate them. Ask staff to rate how well the business delivers on these issues.	AN Other	May	Staff satisfaction.
2	1	Identify action plan to address these issues. Identify those issues within Studios, OB's and Post Production control and those outside our direct control.	AN Other	July	Staff satisfaction.
3	1	Communicate survey results and intended actions.	AN Other	August	Staff satisfaction.
4	1	Implement Communication plan to improve staff understanding about the direction of Res Ltd, recognizes peoples value, encourages a sense of identity.	AN Other	July	
	1	Agree date for follow up audit.	AN Other	August	

Figure 4.8 CSF planning document

for implementation was less than 18 months and the transformation in operations has made Resources Ltd an attractive commercial proposition.

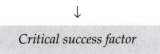

Mission

Two of the statements in a well-known management consultancy's mission are:

Gain and maintain a position as Europe's foremost management consultancy in the development of organizations through management of change.

Provide the consultancy, training and facilitation necessary to assist with making continuous improvement an integral part of our customers' business strategy.

↓

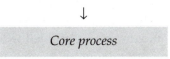

Critical success factor

One of the CSFs which clearly relates to this is:

We need a high level of external awareness of our capabilities.

↓

Core process

One of the core processes which clearly must be done particularly well to achieve this CSF is to:

Promote, advertise, and communicate the company's business capability.

↓

Subprocess

One of the subprocesses which results from a breakdown of this core process is:

Prepare the company's information pack.

↓

Activity

One of the activities which contributes to this subprocess is:

> Prepare **one** of the subject booklets, i.e. 'Strategic Process Management'.

↓

One of the tasks which contributes to this is:

> Write the detailed leaflet for a particular aspect of process management services, e.g. business process re-design.

Individuals, tasks and teams

Having broken down the processes into subprocesses, activities and tasks in this way, it is now possible to link this with the Adair model of action-centered leadership and teamwork (see Chapter 15).

The tasks are clearly performed, at least initially, by individuals. For example, somebody has to sit down and draft out the first version of a leaflet. There has to be an understanding by the individual of the task and its position in the hierarchy of processes. Once the initial task has been performed, the results must be checked against the activity of co-ordinating the promotional booklet. This clearly brings in the team, and there must be interfaces between the needs of the *tasks*, the *individuals* who performed them and the *team* concerned with the *activities*.

Performance measurement and metrics

Once the processes have been analyzed in this way, it should be possible to develop metrics for measuring the performance of the processes, subprocesses, activities, and tasks. These must be meaningful in terms of the *inputs* and *outputs* of the processes, and in terms of the *customers* and of *suppliers* to the processes (Figure 4.5).

At first thought, this form of measurement can seem difficult for processes such as preparing a sales brochure or writing leaflets promoting consultancy, but, if we think carefully about the *customers* for the leaflet-writing tasks, these will include the *internal* ones, i.e. the consultants, and we can ask whether the output meets their requirements. Does it really say what the service is about, what its objectives are and what a project might be? Clearly, one of the 'measures' of the leaflet-writing task could be the number of typing errors in it, but is this a *key* measure of the performance of the process? Only in the context of office management is this an important measure. Elsewhere it is not.

The same goes for the *activity* of preparing the subject booklet. Does it tell the 'customer' what process management is and how the consultancy can help? For the *subprocess* of preparing the company brochure, does it inform people about the company and does it bring in enquiries from which customers can be developed? Clearly, some of these measures require *external market research*, and some of them *internal research*. The main point is that metrics must be developed and used to reflect the *true performance* of the processes, subprocesses, activities and tasks. These must involve good contact with external and internal customers of the processes. The metrics may be quoted as *ratios*, e.g. numbers of customers derived per number of brochures distributed. Good data collection, record keeping, and analysis are clearly required.

It is hoped that this illustration will help the reader to:

- Understand the breakdown of processes into subprocesses, activities and tasks.
- Understand the links between the process breakdowns and the task, individual and team concepts.
- Link the hierarchy of processes with the hierarchy of quality teams.
- Begin to assemble a cascade of flowcharts representing the process breakdowns, which can form the basis of the quality management system and communicate what is going on throughout the business.
- Understand the way in which metrics may be developed to measure the true performance of the processes, and their links with the customers, suppliers, inputs and outputs of the processes.

The changed patterns of co-ordination, driven by the process maps, should increase collaboration and information sharing. Clearly the senior and middle managers need to provide the right support. Once employees, at all levels, identify what kinds of new skills are needed, they will ask for the formal training programs in order to develop those skills further. This is a key area, because teamwork around the processes will ask more of employees, so they will need increasing support from their managers.

This has been called 'just-in-time' training, which describes very well the nature of the training process required. This contrasts with the blanket or carpet bombing training associated with many unsuccessful change programs, which targets competencies or skills, but does not change the organization's patterns of collaboration and co-ordination.

Case study ■■■

Quality improvement is embedded in BT Retail's strategy. BT Retail's strategic goals are to 'Delight our customers, motivate our people and increase shareholder value'. This is being achieved by setting seven very clear strategic objectives for: improving the customer experience; optimizing transaction economics; achieving operational excellence; reducing the cost of failure; defending core revenues; creating new revenue streams; and creating the place to work for our employees. The leadership team has placed considerable emphasis on communication of the strategy and objectives to both employees and the City.

The strategy is deployed through two key mechanisms. All senior managers have a balanced scorecard, which reflects their key objectives, and all managers have objectives aligned to these scorecards. BT Retail has also established a clear set of key change programs which drives the major change required to deliver BT's longer-term strategic objectives. Delivery of key programs is also included in senior manager's scorecards and the benefits from the programs forming an integral part of the budget process.

Step 6 Ensure process and people alignment through a policy deployment or goal translation process

One of the keys to integrating quality into the business strategy is a formal 'goal translation' or 'policy deployment' process. If the mission and measurable goals have been analyzed in terms of critical success factors and core processes, then the organization has begun to understand how to achieve the mission. Goal translation ensures that the 'whats' are converted into 'hows', passing this right down through the organization, using a quality function deployment (QFD) type process, Figure 4.9 – See Chapter 6. The method is best described by an example.

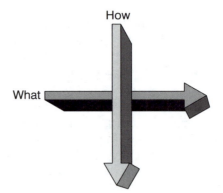

■ **Figure 4.9**
The goal translation process

At the top of an organization in the petro-chemical process industries, five measurable goals have been identified. These are listed under the heading 'What' in Figure 4.10. The top team listens to the 'voice of the customer' and tries to understand *how* these business goals will be achieved. They realize that product consistency, on-time delivery, and speed or quality of response are the keys. These CSFs are placed along the first row of the matrix and the relationships between the *what* and the *how* estimated as strong, medium or weak. A measurement target for the hows is then specified.

The *how* becomes the *what* for the next layer of management. The top team share their goals with their immediate reports and ask them to determine their *hows*, indicate the relationship and set measurement targets. This continues down the organization through a 'catch-ball' process until the senior management goals have been translated through the *what/how* → *what/how* → *what/how* matrices to the individual tasks within the organization. This provides a good discipline to support the breakdown and understanding of the business process mapping described in Chapter 10.

A successful approach to policy/goal deployment and strategic planning in an organization with several business units or divisions is that mission, CSFs with KPOs and targets, and core processes are determined at the corporate level, typically by the board. While there needs to be some flexibility about exactly how this is translated into the business units, typically it would be expected that the process is repeated with the senior team in each business unit or division. Each business unit head should be part of the top team that did the work at the corporate level, and each of them would develop a version of the same process with which they feel comfortable.

Each business unit would then follow a similar series of steps to develop their own mission (perhaps) and certainly their own CSFs and KPOs with targets. A matrix for each business unit showing the impact of achieving the business unit CSFs on the corporate CSFs would be developed. In other words, the first deployment of the corporate 'whats' CSFs is into the 'hows' – the business unit CSFs (Figure 4.2).

If each business unit follows the same pattern, the business unit teams will each identify unit CSFs, KPOs with targets and core processes, which are interlinked with the ones at corporate level. Indeed the core processes at corporate and business unit level may be the same, with any specific additional processes identified at business unit level to catch the flavor and business needs of the unit. It cannot be overemphasized how much ownership there needs to be at the business unit management level for this to work properly.

What

	Generic product consistency	On-time delivery	Speed/quality of response/service
Be the preferred lowest cost supplier by end Yr 1	◎	◎	◎
Double sales by Yr 2	◎	◎	◎
Achieve 12% market position in Europe by Yr 1	◎	◎	◎
Increase value added over GP consistent with our capabilities	◎	◎	◎
Achieve the highest organizational productivity in the industry	◎	◎	◎
Attain at least 12% integrated trend line return			
Target/measurement	All reactors on statistical control. '0' defects from post reactor to customer	100% delivery on time, on customer-agreed date for target customers	Leadership in customer satisfaction survey

Relationship ◎ Strong ○ Medium △ Weak

How

What

Generic product consistency

On-time delivery

Speed/quality of response/service

Targets/measures

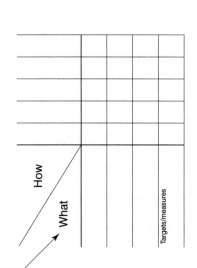

How

What

Targets/measures

Figure 4.10 The goal translation process in practice

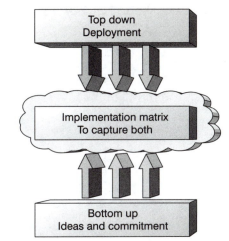

Implementation: top-down and
bottom-up approach

With regard to core processes, each business unit or function will begin to map these at the top level. This will lead to an understanding of the purpose, scope, inputs, control, and resources for each process and provide an understanding of how the subprocesses are linked together. Flowcharting showing connections with procedures will then allow specific areas for improvement to be identified so that the continuous improvement, 'bottom-up' activities can be deployed, and benefit derived from the process improvement training to be provided (Figure 4.11).

It is important to get clarity at the corporate and business unit management levels about the whats/hows relationships, but the ethos of the whole process is one of involvement and participation in goal/target setting, based on a good understanding of processes – so that it is known and agreed what can be achieved and what needs measuring and targeting at the business unit level.

Senior management may find it useful to monitor performance against the CSFs, KPOs and targets, and to keep track of processes using a reporting matrix, perhaps at their monthly meetings. A simplified version of this, developed for use in a small company, is shown in Figure 4.12. The frequency of reporting for each CSF, KPO, and process can be determined in a business planning calendar.

As previously described, in a larger organization, this approach may be used to deploy the goals from the corporate level through divisions to site/departmental level, Figure 4.13. This form of implementation should ensure the top-down *and* bottom-up approach to the deployment of policies and goals.

Core processes

	CSFs: We must have	Measures	Year targets	Target CSF owner	Manage people	Develop products	Develop new business	Manage our accounts	Manage financials	Manage int. systems	Conduct research
	Satisfactory financial and non-financial performance	Sales volume. Profit. Costs versus plan. Shareholder return Associate/employee utilisation figures	Turnover £2m. Profit £200k. Return for shareholders. Days/month per person		X		X	X	X	X	X
	A growing base of satisfied customers	Sales/customer Complaints/recommendations Customer satisfaction	>£200k=1 client. £100k-£200k=5 clients. £50k-£100k=6 clients <£50k=12 clients		X	X	X	X		X	
	A sufficient number of committed and competent people	No. of employed staff/associates Gaps in competency matrix. Appraisal results Perceptions of associates and staff	15 employed staff 10 associates including 6 new by end of year		X				X	X	
	Research projects properly completed and published	Proportion completed on time, in budget with customers satisfied. Number of publications per project	3 completed on time, in budget with satisfied customers					X			X
	** = Priority for improvement				**		**	**			
	Process owner										
	Process performance										
	Measures and targets										

Figure 4.12 CSF/core process reporting matrix

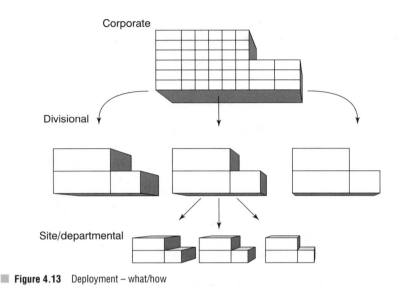

Corporate

Divisional

Site/departmental

Figure 4.13 Deployment – what/how

Deliverables

The deliverables after one planning cycle of this process in a business will be:

1 An agreed framework for policy/goal deployment through the business.
2 Agreed mission statement for the business and, if required, for the business units/division.
3 Agreed critical success factors (CSFs) with ownership at top team level for the business and business units/divisions.
4 Agreed key performance outcomes (KPOs) with targets throughout the business.
5 Agreed core business processes, with sponsorship at top team level.
6 A corporate CSF/business unit CSF matrix showing the impacts and the first 'whats/hows' deployment.
7 A what/how (CSF/process) matrix approach for deploying the goals into the organization through process definition, understanding, and measured improvement at the business unit level.
8 Focused business improvement, linked back to the CSFs, with prioritized action plans and involvement of employees.

Strategic and operational planning

Changing the culture of an organization to incorporate a sustainable ethos of continuous improvement and responsive business planning will come about only as the result of a carefully planned and managed process. Clearly many factors are involved, including:

- identifying strategic issues to be considered by the senior management team;
- balancing the present needs of the business against the vital needs of the future;
- concentrating finite resources on important things;
- providing awareness of impending changes in the business environment in order to adapt more rapidly and more appropriately.

Strategic planning is the continuous process by which any organization will describe its destination, assess barriers standing in the way of reaching that destination, and select approaches for dealing with those barriers and moving forward. Of course the real contributors to a successful strategic plan are the participants.

The strategic and operational planning process described in this chapter will:

- Provide the senior management team with the means to manage the organization and strengths and weaknesses through the change process.
- Allow the senior management team members to have a clear understanding and to achieve agreement on the strategic direction, including vision and mission.
- Identify and document those factors critical to success (CSFs) in achieving the strategic direction and the means by which success will be measured (KPOs) and targeted.
- Identify, document and encourage ownership of the core processes that drive the business.
- Reach agreement on the priority processes for action by process improvement teams, incorporating current initiatives into an overall, cohesive framework.
- Provide a framework for successfully deploying all goals and objectives through all organizational levels through a two-way 'catch-ball' process.
- Provide a mechanism by which goals and objectives are monitored, reviewed, and appropriate actions taken, at appropriate frequencies throughout the operational year.
- Transfer the skills and knowledge necessary to sustain the process.

The components outlined above will provide a means of effectively deploying a common vision and strategy throughout the organization. They will also allow for the incorporation of all change projects, as well as 'business as usual' activities, into a common framework which will form the basis of detailed operating plans.

■ The development of policies and strategies

Let us assume that a management team are to develop the policies and strategies based on stakeholder needs and the organization's capabilities, and that it wants to ensure these are communicated, implemented, reviewed and updated. Clearly a detailed review is required of the major stakeholders' needs, the performance of competitors, the state of the market and industry/sector conditions. This can then form the basis of top level goals, planning activities and setting of objectives and targets.

How individual organizations do this varies greatly, of course, and some of this variation can be seen in the many organizations that have adopted this approach. However, some common themes emerge under six headings:

Customer/market

- Data collected, analyzed and understood in terms of where the organization will operate.
- Customers' needs and expectations understood, now and in the future.
- Developments anticipated and understood, including those of competitors and their performance.
- The organization's performance in the market place known.
- Benchmarking against best in class organizations.

Shareholders/major stakeholders

- Shareholders'/major stakeholders' needs and ideas understood.
- Appropriate economic trends/indicators and their impact analyzed and understood.
- Policies and strategies appropriate to shareholder/stakeholder needs and expectations developed.
- Needs and expectations balanced.
- Various scenarios and plans to manage risks developed.

People

- The needs and expectations of the employees understood.
- Data collected, analyzed and understood in terms of the internal performance of the organization.
- Output from learning activities understood.
- Everyone appropriately informed about the policies and strategies.

Processes

- A key process framework to deliver the policies and strategies designed, understood and implemented.
- Key process owners identified.
- Each key process and its major stakeholders defined.
- Key process framework reviewed periodically in terms of its suitability to deliver to organization's requirements.

Partners/resources

- Appropriate technology understood.
- Impact of new technologies analyzed.
- Needs and expectations of partners understood.
- Policies and strategies aligned with those of partners.
- Financial strategies developed.
- Appropriate buildings, equipment and materials identified/sourced.

Society

- Society, legal and environmental issues understood.
- Environment and corporate responsibility policies developed.

The whole field of business policy, strategy development and planning is huge and there are many excellent texts on the subject. It is outside the scope of this book to cover this area in detail, of course, but one of the most widely used and comprehensive texts is *Exploring Corporate Strategy – text and cases*, 6th Edn by Gerry Johnson and Kevan Scholes. This covers strategic positioning and choices, and strategy implementation at all levels. The author is proud to have a case study included in this text – the ST Micro electronics case.

Case study ■■■ ▬▬▬▬▬▬▬▬▬▬▬▬▬▬▬

■
■ Policy deployment at STM

Policy deployment (PD) is the primary method used in STM to make TQM 'the way we manage' rather than something added to operational management. In order to make it effective, STM have simplified the approach, combining as many existing initiatives as possible, to leave only one set of key improvement goals deriving from both internal and external identified needs. In this process the management of STM also provided a mechanism for 'real time' visual follow-up of breakthrough priorities to support very rapid progress.

In STM policy deployment is regarded as:

- The 'backbone' of TQM.
- The way to translate the corporate vision, objectives and strategies into concrete specific goals, plans and actions at the operative level.
- A means to focus everyone's contributions in support of employee empowerment.
- The mechanism for jointly identifying objectives and the actions required to obtain the expected results.
- A vehicle to ensure that the corporate quality, service and cost goals are given superordinate importance in annual operational planning and performance evaluation.
- The method to integrate the entire organization's daily priority activities with its long-term goals.
- A process to focus attention on managing STM's future, rather than the past.

A policy deployment manual, addressed to all managers at any level of ST Microelectronics, was developed as a methodological and operative user guide for those charged with planning and achieving significant improvement goals. Examples, detailed explanations, and descriptions of tools/forms were included in the manual.

Policy deployment operates at two levels: continuous focused improvement and strategic breakthrough – referred to as Level 1 and Level 2. The yearly plan is designed by assembling the budget and improvement plan, but also taking into account the investment plan.

All these elements must be consistent and coherent. Current year business result goals are defined in the budget and the underlying operations and capability improvement goals have to be approached using policy deployment. Among all the improvement goals, a very few (one to three per year) are then selected for a more intensive management. These are the breakthrough goals and must be managed using special attention and techniques. Policy deployment goals have to be consistent with long-term policies, and finally, everything must be consistent with and must be supported by the investment plan.

Continuously improving performance and capabilities, and especially achieving 'break-throughs', i.e. dramatic improvements in short times, was the main task that each manager was asked to face and carry out in his/her activities. Once the importance of achieving dramatic goals was clear, the problem arose of how to identify and prioritize them. To assist, STM fixed four long-term policies (broad and generic objectives):

- become number one in service;
- be among the top three suppliers in quality;
- have world class manufacturing capabilities;
- become a leader in TQM in the Western business world.

These long-term policies reflected the need to improve **strategic capabilities**. They were implemented progressively by achieving sequential sets of shorter-term goals focused on **operational capabilities, operational performance**, and urgent requirements, as illustrated in Figure 4.14. STM recognized that a successful enterprise ensures consistency between its short-term efforts and long-term goals.

URGENCIES	SHORT TERM	MEDIUM TERM	LONG TERM
• Removal of problems	• Improvement of operational performance	• Improvement of operational capabilities	• Improvement of strategic capabilities
• Catching of opportunities Examples:	Examples:	Examples:	Examples:
– Quality problem	– JIT	– Concurrent engineering	– Human Resources
– Process breakdown	– Cycle time	– TPM	capabilities
– Customer complaint	– Inventory turns	– Logistics	– Technological
– Important opportunity	– Defectiveness	– Self-managing teams	breakthroughs
	– Yield	– Planning/Scheduling	– Multi-processing
	– Productivity		– Mixed design know-how
			– Time to market

■ **Figure 4.14** STM 1 Example of objectives by different horizon

The yearly plan comprises all the goals and the performances the company have to reach during the year. Goals related to sales volume, profit and loss, inventories, standard costs, expenses, etc. are generally managed by management control through the budget. In order to be more and more competitive, however, more challenging goals have to be identified each year and these goals – the ones that constitute the improvement plan – need 'special management' through a specific approach. This approach is policy deployment, in which a policy can be fully defined as the combination of goals/targets and means.

Policy deployment applies both to 'What' goals, i.e. mainly results oriented, and 'How' goals that are more related to operational, technological, organizational and behavioral aspects, mainly process oriented.

■ Reference

Johnson, G. and Scholes, K. *Exploring Corporate Strategy, Text and Cases* (6th edn), Prentice-Hall, London, 2002

Chapter highlights
■ ■ ■

Integrating quality into the policy and strategy

- Policy and strategy is concerned with how the organization implements its mission and vision in a clear stakeholder-focused strategy supported by relevant policies, plans, objectives, targets and processes.
- Senior management may begin the task of alignment through six steps:
  - develop a shared vision and mission;
  - develop the critical success factors;
  - define the key performance outcomes (balanced scorecard);
  - understand the core process and gain ownership;

- break down the core processes into subprocesses, activities and tasks;
- ensure process and people alignment through a policy deployment or goal translation process.

■ The deliverables after one planning cycle will include: an agreed policy/goal deployment framework; agreed mission statements; agreed CSFs and owners; agreed KPOs and targets; agreed core processes and sponsors; whats/hows deployment matrices; focused business improvement plans.

The development of policies and strategies

■ The development of policies and strategies requires a detailed review of the major stakeholders' needs, the performance of competitors, the market/industry/sector conditions to form the basis of top level goals, planning activities and setting of objectives and targets.

■ The common themes for planning strategies may be considered under the headings of customers/market, shareholders/major stakeholders, people, processes, partners, resources and society.

■ The field of policy and strategy development is huge and the text by Johnson and Scholes is recommended reading.

Chapter 5

■ Partnerships and resources

■ Partnering

In recent years business, technologies and economies have developed in such a way that organizations recognize the increasing needs to establish mutually beneficial relationships with other organizations, often called 'partners'. The philosophies behind quality management and the excellence models support the establishment of partnerships and lay down principles and guidelines for them.

How companies in the private sector plan and manage their partnerships can mean the difference between success and failure for it is extremely rare to find companies which can sustain a credible business operation now without a network of co-operation between individuals and organizations or parts of them. This extends the internal customer supplier relationship ideas into the supply chain of a company making sure that all the necessary materials, services, equipment, information skills and experience are available in totality to deliver the right products or services to the end customer. Gone are the days, hopefully, of conflict and dispute between a customer and their suppliers. An efficient supply chain process, built on strong confident partnerships, will create high levels of people satisfaction, customer satisfaction and support and, in turn, good business results.

Similarly in public sector organizations, where the involvement of the private sector has been a key development feature in recent times, there

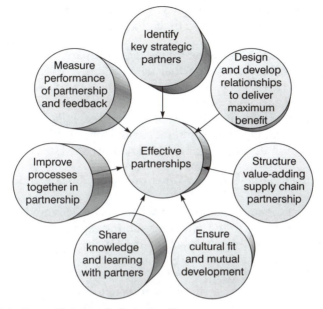

Figure 5.1   The contributors to effective partnerships

is the need to recognize and build strong external partnerships. The bidding and tendering processes of such organizations by necessity will always be different to those in the private sector. Nevertheless, government departments, the health service, education, police, the armed forces, tax collection bodies and local authorities need to understand, develop and deliver strong external partnerships if they are to achieve the performance levels and targets that the general public in any country desires.

How an organization plans and manages the external partnerships must be in line with its overall policies and strategies, being designed and developed to support the effective operation of its processes (Figure 5.1). A key part of this, of course, is identifying with whom those key strategic partnerships will be formed. Whether it is working with key suppliers to deliver materials or components to the required quality, plan (lead times) and costs, or the supply of information technology, transport, broadcasting or consultancy services, the quality of partnerships has been recognized throughout the world as a key success criterion.

There are various ways of ensuring the partnership processes work well for an organization. These range from the use of quality management system audits and reviews, through certificates of competence, to performance reviews and joint action plans. Another key aspect of successful partnerships is good communications and exchange of information. This supports learning between two organizations and often leads to innovative solutions to problems that have remained unsolved in the separate organizations, prior to their close collaboration.

One example of this in a medium-sized company involved a close working relationship with a partner who had been responsible for producing the written content of the company's work-based learning materials. The customer of this service had shared their ideas for the future in order to exploit areas of possible mutual benefit. Arising from this, and other similar partnership relationships in the company, was improved customer satisfaction and refined processes.

When establishing partnerships, attention should be given to:

■ maximizing the understanding of what is to be delivered by the partnership – the needs of the customer and the capability of the supplier must match perfectly if satisfaction and loyalty are to be the result;
■ understanding what represents value for money – getting the commercial relationship right;
■ understanding the respective roles and ensuring an appropriate allocation of responsibilities – to the party best able to manage them;
■ working in a supportive, constructive and a team-based relationship;
■ having solid programs of work, comprising agreed plans, timetables, targets, key milestones and decision points;
■ structuring the resolution of complaints, concerns or disputes rapidly and at the lowest practical level;
■ enabling the incorporation of knowledge transfer and making sure this adds value;
■ developing a stronger and stronger working relationship geared to delivering better and better products or services to the end customer – based on continuous improvement principles.

■ The role of purchasing in partnerships

A company selling wooden products had a very simple purchasing policy: it bought the cheapest wood it could find anywhere in the world. Down in the workshops they were scrapping doors and window frames as if they were going out of fashion – warping, knots in the wood, 'flaking', cracking, splits, etc. When the purchasing manager was informed, he visited the workshops and explained to the supervisors how cheap the wood was and instructed them to 'do the best you can – the customer will never notice'. On challenging this policy, the author was told that it would not change until someone proved to the purchasing manager, in a quantitative way, that the policy was wrong. That year the company 'lost' $1 million worth of wood – in scrap and rework. You can go out of business waiting for such proof.

Very few organizations are self-contained to the extent that their products and services are all generated at one location, from basic materials. Some materials or services are usually purchased from outside organizations, and the primary objective of purchasing is to obtain the correct equipment, materials, and services in the right quantity, of the right quality, from the right origin, at the right time and cost. Purchasing can also play a vital role as the organization's 'window-on-the-world', providing information on any new products, processes, materials and services that become available. It can also advise on probable prices, deliveries, and performance of products under consideration by the research, design and development functions. In other words it should support any partnership in the supply chain.

Although purchasing is clearly an important area of managerial activity, it is often neglected by both manufacturing and service industries. The separation of purchasing from selling has, however, been removed in many large retail organizations, which have recognized that the purchasing or 'merchandising' must be responsible for the whole 'product line' – its selection, quality, specification, delivery, price, acceptability, and reliability. If any part of this chain is wrong, customer satisfaction will suffer. This concept is clearly very appropriate in retailing, where transformation activities on the product itself, between purchase and sale, are small or zero, but it shows the need to include market information in the buying decision processes in all organizations.

The purchasing or procurement system should be documented and include:

1 Assigning responsibilities for and within the purchasing procurement function.
2 Defining the manner in which suppliers are selected, to ensure that they are continually capable of supplying the requirements.
3 Specifying the purchasing documentation – written orders, specifications, etc. – required in any modern procurement activity.

Historically many organizations, particularly in the manufacturing industries, have operated an inspection-oriented quality system for bought-in parts and materials. Such an approach has many disadvantages. It is expensive, imprecise, and impossible to apply evenly across all material and parts, which all lead to variability in the degree of appraisal. Many organizations, such as Ford, have found that survival and future growth in both volume and variety demand that changes be made to this approach.

The prohibitive cost of holding large stocks of components and raw materials also pushed forward the 'just-in-time' (JIT) concept. As this requires that suppliers make frequent, on time, deliveries of small quantities of material, parts, components, etc., often straight to the point of use, in order that stocks can be kept to a minimum, the approach requires an effective supplier network – one producing goods and services that can be trusted to conform to the real requirements with a high degree of confidence.

Commitment and involvement

The process of improving supplier performance is complex and clearly relies very heavily on securing real commitment from the senior management of both organizations to a partnership. This may be aided by presentations made to groups of directors of the suppliers brought together to share the realization of the importance of their organizations' performance in the quality chains. The synergy derived from members of the partnership meeting together, being educated, and discussing mutual problems, will be tremendous. If this can be achieved, within the constraints of business and technical confidentiality, it is always a better approach than the arm's-length method of purchasing still used by many companies.

The author recalls the benefits that accrued from bringing together suppliers of a photocopier, paper, and ring binders to explain to them the way their inputs were used to generate executive development-course materials and how they in turn were used during the courses themselves. The suppliers were able to understand the business in which their customers were engaged, and play their part in the whole process. A supplier of goods *or* services that has received such attention, education, and training, and understands the role its inputs play, is less likely knowingly to offer non-conforming materials and services, and more likely to alert customers to potential problems.

Policy

One of the first things to communicate to any external supplier is the purchasing organization's policy on quality of incoming goods and services. This can include such statements as:

■ It is the policy of this company to ensure that the quality of all purchased materials and services meets its requirements.
■ Suppliers who incorporate a quality management system into their operations will be selected. This system should be designed, implemented and operated according to the International Standards Organization (ISO) 9000 series (see Chapter 12).

- Suppliers who incorporate statistical process control (SPC) and continuous improvement methods into their operations (see Chapter 13) will be selected.
- Routine inspection, checking, measurement and testing of incoming goods and services will *not* be carried out by this company on receipt.
- Suppliers will be audited and their operating procedures, systems, and SPC methods will be reviewed periodically to ensure a never-ending improvement approach.
- It is the policy of this company to pursue uniformity of supply, and to encourage suppliers to strive for continual reduction in variability (this may well lead to the narrowing of specification ranges).

Quality management system assessment certification

Many customers examine their suppliers' quality management systems themselves, operating a second party assessment scheme (see Chapters 8 and 12). This can lead to high costs and duplication of activity, for both the customer and supplier. If a qualified, independent third party is used instead to carry out the assessment, attention may be focused by the customer on any special needs and in developing closer partnerships with suppliers. Visits and dialog across the customer/supplier interface are a necessity for the true requirements to be met, and for future growth of the whole business chain. Visits should be concentrated, however, on improving understanding and capability, rather than on close scrutiny of operating procedures, which is best left to experts, including those within the supplier organizations charged with carrying out internal system audits and reviews.

Supplier approval and single sourcing

Some organizations have as an objective to obtain at least two 'approved' suppliers for each material or service purchased on a regular basis. It may be argued, however, that single sourcing – the development of an extremely close relationship with just one supplier for each item or service – encourages greater commitment and a true partnership to be created. This clearly needs careful management, but it is a sound policy, based on the premise that it is better to work together with a supplier to remove problems, improve capability, and generate a mutual understanding of the *real* requirements, than to hop from one supplier to another and thereby experience a different set of problems each time.

To become an 'approved supplier' or partner, it is usually necessary to pass through a number of stages:

1 *Technical approval* – largely to determine if the product/service meets the technical requirements. This stage should be directed at agreeing a specification that is consistent with the supplier's process capability.
2 *Conditional approval* – at this stage it is known that the product/service meets the requirements, following customer in-process trials, and there is a good commercial reason for purchase.
3 *Full approval* – when all the requirements are being met, including those concerning the operation of the appropriate management systems, and the commercial arrangements have been agreed.

It is, of course, normal for organizations to carry out audits of their suppliers and to review periodically their systems and process capabilities. This is useful in developing the partnership and ensuring that the customer needs continue to be met.

■ Just-in-time (JIT) management

There are so many organizations throughout the world now that are practicing just-in-time (JIT) management principles that the probability of encountering it is very high. JIT, like many modern management concepts, is credited to the Japanese, who developed and began to use it in the late 1950s. It took approximately 20 years for JIT methods to reach Western hard goods industries and a further ten years before businesses realized the generality of the concepts.

Basically JIT is a program directed towards ensuring that the right quantities are purchased or produced at the right time, and that there is no waste. Anyone who perceives it purely as a material-control system, however, is bound to fail with JIT. JIT fits well under the quality management umbrella, for many of the ideas and techniques are very similar and, moreover, JIT will not work without good management of quality. Writing down a definition of JIT for all types of organization is extremely difficult, because the range of products, services and organization structures leads to different impressions of the nature and scope of JIT. It is essentially:

■ A series of operating concepts that allows systematic identification of operational problems.
■ A series of technology-based tools for correcting problems following their identification.

An important outcome of JIT is a disciplined program for improving productivity and reducing waste. This program leads to cost-effective production or operation and delivery of only the required goods or services, in the correct quantity, at the right time and place. This is achieved with the minimum amount of resources – facilities, equipment, materials, and people. The successful operation of JIT is dependent upon a balance between the suppliers' flexibility and the users' stability, and of course requires total management and employee commitment and teamwork.

Aims of JIT

The fundamental aims of JIT are to produce or operate to meet the requirements of the customer exactly, without waste, immediately on demand. In some manufacturing companies JIT has been introduced as 'continuous flow production', which describes very well the objective of achieving conversion of purchased material or service receipt to delivery, i.e. from supplier to customer. If this extends into the supplier and customer chains, all operating with JIT a perfectly continuous flow of material, information or service will be achieved. JIT may be used in non-manufacturing, in administration areas, for example, by using external standards as reference points.

The JIT concepts identify operational problems by tracking the following:

1 *Material movements* – when material stops, diverts or turns backwards, these always correlate with an aberration in the 'process'.
2 *Material accumulations* – these are there as a buffer for problems, excessive variability, etc., like water covering up 'rocks'.
3 *Process flexibility* – an absolute necessity for flexible operation and design.
4 *Value-added efforts* – where much of what is done does not add value, the customer will not pay for it.

The operation of JIT

The tools to carry out the monitoring required are familiar quality and operations management methods, such as:

- Flowcharting.
- Process study and analysis.
- Preventive maintenance.
- Plant layout methods.
- Standardized design.
- Statistical process control.
- Value analysis and value engineering.

But some techniques are more directly associated with the operation of JIT systems:

1 Batch or lot size reduction.
2 Flexible workforce.
3 Kanban or cards with material visibility.
4 Mistake-proofing.
5 Pull-scheduling.
6 Set-up time reduction.
7 Standardized containers.

In addition, joint development programs with suppliers and customers will be required to establish long-term relationships and develop single sourcing arrangements that provide frequent deliveries in small quantities. These can only be achieved through close communications and meaningful certified quality.

There is clear evidence that JIT has been an important component of business success in the Far East and that it is used by Japanese companies operating in the West. Many European and American companies that have adopted JIT have made spectacular improvements in performance. These include:

■ Increased flexibility (particularly of the workforce).
■ Reduction in stock and work-in-progress, and the space it occupies.
■ Simplification of products and processes.

These programs are **always** characterized by a real commitment to continuous improvement. Organizations have been rewarded, however, by the low cost, low risk aspects of implementation provided a sensible attitude prevails. The golden rule is to never remove resources – such as stock – before the organization is ready and able to correct the problems that will be exposed by doing so. Reduction of the water level to reveal the rocks, so that they may be demolished, is fine, provided that we can quickly get our hands back on the stock while the problem is being corrected.

The Kanban system

Kanban is a Japanese word meaning 'visible record', but in the West it is generally taken to mean a 'card' that signals the need to deliver or produce more parts or components. In manufacturing, various types of records, e.g. job orders or route information, are used for ordering more parts in a *push* type, schedule-based system. In a push system a multi-period master production schedule of future demands is prepared, and a computer explodes this into detailed schedules for producing or purchasing the appropriate parts or materials. The schedules then *push* the production of the part or components out and onward. These systems, when computer-based, were originally called 'Material Requirements Planning' (MRP) but have been extended in many organizations to 'Enterprise Resource Planning' (ERP) systems. The main feature of the Kanban system is that it *pulls* parts and components through the production processes when they are needed. Each material, component or part has its own special container designed to hold a precise, preferably small, quantity. The number of containers for each part is a carefully considered management decision. Only standard containers are used, and they are always filled with the prescribed quantity.

A Kanban system provides parts when they are needed but without guesswork, and therefore without the excess inventory that results from bad guesses. The system will only work well, however, within the context of a JIT system in general, and the reduction of set-up times and lot sizes in particular. A JIT program can succeed without a Kanban-based operation, but Kanban will not function effectively independently of JIT.

Just-in-time in partnerships and the supply chain

The development of long-term partnerships with a few suppliers, rather than short-term ones with many, leads to the concept of *co-producers* in networks of trust providing dependable quality and delivery of goods and services. Each organization in the chain of supply is often encouraged to extend JIT methods to its suppliers. The requirements of JIT mean that suppliers are usually located near the purchaser's premises, delivering small quantities, often several times per day, to match the usage rate. Administration is kept to a minimum and standard quantities in standard containers are usual. The requirement for suppliers to be located near the buying organization, which places those at some distance at a competitive disadvantage, causes lead times to be shorter and deliveries to be more reliable.

It can be argued that JIT purchasing and delivery are suitable mainly for assembly line operations, and less so for certain process and service industries, but the reduction in the inventory and transport costs that it brings should encourage innovations to lead to its widespread adoption. The main point is that there must be recognition of the need to develop closer relationships and to begin the dialog – the sharing of information and problems – that leads to the product or service of the right quality, being delivered in the right quantity, at the right time.

■ Resources

All organizations assemble resources, other than human, to support the effective operation of the processes that hopefully will deliver the strategy. These come in many forms but certainly include financial resources, buildings, equipment, materials, technology, information and knowledge. How these are managed will have a serious effect on the effectiveness, efficiency and quality of any establishment, whether it be in manufacturing, service provision, or the public sector.

Financial resources

Investment is key for the future development and growth of business. The ability to attract investment often determines the strategic direction of commercial enterprises. Similarly the acquisition of funding will affect the ability of public sector organizations in health, education or law establishments to function effectively. The development and implementation of appropriate financial strategies and processes will, therefore, be driven by the financial goals and performance of the business. Focus on, for example, improving earnings before interest and tax (EBIT – a measure of profitability) and economic value added (EVA – a measure of the degree to which the returns generated exceed the costs of financing the assets used) can in a private company be the drivers for linking the strategy to action. The construction of plans for the allocation of financial resources in support of the policies and strategies should lead to the appropriate and significant activities being carried out within the business to deliver the strategy.

Consolidation of these plans, coupled with an iterative review and approval, provides a mechanism of providing the best possible chance for success. Use of a 'balanced scorecard' approach (see Chapters 2, 7 and 8) can help in ensuring that the long-term impact of financial decisions on processes, innovation and customer satisfaction is understood and taken into account. The extent to which financial resources are being used to support strategy needs to be subject to continuous appraisal – this will include evaluating investment in the tangible and non-tangible assets, such as knowledge.

In the public sector, of course, financial policies are often derived from legislation and public accountability. In such situations a 'Director of Finance' often supports the organization's financial system, which is subject to independent review by appropriate authorities. Whatever the system, it is important to ensure alignment with policy and strategy, and that objectives are agreed through incorporation of targets, budgets and accounts, and that the risks to financial resources are managed.

In small- and medium-sized enterprises it is even more important that the financial strategy forms a key part of the strategic planning system, and that key financial goals are identified, deployed, and regularly scrutinized.

Other resources

Many different types of resources are deployed by different types of organizations. Most organizations are established in some sort of building, use equipment and consume materials. In these areas directors and managers must pay attention to:

- utilization of these resources;
- security of the assets;
- maintenance of building and equipment;
- managing material inventories and consumption (see earlier sections on purchasing and JIT);
- waste reduction and recycling;
- environmental aspects, including conservation of non-renewable resources and adverse impact of products and processes.

Technology is a splendid and vital resource in the modern age. Exciting alternative and emerging technologies need to be identified, evaluated and appropriately deployed in the drive towards achieving organizational goals. This will include managing the replacement of 'old technologies' and the innovations which will lead to the adoption of new ones. There are clear links here, of course, with process redesign and re-engineering (see Chapter 11). It is not possible to create the 'paperless' courtroom, for example, without consideration of the processes involved. A murder trial typically involves a million pieces of paper, which are traditionally wheeled into courtrooms on trolleys. To replace this with computer systems and files on disks requires more than just a flick of a switch. The whole end-to-end process of the criminal justice system may come under scrutiny in order to deliver the paper-free trial, and this will involve many agencies in the process – police, prosecution service, courts, probation services, and the legal profession. Their involvement in the end-to-end process design will be vital if technology solutions are to add value and deliver the improvement in justice and reductions in costs that the systems in most countries clearly need.

Most organizations' strategies these days have some if not considerable focus on technology and information systems, as these play significant roles in how they supply products and services to and communicate with customers. They need to identify technology requirements through business planning processes and work with technology partners and IT system providers to exploit technology to their best advantage, improve processes and meet business objectives. Whether this requires a dedicated IT team to develop the strategy will depend on the size and nature of the business but it will always be necessary to assess information resource requirements, provide the right balance, and ensure quality and value for money is provided. This is often a tall order it seems in the provision of IT services! Close effective partnerships that deliver in this area are often essential.

In the piloting and evaluation of new technology the impact on customers and the business itself should be determined. The rollout of any new systems involves people across the organization, and communication cycles need to be used to identify any IT issues and feedback to partners (see Chapter 16). IT support should be designed in collaboration with users to confirm business processes, functionality and the expected utilization and availability. Responsibilities and accountabilities are important here, of course, and in smaller organizations this usually falls on line management.

Like any other resource, knowledge and information need managing and this requires careful consideration in its own right. Chapter 16 on communication, innovation and learning covers this in some detail.

In the design of quality management systems, resource management is an important consideration and is covered by the detail to be found in the ISO 9000:2000 family of standards (see Chapter 12).

Chapter highlights

Partnering

- Organizations increasingly recognize the need to establish mutually beneficial relationships in partnerships. The philosophies behind quality management and the 'Excellence Models' lay down principles and guidelines to support them.
- How partnerships are planned and managed must be in line with overall policies and strategies and support the operation of the processes.
- Establishing effective partnerships requires attention to identification of key strategic partners, design/development of relationships, structured value-adding supply chains, cultural fit and mutual development, shared knowledge and learning, improved processes, measured performance and feedback.

The role of purchasing in partnerships

- The prime objective of purchasing is to obtain the correct equipment, materials, and services in the right quantity, of the right quality, from the right origin, at the right time and cost. Purchasing also acts as a 'window-on-the-world'.
- The separation of purchasing from selling has been eliminated in many retail organizations, to give responsibility for a whole 'product line'. Market information must be included in *any* buying decision.
- The purchasing system should be documented and assign responsibilities, define the means of selecting suppliers, and specify the documentation to be used.
- Improving supplier performance requires commitment, education, a policy, an assessed quality system, and supplier approval from the suppliers' senior management.
- Single sourcing – the close relationship with one supplier for each item or service – depends on technical, conditional, and full stages of approval.

Just-in-time (JIT) management

- JIT fits well under the quality umbrella and is essentially a series of operating concepts that allow the systematic identification of problems, *and* tools for correcting them.
- JIT aims to produce or operate, in accordance with customer requirements, without waste, immediately on demand. Some of the direct techniques associated with JIT are batch or lot size reduction, flexible workforce, Kanban cards, mistake-proofing, set-up time reduction, and standardized containers.
- The development of long-term relationships with a few suppliers or 'co-producers' is an important feature of JIT. These exist in a network of trust to provide quality goods and services.

Resources

- All organizations assemble resources to support operation of the processes and deliver the strategy. These include finance, buildings, equipment, materials, technology, information and knowledge.
- Investment and/or funding is key for future development of all organizations and often determines strategic direction. Financial goals and performance will, therefore, drive strategies and processes. Use of a 'balanced scorecard' approach with continuous appraisal helps in understanding the long-term impact of financial decisions.

- In the management of buildings, equipment and materials, attention must be given to utilization, security, maintenance, inventory, consumption, waste and environmental aspects.
- Technology plays a key role in most organizations and management of existing alternative and emerging technologies need to be identified, evaluated and deployed to achieve organizational goals.
- There are clear links between the introduction of new or the replacement of old technologies and process redesign/engineering. The rollout of any new systems also involves people across the organization and good communications are vital.

Chapter 6

Design for quality

■ Design, innovation and improvement

Products, services and processes may be designed, both to add value to customers and to become more profitable. But leadership and management style is also designed, perhaps through internal communication methods and materials. Almost all areas of organizations have design issues inherent within them.

Design can be used to gain and hold on to competitive edge, save time and effort, deliver innovation, stimulate and motivate staff, simplify complex tasks, delight clients and stakeholders, dishearten competitors, achieve impact in a crowded market, and justify a premium price. Design can be used to take the drudgery out of the mundane and turn it into something inspiring, or simply make money. Design can be considered as a management function, a cultural phenomenon, an art form, a problem-solving process, a discrete activity, an end product or a service.

In the Collins *Cobuild English Language Dictionary*, design is defined as: 'the way in which something has been planned and made, including what it looks like and how well it works'. Using this definition, there is very little of an organization's activities that is not covered by 'planning' or 'making'. Clearly the consideration of what it looks like and how well it works in the eyes of the customer determines the success of products or services in the market place.

All organizations need to update their products, processes and services periodically. In markets such as electronics, audio and visual goods, and office automation, new variants of products are offered frequently – almost like fashion goods. While in other markets the pace of innovation may not be as fast and furious there is no doubt that the rate of change for product, service, technology and process design has accelerated on a broad front.

Innovation entails both the invention and design of radically new products and services, embodying novel ideas, discoveries and advanced technologies, *and* the continuous development and improvement of existing products, services, and processes to enhance their performance and quality. It may also be directed at reducing costs of production or operations throughout the life cycle of the product or service system.

In many organizations innovation is predominantly either technology-led, e.g. in some information and communications industries, or marketing-led, e.g. in some food companies. What is always striking about leading product or service innovators is that their developments are market-led, which is different from marketing-led. The latter means that the marketing function takes the lead in product and service developments. But most leading innovators identify and set out to meet the existing and potential demands profitably and, therefore, are market-led constantly striving to meet the requirements even more effectively through appropriate experimentation.

Everything we experience in or from an organization is the result of a design decision, or lack of one. This applies not just to the tangible things like products and services, but the intangibles too: the systems and processes which affect the generation of products and delivery of services. Design is about combining function and form to achieve fitness for purpose, be it an improvement to a supersonic aircraft, the synthesis of a new drug, a staff incentive scheme or this book.

Once fitness for purpose has been achieved, of course, the goal posts change. Events force a reassessment of needs and expectations and customers want something different. In such a changing world, design is an ongoing activity, dynamic not static, a verb not a noun – *design is a process*.

■ The design process

Commitment in the most senior management helps to build quality throughout the *design process* and to ensure good relationships and communication between various groups and functional areas. Designing customer satisfaction and loyalty into products and services contributes greatly to competitive success. Clearly, it does not guarantee it, because the conformance aspect of quality must be present and the operational processes must be capable of producing to the design. As in the marketing/operations interfaces, it is never acceptable to design a product, service, system or process that the customer wants but the organization is incapable of achieving.

The design process often concerns technological innovation in response to, or in anticipation of, changing market requirements and trends in technology. Those companies with impressive records of product- or service-led growth have demonstrated a state-of-the-art approach to innovation based on three principles:

- *Strategic balance* to ensure that both old and new product service developments are important. Updating old products, services and processes, ensures continuing cash generation from which completely new products may be funded.
- *Top management approach* to design to set the tone and ensure that commitment is the common objective by visibly supporting the design effort. Direct control should be concentrated on critical decision points, since overmeddling by very senior people in day-to-day project management can delay and demotivate staff.
- *Teamwork*, to ensure that once projects are under way, specialist inputs, e.g. from marketing and technical experts, are fused and problems are tackled simultaneously. The teamwork should be urgent yet informal, for too much formality will stifle initiative, flair and the fun of design.

The extent of the design process should not be underestimated, but it often is. Many people associate design with *styling* of products, and this is certainly an important aspect. But for certain products and many service operations the *secondary design* considerations are vital. Anyone who has bought an 'assemble-it-yourself' kitchen unit will know the importance of the design of the assembly instructions, for example. Aspects of design that affect quality in this way are packaging, customer-service arrangements, maintenance routines, warranty details and their fulfillment, spare-part availability, etc.

An industry that has learned much about the secondary design features of its products is personal computers. Many of the problems of customer dissatisfaction experienced in this market have not been product design features but problems with user manuals, availability and loading of software, and applications. For technically complex products or service systems, the design and marketing of after-sales arrangements are an essential component of the design activity. The design of production equipment and its layout to allow ease of access for repair and essential maintenance, or simple use as intended, widens the management of design quality into suppliers and contractors and requires their total commitment.

Proper design of plant and equipment plays a major role in the elimination of errors, defectives, and waste. Correct initial design also obviates the need for costly and wasteful modifications to be carried out after the plant or equipment has been constructed. It is at the plant design stage that such important matters as variability, reproducibility, ease of use in operation, maintainability, should receive detailed consideration.

Designing

If design quality is taking care of all aspects of the customer's requirements, including cost, production, safe and easy use, and maintainability of products and services, then *designing* must take place in all aspects of:

- Identifying the need (including need for change).
- Developing that which satisfies the need.
- Checking the conformance to the need.
- Ensuring that the need is satisfied.

Designing covers every aspect, from the identification of a problem to be solved, usually a market need, through the development of design concepts and prototypes to the generation of detailed specifications or instructions required to produce the artefact or provide the service. It is the process of presenting needs in some physical form, initially as a solution, and then as a specific configuration or arrangement of materials resources, equipment, and people. Design permeates strategically and operationally through many areas of an organization and, while design professionals may control detailed product styling, decisions on design involve many people from other functions. Quality management supports such cross-functional interpretation of design.

Design, like any other activity, must be carefully managed. A flowchart of the various stages and activities involved in the design and development process appears in Figure 6.1.

By structuring the design process in this way, it is possible to:

- Control the various stages.
- Check that they have been completed.

- Decide which management functions need to be brought in and at what stage.
- Estimate the level of resources needed.

The control of the design process must be carefully handled to avoid stifling the creativity of the designer(s), which is crucial in making design solutions a reality. It is clear that the design process requires a range of specialized skills, and the way in which these skills are managed, the way they interact, and the amount of effort devoted to the different stages of the design and development process is fundamental to the quality, producibility, and price of the service or final product. A team

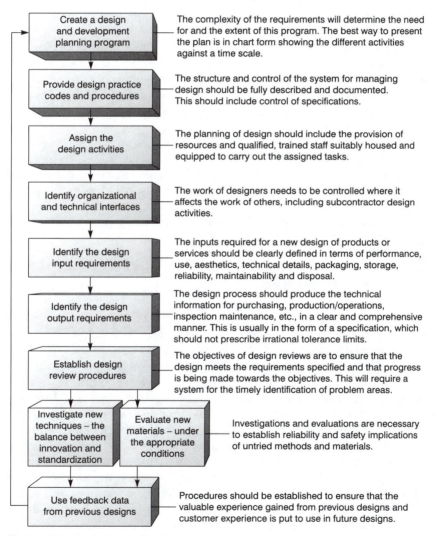

Figure 6.1 The design and development process

approach to the management of design can play a major role in the success of a project.

It is never possible to exert the same tight control on the design effort as on other operational efforts, yet the cost and the time used are often substantial, and both must appear somewhere within the organization's budget.

Certain features make control of the design process difficult:

1 No design will ever be 'complete' in the sense that, with effort, some modification or improvement cannot be made.
2 Few designs are entirely novel. An examination of most 'new' products, services or processes will show that they employ existing techniques, components or systems to which have been added novel elements.
3 The longer the time spent on a design, the less the increase in the value of the design tends to be, unless a technological breakthrough is achieved. This diminishing return from the design effort must be carefully managed.
4 External and/or internal customers will impose limitations on design time and cost. It is as difficult to imagine a design project whose completion date is not implicitly fixed, either by a promise to a customer, the opening of a trade show or exhibition, a seasonal 'deadline', a production schedule or some other constraint, as it is to imagine an organization whose funds are unlimited, or a product whose price has no ceiling.

Total design processes

Quality of design, then, concerns far more than the product or service design and its ability to meet the customer requirements. It is also about the activities of design and development. The appropriateness of the actual *design process* has a profound influence on the performance of any organization, and much can be learned by examining successful companies and how their strategies for research, design, and development are linked to the efforts of marketing and operations. In some quarters this is referred to as 'total design', and the term 'simultaneous engineering' has been used. This is an integrated approach to a new product or service introduction, similar in many ways to quality function deployment (QFD – see next section) in using multifunction teams or task forces to ensure that research, design, development, manufacturing, purchasing, supply, and marketing all work in parallel from concept through to the final launch of the product or service into the market place, including servicing and maintenance.

■ Quality function deployment (QFD) – the house of ■ quality

The 'house of quality' is the framework of the approach to design management known as quality function deployment (QFD). It originated in Japan in 1972 at Mitsubishi's Kobe shipyard, but it has been developed in numerous ways by Toyota and its suppliers, and many other organizations. The house of quality (HoQ) concept, initially referred to as quality tables, has been used successfully by manufacturers of integrated circuits, synthetic rubber, construction equipment, engines, home appliances, clothing, and electronics, mostly Japanese. Ford and General Motors use it, and other organizations, including AT&T, Bell Laboratories, Dell, Hewlett-Packard, Procter & Gamble, ITT, Rank Xerox, and Jaguar (now Ford), have applications. In Japan its design applications include public services, retail outlets, and apartment layout.

QFD is a 'system' for designing a product or service, based on customer requirements, with the participation of members of all functions of the supplier organization. It translates the customer's requirements into the appropriate technical requirements for each stage. The activities included in QFD are:

1 Market research.
2 Basic research.
3 Innovation.
4 Concept design.
5 Prototype testing.
6 Final-product or service testing.
7 After-sales service and troubleshooting.

These are performed by people with different skills in a team whose composition depends on many factors, including the products or services being developed and the size of the operation. In many industries, such as cars, digital equipment, electronics, and computers, 'engineering' designers are seen to be heavily into 'designing'. But in other industries and service operations designing is carried out by people who do not carry the word 'designer' in their job title. The failure to recognize the design inputs they make, and to provide appropriate training and support, will limit the success of the design activities and result in some offering that does not satisfy the customer. This is particularly true of internal customers.

The QFD team in operation

The first step of a QFD exercise is to form a cross-functional QFD team. Its purpose is to take the needs of the market and translate them into such a form that they can be satisfied within the operating unit and delivered to the customers.

As with all organizational problems, the structure of the QFD team must be decided on the basis of the detailed requirements of each organization. One thing, however, is clear – close liaison must be maintained at all times between the design, marketing and operational functions represented in the team.

The QFD team must answer three questions – *WHO, WHAT* and *HOW,* i.e.:

WHO are the customers?
WHAT does the customer need?
HOW will the needs be satisfied?

WHO may be decided by asking 'Who will benefit from the successful introduction of this product, service, or process? Once the customers have been identified, *WHAT* can be ascertained through interview/ questionnaire/focus group processes, or from the knowledge and judgement of the QFD team members. *HOW* is more difficult to determine, and will consist of the attributes of the product, service, or process under development. This will constitute many of the action steps in a 'QFD strategic plan'.

WHO, WHAT and *HOW* are entered into a QFD matrix or grid of 'house of quality' (HoQ), which is a simple 'quality table'. The *WHATs* are recorded in rows and the *HOWs* are placed in the columns.

The house of quality provides structure to the design and development cycle, often likened to the construction of a house, because of the shape of matrices when they are fitted together. The key to building the house is the focus on the customer requirements, so that the design and development processes are driven more by what the customer needs than by innovations in technology. This ensures that more effort is used to obtain vital customer information. It may increase the initial planning time in a particular development project, but the overall time, including design and redesign, taken to bringing a product or service to the market will be reduced.

This requires that marketing people, design staff (including engineers), and production/operations personnel work closely together from the time the new service, process, or product is conceived. It will need to replace in many organizations the 'throwing it over the wall' approach, where a solid wall exists between each pair of functions (Figure 6.2).

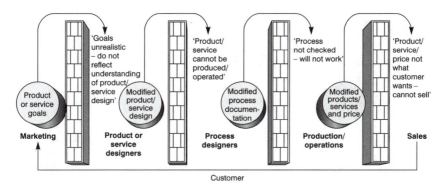

Figure 6.2 'Throw it over the wall.' The design and development process is sequential and walled into separate functions

The HoQ provides an organization with the means for interdepartmental or interfunctional planning and communications, starting with the so-called customer attributes (CAs). These are phrases customers use to describe product, process, and service characteristics.

A complete QFD project will lead to the construction of a sequence of house of quality diagrams, which translate the customer requirements into specific operational process steps. For example, the 'feel' that customers like on the steering wheel of a motor car may translate into a specification for 45 standard degrees of synthetic polymer hardness, which in turn translates into specific manufacturing process steps, including the use of certain catalysts, temperatures, processes, and additives.

The first steps in QFD lead to a consideration of the product as a whole and subsequent steps to consideration of the individual components. For example, a complete hotel service would be considered at the first level, but subsequent QFD exercises would tackle the restaurant, bedrooms and reception. Each of the subservices would have customer requirements, but they all would need to be compatible with the general service concept.

The QFD or house of quality tables _____

Figure 6.3 shows the essential components of the quality table or HoQ diagram. The construction begins with the *customer requirements*, which are determined through the 'voice of the customer' – the marketing and market research activities. These are entered into the blocks to the left of the central relationship matrix. Understanding and prioritizing the customer requirements by the QFD team may require the use of competitive and compliant analysis, focus groups, and the analysis of market

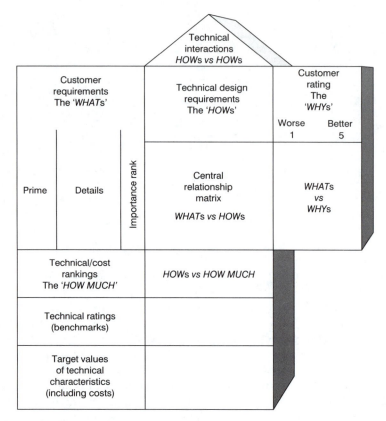

potential. The prime or broad requirements should lead to the detailed
WHATs.

Once the customer requirements have been determined and entered
into the table, the *importance* of each is rated and rankings are added.
The use of the 'emphasis technique' or paired comparison may be helpful here (see Chapter 13).

Each customer requirement should then be examined in terms of customer rating; a group of customers may be asked how they perceive the
performance of the organization's product or service versus those of
competitors. These results are placed to the right of the central matrix.
Hence the customer requirements' importance rankings and competition ratings appear from left to right across the house.

The *WHATs* must now be converted into the *HOWs*. These are called
the *technical design requirements* and appear on the diagram from top to
bottom in terms of requirements, rankings (or costs) and ratings against

competition (technical benchmarking, see Chapter 9). These will provide the 'voice of the process'.

The technical design requirements themselves are placed immediately above the central matrix and may also be given a hierarchy of prime and detailed requirements. Immediately below the customer requirements appear the rankings of technical difficulty, development time, or costs. These will enable the QFD team to discuss the efficiency of the various technical solutions. Below the technical rankings on the diagram comes the benchmark data, which compares the technical processes of the organization against its competitors'.

The *central relationship matrix* is the working core of the house of quality diagram. Here the WHATs are matched with the HOWs, and each customer requirement is systematically assessed against each technical design requirement. The nature of any relationship – strong positive, positive, neutral, negative, strong negative – is shown by symbols in the matrix. The QFD team carries out the relationship estimation, using experience and judgement, the aim being to identify HOW the WHATs may be achieved. All the HOWs listed must be necessary and together sufficient to achieve the WHATs. Blank rows (customer requirement not met) and columns (redundant technical characteristics) should not exist.

The roof of the house shows the interactions between the technical design requirements. Each characteristic is matched against the others, and the diagonal format allows the nature of relationships to be displayed. The symbols used are the same as those in the central matrix.

The complete QFD process is time-consuming, because each cell in the central and roof matrices must be examined by the whole team. The team must examine the matrix to determine which technical requirement will need design attention, and the costs of that attention will be given in the bottom row. If certain technical costs become a major issue, the priorities may then be changed. It will be clear from the central matrix if there is more than one way to achieve a particular customer requirement, and the roof matrix will show if the technical requirements to achieve one customer requirement will have a negative effect on another technical issue.

The very bottom of the house of quality diagram shows the *target* values of the *technical characteristics*, which are expressed in physical terms. They can only be decided by the team after discussion of the complete house contents. While these targets are the physical output of the QFD exercise, the whole process of information-gathering, structuring, and ranking generates a tremendous improvement in the team's cross-functional understanding of the product/service design delivery system.

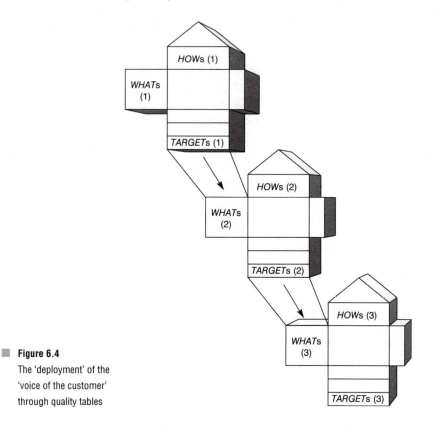

■ **Figure 6.4**
The 'deployment' of the
'voice of the customer'
through quality tables

The target technical characteristics may be used to generate the next level house of quality diagram, where they become the *WHATs*, and the QFD process determines the further details of *HOW* they are to be achieved. In this way the process 'deploys' the customer requirements all the way to the final operational stages. Figure 6.4 shows how the target technical characteristics, at each level, become the input to the next level matrix.

QFD progresses now through the use of the 'seven new planning tools' (see Chapter 13) and other standard techniques such as value analysis,[1] experimental design,[2] statistical process control,[3] and so on.

The benefits of QFD

The aim of the HoQ is to co-ordinate the interfunctional activities and skills within an organization. This should lead to products and services designed, produced/operated, and marketed so that customers will want to purchase them and continue doing so.

The use of competitive information in QFD should help to prioritize resources and to structure the existing experience and information. This allows the identification of items that can be acted upon.

There should be reductions in the number of midstream design changes, and these in turn will limit post-introduction problems and reduce implementation time. Because QFD is consensus based, it promotes teamwork and creates communications at functional interfaces, while also identifying required actions. It should lead to a 'global view' of the development process, from a consideration of all the details.

If QFD is introduced systematically, it should add structure to the information, generate a framework for sensitivity analysis, and provide documentation, which is 'living' and adaptable to change. In order to understand the full impact of QFD it is necessary to examine the changes that take place in the team and the organization during the design and development process.

The main benefit of QFD is of course the increase in customer satisfaction and loyalty, which may be measured in terms of, for example, reductions in warranty claims, and repeat business.

■ Specifications and standards

There is a strong relationship between standardization and specification. To ensure that a product or a service is *standardized* and may be repeated a large number of times in exactly the manner required, *specifications* must be written so that they are open to only one interpretation. The requirements, and therefore the quality, must be built into the design specification. There are national and international standards which, if used, help to ensure that specifications will meet certain accepted criteria of technical or managerial performance, safety, etc.

Standardization does not guarantee that the best design or specification is selected. It may be argued that the whole process of standardization slows down the rate and direction of technological development, and affects what is produced. If standards are used correctly, however, the process of drawing up specifications should provide opportunities to learn more about particular innovations and to change the standards accordingly.

It is possible to strike a balance between innovation and standardization. Clearly, it is desirable for designers to adhere where possible to past-proven materials and methods, in the interests of reliability, maintainability and variety control. Hindering designers from using recently developed materials, components, or techniques, however, can cause the design process to stagnate technologically. A balance must be achieved by analysis of materials, products and processes proposed in the design, against the background of their known reproducibility and reliability. If breakthrough innovations are proposed, then analysis or testing should objectively justify their adoption in preference to the established alternatives.

It is useful to define a specification. The International Standards Organization (ISO) defines it in ISO 8402 (1986) as 'The document that prescribes the requirements with which the product or service has to conform'. A document not giving a detailed statement or description of the requirements to which the product, service or process must comply cannot be regarded as a specification, and this is true of much sales literature.

The specification conveys the customer requirements to the supplier to allow the product or service to be designed, engineered, produced, or operated by means of conventional or stipulated equipment, techniques, and technology. The basic requirements of a specification are that it gives the:

- Performance requirements of the product or service.
- Parameters – such as dimensions, concentration, turn-round time – which describe the product or service adequately (these should be quantified and include the units of measurement).
- Materials to be used by stipulating properties or referring to other specifications.
- Method of production or delivery of the service.
- Inspection/testing/checking requirements.
- References to other applicable specifications or documents.

To fulfill its purpose the specifications must be written in terminology that is readily understood, and in a manner that is unambiguous and so cannot be subject to differing interpretation. This is not an easy task, and one which requires all the expertise and knowledge available. Good specifications are usually the product of much discussion, deliberation and sifting of information and data, and represent tangible output from a QFD team.

■ Design for quality in the service sector

The emergence of the services sector has been suggested by economists to be part of the natural progression in which economic dominance changes first from agriculture to manufacturing and then to services. It is argued that if income elasticity of demand is higher for services than it is for goods, then as incomes rise, resources will shift toward services. The continuing growth of services verifies this, and is further explained by changes in culture, fitness, safety, demography and lifestyles.

In considering the design of services it is important to consider the differences between goods and services. Some authors argue that the marketing and design of goods and services should conform to the same fundamental rules, whereas others claim that there is a need for a different approach to service because of the recognizable differences between the goods and services themselves.

In terms of design, it is possible to recognize three distinct elements in the service package – the physical elements or facilitating goods, the explicit service or sensual benefits, and implicit service or psychological benefits. In addition, the particular characteristics of service delivery systems may be itemized:

- Intangibility.
- Perishability.
- Simultaneity.
- Heterogeneity.

It is difficult, if not impossible, to design the intangible aspects of a service, since consumers often must use experience or the reputation of a service organization and its representatives to judge quality.

Perishability is often an important issue in services, since it is often impossible or undesirable to hold stocks of the explicit service element of the service package. This aspect often requires that service operation and service delivery must exist simultaneously.

Simultaneity occurs because the consumer must be present before many services can take place. Hence, services are often formed in small and dispersed units, and it is difficult to take advantage of economies of scale. The rapid developments in computing and communications technologies are changing this in sectors such as banking, but contact continues to be necessary for many service sectors. Design considerations here include the environment and the systems used. Service facilities, procedures, and systems should be designed with the customer in mind, as well as the 'product' and the human resources. Managers need a picture of the total span of the operation, so that factors which are crucial to success are not neglected. This clearly means that the functions of marketing, design, and operations cannot be separated in services, and this must be taken into account in the design of the operational controls, such as the diagnosing of individual customer expectations. A QFD approach here is most appropriate.

Heterogeneity of services occurs in consequence of explicit and implicit service elements relying on individual preferences and perceptions. Differences exist in the outputs of organizations generating the same service, within the same organization, and even the same employee on different occasions. Clearly, unnecessary variation needs to be controlled, but the variation attributed to estimating, and then matching, the consumers' requirements is essential to customer satisfaction and loyalty and must be designed into the systems. This inherent variability does, however, make it difficult to set precise quantifiable standards for all the elements of the service.

In the design of services it is useful to classify them in some way. Several sources from the literature on the subject help us to place services in one of five categories:

■ Personal services.
■ Service shop.
■ Professional services.
■ Mass services.
■ Service factory.

Several service attributes have particular significance for the design of service operations:

1 *Labor intensity* – the ratio of labor costs incurred to the value of assets and equipment used (people versus equipment-based services).
2 *Contact* – the proportion of the total time required to provide the service for which the consumer is present in the system.
3 *Interaction* – the extent to which the consumer actively intervenes in the service process to change the content of the service; this includes customer participation to provide information from which needs can be assessed, and customer feedback from which satisfaction levels can be inferred.
4 *Customization* – which includes *choice* (providing one or more selections from a range of options, which can be single or *fixed*) and *adaptation* (the interaction process in which the requirement is decided, designed and delivered to match the need).
5 *Nature of service act* – either tangible, i.e. perceptible to touch and can be owned, or intangible, i.e. insubstantial.
6 *Recipient of service* – either people or things.

Table 6.1 gives a list of some services with their assigned attribute types and Table 6.2 shows how these may be used to group the services under the various classifications.

Table 6.1 A classification of selected services

Service	Labour intensity	Contact	Interaction	Customization	Nature of act	Recipient of service
Accountant	High	Low	High	Adapt	Intangible	Things
Architect	High	Low	High	Adapt	Intangible	Things
Bank	Low	Low	Low	Fixed	Intangible	Things
Beautician	High	High	High	Adapt	Tangible	People
Bus/coach service	Low	High	High	Choice	Tangible	People
Cafeteria	Low	High	High	Choice	Tangible	People
Cleaning firm	High	Low	Low	Fixed	Tangible	People
Clinic	Low	High	High	Adapt	Tangible	People
Coaching	High	High	High	Adapt	Intangible	People
College	High	High	Low	Fixed	Intangible	People
Courier firm	High	Low	Low	Adapt	Tangible	Things
Dental practice	High	High	High	Adapt	Tangible	Things
Driving school	High	High	High	Adapt	Intangible	People
Equipment hire	Low	Low	Low	Choice	Tangible	Things
Finance consultant	High	Low	High	Adapt	Intangible	People
Hairdresser	High	High	High	Adapt	Tangible	People
Hotel	High	High	Low	Choice	Tangible	People
Leisure center	Low	High	High	Choice	Tangible	People
Maintenance	Low	Low	Low	Choice	Tangible	Things
Management consultant	High	High	High	Adapt	Intangible	People
Nursery	High	Low	Low	Fixed	Tangible	People
Optician	High	Low	High	Adapt	Tangible	People
Postal service	Low	Low	Low	Adapt	Tangible	Things
Rail service	Low	High	Low	Choice	Tangible	People
Repair firm	Low	Low	Low	Adapt	Tangible	Things
Restaurant	High	High	Low	Choice	Tangible	People
Service station	Low	High	High	Choice	Tangible	People
Solicitors	High	Low	High	Adapt	Intangible	Things
Takeaway	High	Low	Low	Choice	Tangible	People
Veterinary	High	Low	High	Adapt	Tangible	Things

It is apparent that services are part of almost all organizations and not confined to the service sector. What is clear is that the service classifications and different attributes must be considered in any service design process.

The author is grateful to the contribution made by John Dotchin to this section of Chapter 6.

Table 6.2 Grouping of similar services

Personal services

Driving school	Sports coaching
Beautician	Dental practice
Hairdresser	Optician

Service shop

Clinic	Cafeteria
Leisure center	Service station

Professional services

Accountant	Architect
Finance consultant	Management consultant
Solicitor	Veterinary

Mass services

Hotel	Restaurant
College	Bus service
Coach service	Rail service
Takeaway	Nursery
Courier firm	

Service factory

Cleaning firm	Postal service
Repair firm	Equipment hire
Maintenance	Bank

Failure mode, effect and criticality analysis (FMECA)

In the design of products, services and processes it is possible to determine potential modes of failure and their effects on the performance of the product or operation of the process or service system. Failure mode and effect analysis (FMEA) is the study of potential failures to determine their effects. If the results of an FMEA are ranked in order of seriousness, then the word criticality is added to give FMECA. The primary objective of an FMECA is to determine the features of product design, production or operation and distribution that are critical to the various modes of failure, in order to reduce failure. It uses all the available experience and expertise, from marketing, design, technology, purchasing, production/operation, distribution, service, etc., to identify the importance levels or criticality of potential problems and stimulate action to reduce these levels. FMECA should be a major consideration at the design stage of a product or service.

The elements of a complete FMECA are:

- *Failure mode* – the anticipated conditions of operation are used as the background to study the most probable failure mode, location and mechanism of the product or system and its components.
- *Failure effect* – the potential failures are studied to determine their probable effects on the performance of the whole product, process, or service, and the effects of the various components on each other.
- *Failure criticality* – the potential failures on the various parts of the product or service system are examined to determine the severity of each failure effect in terms of lowering of performance, safety hazard, total loss of function, etc.

FMECA may be applied to any stage of design, development, production/ operation or use, but since its main aim is to prevent failure, it is most suitably applied at the design stage to identify and eliminate causes. With more complex product or service systems, it may be appropriate to consider these as smaller units or subsystems, each one being the subject of a separate FMECA.

Special FMECA pro formas are available (for example see Table 6.3) and they set out the steps of the analysis as follows:

1 Identify the product or system components, or process function.
2 List all possible failure modes of each component.
3 Set down the effects that each mode of failure would have on the function of the product or system.
4 List all the possible causes of each failure mode.
5 Assess numerically the failure modes on a scale from 1 to 10. Experience and reliability data should be used, together with judgement, to determine the values, on a scale 1–10, for:

   P the probability of each failure mode occurring (1 = low, 10 = high).
   S the seriousness or criticality of the failure (1 = low, 10 = high).
   D the difficulty of detecting the failure before the product or service is used by the consumer (1 = easy, 10 = very difficult). See Table 6.4.

6 Calculate the product of the ratings, $C = P \times S \times D$, known as the criticality index or risk priority number (RPN) for each failure mode. This indicates the relative priority of each mode in the failure prevention activities.
7 Indicate briefly the corrective action required and, if possible, which department or person is responsible and the expected completion date.

When the criticality index has been calculated, the failures may be ranked accordingly. It is usually advisable, therefore, to determine the value of C for each failure mode before completing the last columns. In this way the action required against each item can be judged in the light of the ranked severity and the resources available.

Table 6.3 Failure mode, effect and criticality analysis (FMECA)

Part name: emission assembly

Process/ function (1)	Possible failure mode (2)	Effect(s) of failure (3)	Possible cause(s) of failure (4)	P	S	D	C (6)	Corrective action (7)
				(5)				
Inspection of inwards goods	Base material incorrect	Early failure in service	Wrong selection of material by supplier	1	8	9	72	None
	Dimensions incorrect	Loose or tight fit on spigot and in molding	Extrusion process out of statistical control	8	5	7	280	Supplier to certify and is applying in process controls
	Label incorrect regarding size	Wrong part supplied to customer	Specification incorrect	2	5	7	70	None
	Faulty molding	Inability to assemble.	Incorrect inspection	2	5	5	50	None
		Early failure in service		2	8	6	96	
Washing	Not washed sufficiently	Line marking becomes indistinct and adhesion values reduced	Detergent not added to washer	3	2	6	36	None
			Water not changed	2	6	6	72	None
Air blow	Omitted	Blocked hose.	Operator error	2	8	6	96	Positive release system of passed tubing being investigated
		Wet hose could affect sintered disc		2	8	6	96	

Process	Failure mode	Effect	Cause					Action
Storage	Not stored correctly	Wet hose could affect sintered disc	Incorrect packaging	2	8	6	96	Drying facility under investigation Sept.
	Dirt in component	Incorrect operation	Bad storage	1	8	9	72	None
			Human error operator	2	8	6	96	None
Stripe application	Wrong color	Incorrect fitment in assembly	Incorrectly planned	2	8	6	96	None
	Incomplete	Cosmetic rejection	Machine malfunction	8	2	6	96	Co-extruded stripe under investigation Sept. Target date
Cut to length	Short	Cannot assemble at customer or in plant	Machine capability incorrect	5	9	7	315	New design cutting machine to be evaluated Sept.
	Long	Fouls on fitment; in-house difficulty to perform next operation		4	3	7	84	
Storage awaiting assembly	Mixed parts	Incorrect length supplied to assembler	Parts mixed during transit	2	5	7	70	None
	Incorrect labeling	Incorrect length supplied to assembler	Operator error	2	8	6	96	Assembly boards and attribute charts to be introduced
Subassembly of moldings	Incorrect units assembled.	Customer cannot fit unit	Incorrect selection	2	8	6	96	None
	Bonding	Leaks. Failing vacuum requirement affects driveability	Incorrect bonding agent preparation, overage material	2	8	7	112	Improved adhesive application under evaluation
	Fit incorrect length of tubing	Long and short lengths can result in customer being unable to assemble	Incorrect selection	2	8	6	96	Increased use of inspection boards as per sampling plan
				2	3	6	36	

(continued)

■ Table 6.3 (continued)

Process/function (1)	Possible failure mode (2)	Effect(s) of failure (3)	Possible cause(s) of failure (4)	(5) P	S	D	(6) C	Corrective action (7)
Assembly	Assembly fitted circuit broken	Short length unable to join components		2	8	7	112	
	Missed parts	Failure of unit to function on vehicle	Operator error	6	6	6	216	Increased use of inspection boards in hand
	Valve orientation incorrect	Failure of unit to function on vehicle	Operator error	2	7	6	84	Increased use of inspection boards in hand
	Fuel trap orientation incorrect	Failure of unit to function on vehicle	Operator error	2	8	6	96	Increased use of inspection boards in hand
	Rubber elbow orientation incorrect	Will not fit on vehicle	Operator error	2	6	6	72	Increased use of inspection boards in hand
Packaging	Label omitted	Wrong part supplied	Operator error	2	8	6	96	Increased use of inspection boards in hand
	Incorrect label	Parts sent to wrong destination	Human error, incorrect data	5	7	9	315	Improved labeling system is being introduced
	Incorrect container	Rejected at customer	Human error, incorrect data	2	5	6	60	None

Table 6.4 Probability and seriousness of failure and difficulty of detection

Value	1	2	3	4	5	6	7	8	9	10
P	low chance of occurrence _____ almost certain to occur									
S	not serious, minor nuisance _____ total failure, safety hazard									
D	easily detected _____ unlikely to be detected									

Moments of truth

Moments of truth (MoT) is a concept that has much in common with FMEA. The idea was created by Jan Carlzon,[4] CEO of Scandinavian Airlines (SAS), and was made popular by Albrecht and Zemke.[5] An MoT is the moment in time when a customer first comes into contact with the people, systems, procedures, or products of an organization, which leads to the customer making a judgement about the quality of the organization's services or products.

In MoT analysis the points of potential dissatisfaction are identified proactively, beginning with the assembly of process flow chart-type diagrams. Every small step taken by a customer in his/her dealings with the organization's people, products, or services is recorded. It may be difficult or impossible to identify all the MoTs, but the systematic approach should lead to a minimalization of the number and severity of unexpected failures, and this provides the link with FMEA.

The links between good design and managing the business

Research carried out by the European Centre for Business Excellence[6] has led to a series of specific aspects that should be addressed to integrate design into the business or organization. These are presented under various business criteria below.

Leadership and management style

- 'Listening' is designed into the organization.
- Management communicates the importance of good design in good partnerships and vice versa.
- A management style is adopted that fosters innovation and creativity, and that motivates employees to work together effectively.

Customers, strategy and planning

- The customer is designed into the organization as a focus to shape policy and strategy decisions.
- Designers and customers communicate directly.
- Customers are included in the design process.
- Customers are helped to articulate and participate in the understanding of their own requirements.
- Systems are in place to ensure that the changing needs of the customers inform changes to policy and strategy.
- Design and innovation performance measures are incorporated into policy and strategy reviews.
- The design process responds quickly to customers.

People – their management and satisfaction

- People are encouraged to gain a holistic view of design within the organization.
- There is commitment to design teams and their motivation, particularly in cross-functional teamwork (e.g. quality function deployment teams).
- The training program is designed, with respect to design, in terms of people skills training (e.g. interpersonal, management, teamwork) and technical training (e.g. resources, software).
- Training helps integrate design activities into the business.
- Training impacts on design (e.g. honing creativity and keeping people up to date with design concepts and activity).
- Design activities are communicated (including new product or service concepts).
- Job satisfaction is harnessed to foster good design.
- The results of employee surveys are fed back into the design process.

Resource and partnership management

- Knowledge is managed proactively, including investment in technology.
- Information is shared in the organization.
- Past experience and learning are captured from design projects and staff.
- Information resources are available for planning design projects.
- Suppliers contribute to innovation, creativity and design concepts.
- Concurrent engineering and design is integrated through the supply chains.

Process management

- Design is placed at the center of process planning to integrate different functions within the organization and form partnerships outside the organization.
- Process management is used to resolve design problems and foster teamwork within the organization and with external partners.

Society and business performance

- Consideration is given to how the design of a product or service impacts on:
  - the environment;
  - the recyclability and disposal of materials;
  - packaging and wastage of resources;
  - the (local) economy (e.g. effects on labor requirements);
  - the business results, both financial and non-financial.

This same research showed that strong links exist between good design and proactive flexible deployment of business policies and strategies. These can be used to further improve design by encouraging the sharing of best practice within and across industries, by allowing designers and customers to communicate directly, by instigating new product/service introduction policies, project audits and design/innovation measurement policies and by communicating the strategy to employees. The findings of this work may be summarized by thinking in terms of the design process acting across the 'value chain', as shown in Figure 6.5.

Effective people management skills are essential for good design – these include the ability to listen and communicate, to motivate employees and encourage teamwork, as well as the ability to create an organizational climate which is conducive to creativity and continuous innovation.

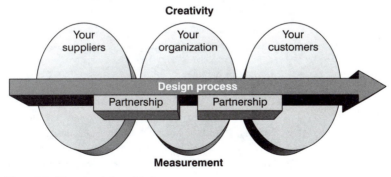

Figure 6.5   The value chain and design process

The only way to ensure that design actively contributes to business performance is to make sure it happens 'by design', rather than by accident. In short, it needs co-ordinating and managing right across the organization.

■ References

1. Lockyer, K. G., Muhlemann, A. P. and Oakland, J. S., *Production and Operations Management*, 6th edition, Pitman, 1992.
2. Caulcutt, R., *Statistics in Research and Development*, 2nd edition, Chapman and Hall, 1991.
3. Oakland, J. S., *Statistical Process Control*, 5th edition, Butterworth-Heinemann, 2002.
4. Carlzon, J., *Moments of Truth*, Harper & Row, 1987.
5. Albrecht, K. and Zemke, R., *Service America! – Doing Business in the New Economy*, Dow Jones-Irwin, Homewood, Ill. (USA), 1985.
6. *Designing Business Excellence*, European Centre for Business Excellence (the Research and Education Division of Oakland Consulting plc, www.ecforbe.com)/British Quality Foundation/Design Council, 1998.

Chapter highlights

■ ■ ■

Design, innovation and improvement

- Design is a multifaceted activity which covers many aspects of an organization.
- All businesses need to update their products, processes and services.
- Innovation entails both invention and design, *and* continuous improvement of existing products, services, and processes.
- Leading product/service innovations are market-led, not marketing-led.
- Everything in or from an organization results from design decisions.
- Design is an ongoing activity, dynamic not static, a verb not a noun – design is a process.

The design process

- Commitment at the top is required to building in quality throughout the design process. Moreover, the operational processes must be capable of achieving the design.
- State-of-the-art approach to innovation is based on a strategic balance of old and new, top management approach to design, and teamwork.

- The 'styling' of products must also be matched by secondary design considerations, such as operating instructions and software support.
- Designing takes in all aspects of identifying the need, developing something to satisfy the need, checking conformance to the need and ensuring the need is satisfied.
- The design process must be carefully managed and can be flowcharted, like any other process, into: planning, practice codes, procedures, activities, assignments, identification of organizational and technical interfaces, and design input requirements, review investigation and evaluation of new techniques and materials, and use of feedback data from previous designs.
- Total design or 'simultaneous engineering' is similar to quality function deployment and uses multifunction teams to provide an integrated approach to product or service introduction.

Quality function deployment (QFD) – the house of quality

- The 'house of quality' is the framework of the approach to design management known as quality function deployment (QFD). It provides structure to the design and development cycle, which is driven by customer needs rather than innovation in technology.
- QFD is a system for designing a product or service, based on customer demands, and bringing in all members of the supplier organization.
- A QFD team's purpose is to take the needs of the market and translate them into such a form that they can be satisfied within the operating unit.
- The QFD team answers the following questions. *WHO* are the customers? *WHAT* do the customers need? *HOW* will the needs be satisfied?
- The answers to the *WHO*, *WHAT* and *HOW* questions are entered into the QFD matrix or quality table, one of the seven new tools of planning and design.
- The foundations of the house of quality are the customer requirements; the framework is the central planning matrix, which matches the 'voice of the customer' with the 'voice of the processes' (the technical descriptions and capabilities); and the roof is the inter-relationships matrix between the technical design requirements.
- The benefits of QFD include customer-driven design, prioritizing of resources, reductions in design change and implementation time, and improvements in teamwork, communications, functional interfaces, and customer satisfaction.

Specifications and standards

- There is a strong relationship between standardization and specifications. If standards are used correctly, the process of drawing up specifications should provide opportunities to learn more about innovations and change standards accordingly.
- The aim of specifications should be to reflect the true requirements of the product/service that are capable of being achieved.

Design for quality in the service sector

- In the design of services three distinct elements may be recognized in the service package: physical (facilitating goods), explicit service (sensual benefits), and implicit service (psychological benefits). Moreover, the characteristics of service delivery may be itemized as intangibility, perishability, simultaneity, and heterogeneity.
- Services may be classified generally as personal services, service shop, professional services, mass services, and service factory. The service attributes that are important in designing services include labour intensity, contact interaction, customerization, nature of service act, and the direct recipient of the act.
- Use of this framework allows services to be grouped under the five classifications.

Failure mode, effect and criticality analysis (FMECA)

- FMEA is the study of potential product, service or process failures and their effects. When the results are ranked in order of criticality, the approach is called FMECA. Its aim is to reduce the probability of failure.
- The elements of a complete FMECA are to study failure mode, effect and criticality. It may be applied at any stage of design, development, production/operation or use.
- Moments of truth (MoT) is a similar concept to FMEA. It refers to the moments in time when customers first come into contact with an organization, leading to judgements about quality.

The links between good design and managing the business

- Research has led to a series of specific aspects to address in order to integrate design into an organization.
- The aspects may be summarized under the headings of: leadership and management style; customers, strategy and planning; people – their management and satisfaction; resource management; process management; society and business performance.
- The research shows that strong links exist between good design and proactive flexible deployment of business policies and strategies – design needs co-ordinating and managing right across the organization.

Part 3

Performance

All words, and no performance!

Philip Massinger, 1583–1640,
from 'Parliament of Love', *c.* 1619

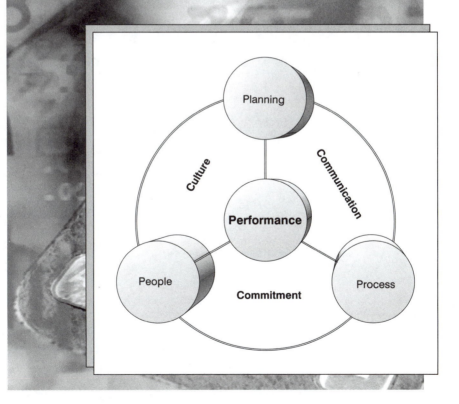

Chapter 7

Performance measurement frameworks

Performance measurement and the improvement cycle

Traditionally, performance measures and indicators have been derived from cost-accounting information, often based on outdated and arbitrary principles. These provide little motivation to support quality management and, in some cases, actually inhibit continuous improvement because they are unable to map onto process performance. In the organization that is to succeed over the long term, performance must begin to be measured by the improvements seen by the customer.

In the cycle of never-ending improvement, measurement plays an important role in:

- Tracking progress against organizational goals.
- Identifying opportunities for improvement.
- Comparing performance against internal standards.
- Comparing performance against external standards.

Measures are used in *process control*, e.g. control charts (see Chapter 13), and in *performance improvement*, e.g. by quality improvement teams (see Chapters 14 and 15), so they should give information about how well processes and people are doing and motivate them to perform better in the future.

The author and his colleagues have seen many examples of so-called performance measurement systems that frustrated improvement efforts. Various problems include systems that:

1 Produce irrelevant or misleading information.
2 Track performance in single, isolated dimensions.
3 Generate financial measures too late, e.g. quarterly, for mid-course corrections or remedial action.
4 Do not take account of the customer perspective, both internal and external.
5 Distort management's understanding of how effective the organization has been in implementing its strategy.
6 Promote behavior that *undermines* the achievement of the strategic objectives.

Typical potentially harmful summary measures of local performance are purchase price, machine or plant efficiencies, direct labour costs, and ratios of direct to indirect labour. These are incompatible with quality and productivity improvement measures such as process and throughput times, supply chain performance, inventory reductions, and increases in flexibility, which are first and foremost *non-financial*. Financial summaries provide valuable information, of course, but they should not be used for control. Effective decision making requires direct measures for operational feedback and improvement.

One example of a 'measure' with these shortcomings is return on investment (ROI). ROI can be computed only after profits have been totaled for a given period. It was designed therefore as a single-period, long-term measure, but it is often used as a short-term one. Perhaps this is because most executive bonus 'packages' in the West are based on short-term measures. ROI tells us what happened, not what is happening or what will happen, and, for complex and detailed projects, ROI is inaccurate and irrelevant.

Many managers have a poor or incomplete understanding of their processes and products or services, and, looking for an alternative stimulus, become interested in financial indicators. The use of ROI, for example, for evaluating strategic requirements and performance can lead to a discriminatory allocation of resources. In many ways the financial indicators used in many organizations have remained static while the environment in which they operate has changed dramatically.

Traditionally, the measures used have not been linked to the processes where the value-adding activities take place. What has been missing is a performance measurement framework that provides feedback to people in all areas of business operations. Of course, quality management stresses the need to start with the process for fulfilling customer needs.

The critical elements of a good performance measurement framework (PMF) are:

- Leadership and commitment.
- Full employee involvement.
- Good planning.
- Sound implementation strategy.
- Measurement and evaluation.
- Control and improvement.
- Achieving and maintaining standards of excellence.

A cycle of continuous improvement based on Deming's ideas – Plan, Do, Check, Act – clearly requires measurement to drive it, and yet it is a useful design aid for the measurement system itself:

Plan: establish performance objectives and standards.
Do: measure actual performance.
Check: compare actual performance with the objectives and standards – determine the gap.
Act: take the necessary actions to close the gap and make the necessary improvements.

Before we use performance measurement in the improvement cycle, however, we should attempt to answer four basic questions:

1 Why measure?
2 What to measure?
3 Where to measure?
4 How to measure?

Why measure?

It has been said often that it is not possible to manage what cannot be measured. Whether this is strictly true or not there are clear arguments for measuring. In a quality-driven, never-ending improvement environment, the following are some of the main reasons *why measurement is needed* and why it plays a key role in quality and productivity improvement:

- To ensure customer requirements *have* been met.
- To be able to set sensible *objectives* and comply with them.
- To provide *standards* for establishing comparisons.
- To provide *visibility* and provide a 'scoreboard' for people to *monitor* their own performance levels.

- To highlight *quality problems* and determine which areas require *priority attention*.
- To give an indication of the *costs of poor quality*.
- To justify the *use of resources*.
- To provide *feedback* for driving the improvement effort.

It is also important to know the impact of improvements efforts on business performance, on sustaining current performance, and perhaps on reducing any decline in performance.

What to measure?

In business performance improvement, process understanding, definition, measurement and management are tied inextricably together. In order to assess and evaluate performance accurately, appropriate measurement must be designed, developed and maintained by people who *own* the processes concerned. They may find it necessary to measure effectiveness, efficiency, quality, impact, and productivity. In these areas there are many types of measurement, including direct output or input figures, the cost of poor quality, economic data, comments and complaints from customers, information from customer or employee surveys, etc., generally continuous variable measures (such as time) or discrete attribute measures (such as absentees).

No one can provide a generic list of what should be measured but, once it has been decided in any one organization what measures are appropriate, they may be converted into indicators. These include ratios, scales, rankings, and financial and time-based indicators. Whichever measures and indicators are used by the process owners, they must reflect the true performance of the process in customer/supplier terms, and emphasize continuous improvement. Time-related measures and indicators often have great value.

Where to measure?

If true measures of the effectiveness of the management of quality are to be obtained, there are three components that must be examined – the human, technical and business components.

The human component is clearly of major importance and the key tests are that, wherever measures are used, they must be:

1 Understood by all the people being measured.
2 Accepted by the individuals concerned.

3 Compatible with the rewards and recognition systems.
4 Designed to offer minimal opportunity for manipulation.

Technically, the measures must be the ones that truly represent the controllable aspects of the processes, rather than simple output measures that cannot be related to process management. They must also be correct, precise and accurate.

The business component requires that the measures are objective, timely, and result-oriented, and above all they must mean something to those working in and around the process, *including the customers.*

How to measure?

Measurement, as any other management system, requires the stages of design, analysis, development, evaluation, implementation and review. The system must be designed to measure *progress*, otherwise it will not engage the improvement cycle. Progress is important in five main areas: effectiveness, efficiency, productivity, quality, and impact.

Effectiveness

Effectiveness may be defined as the percentage actual output over the expected output:

$$\text{Effectiveness} = \frac{\text{Actual output}}{\text{Expected output}} \times 100 \text{ percent}$$

Effectiveness then looks at the *output* side of the process and is about the implementation of the objectives – doing what you said you would do. Effectiveness measures should reflect whether the organization, group or process owner(s) are achieving the desired results, accomplishing the right things. Measures of this may include:

- Quality, e.g. a grade of product, or a level of service.
- Quantity, e.g. tonnes, lots, bedrooms cleaned, accounts opened.
- Timeliness, e.g. speed of response, product lead-times, cycle time.
- Cost/price, e.g. unit costs.

Efficiency

Efficiency is concerned with the percentage resource actually used over the resources that were planned to be used:

$$\text{Efficiency} = \frac{\text{Resources actually used}}{\text{Resources planned to be used}} \times 100 \text{ percent}$$

Clearly, this is a process *input* issue and measures performance of the process system management. It is, of course, possible to use resources 'efficiently' while being *ineffective*, so performance efficiency improvement must be related to certain output objectives.

All process inputs may be subjected to efficiency measurement, so we may use labor/staff efficiency, equipment efficiency (or utilization), materials efficiency, information efficiency, etc. Inventory data and through-put times are often used in efficiency and productivity ratios.

Productivity

Productivity measures should be designed to relate the process outputs to its inputs:

$$\text{Productivity} = \frac{\text{Outputs}}{\text{Inputs}}$$

and this may be quoted as expected or actual productivity:

$$\text{Expected productivity} = \frac{\text{Expected output}}{\text{Resources expected to be consumed}}$$

$$\text{Actual productivity} = \frac{\text{Actual output}}{\text{Resources actually consumed}}$$

There is a vast literature on productivity and its measurement, but simple ratios such as tonnes per man-hour (expected and actual), sales output per telephone operator-day, and many others like this are in use. Productivity measures may be developed for each combination of outputs and inputs, e.g. sales/all employee costs.

Quality

This has been defined elsewhere of course (see Chapter 1). The *non-quality*-related measures include the simple counts of defect or error rates (perhaps in parts per million), percentage outside specification or process capability measures such as Cp/Cpk values; deliveries not on time, or more generally as the costs of poor quality. When the positive costs of prevention of poor quality are included, these provide a balanced measure of the costs of quality (see next section).

The quality measures should also indicate positively whether we are doing a good job in terms of customer satisfaction, implementing the objectives, and whether the designs, systems, and solutions to problems are meeting the requirements. These really are voice-of-the-customer measures.

Impact

Impact measures should lead to key performance indicators for the business or organization, including monitoring improvement over time. Value-added management (VAM) requires the identification and elimination of all non-value-adding wastes, including time. Value added is simply the volume of sales (or other measure of 'turnover') minus the total input costs, and provides a good direct measure of the impact of the improvement activities on the performance of the business. A related ratio, percentage return on value added (ROVA) is another financial indicator that may be used:

$$\text{ROVA} = \frac{\text{Net profits before tax}}{\text{Value added}} \times 100 \text{ percent}$$

Other measures or indicators of impact on the business are *growth* in sales, assets, numbers of passengers/students, etc., and *asset-utilization* measures such as return on investment (ROI) or capital employed (ROCE), earnings per share, etc.

Some of the impact measures may be converted to people productivity ratios, e.g.:

$$\frac{\text{Value added}}{\text{Number of employees (or employee costs)}}$$

Activity-based costing (ABC) is an information system that maintains and processes data on an organization's activities and cost objectives. It is based on the activities performed being identified and the costs being traced to them. ABC uses various 'cost drivers' to trace the cost of activities to the cost of the products or services. The activity and cost-driver concepts are the heart of ABC. Cost drivers reflect the demands placed on activities by products, services or other cost targets. Activities are processes or procedures that cause work and thereby consume resources. This clearly measures impact, both on and by the organization.

■ Costs of quality

Manufacturing a quality product, providing a quality service, or doing a quality job – one with a high degree of customer satisfaction – is not enough. The cost of achieving these goals must be carefully managed, so that the long-term effect on the business or organization is a desirable one. These costs are a true measure of the quality effort. A competitive product or service based on a balance between quality and cost factors is the principal goal of responsible management and may be aided by a competent analysis of the costs of quality (COQ).

The analysis of quality-related costs is a significant management tool that provides:

- A method of assessing the effectiveness of the management of quality.
- A means of determining problem areas, opportunities, savings, and action priorities.

The costs of quality are no different from any other costs. Like the costs of maintenance, design, sales, production/operations, and other activities, they can be budgeted, measured and analyzed.

Having specified the quality of design, the operating units have the task of matching it. The necessary activities will incur costs that may be separated into prevention costs, appraisal costs and failure costs, the so-called P-A-F model first presented by Feigenbaum. Failure costs can be further split into those resulting from internal and external failure.

Prevention costs

These are associated with the design, implementation and maintenance of the quality management system. Prevention costs are planned and are incurred before actual operation. Prevention includes:

Product or service requirements

The determination of requirements and the setting of corresponding specifications (which also takes account of process capability) for incoming materials, processes, intermediates, finished products and services.

Quality planning

The creation of quality, reliability, and operational, production, supervision, process control, inspection and other special plans, e.g. pre-production trials, required to achieve the quality objective.

Quality assurance

The creation and maintenance of the quality system.

Inspection equipment

The design, development and/or purchase of equipment for use in inspection work.

Training

The development, preparation and maintenance of training programs for operators, supervisors, staff, and managers both to achieve and maintain capability.

Miscellaneous

Clerical, travel, supply, shipping, communications and other general office management activities associated with quality.

Resources devoted to prevention give rise to the *'costs of doing it right the first time'*.

Appraisal costs

These costs are associated with the supplier's and customer's evaluation of purchased materials, processes, intermediates, products and services to assure conformance with the specified requirements. Appraisal includes:

Verification

Checking of incoming material, process set-up, first-offs, running processes, intermediates and final products, including product or service performance appraisal against agreed specifications.

Quality audits

To check that the quality system is functioning satisfactorily.

Inspection equipment

The calibration and maintenance of equipment used in all inspection activities.

Vendor rating

The assessment and approval of all suppliers, of both products and services.

Appraisal activities result in the *'costs of checking it is right'*.

Internal failure costs

These costs occur when the results of work fail to reach designed quality standards and are detected before transfer to the customer takes place. Internal failure includes the following:

Waste

The activities associated with doing unnecessary work or holding stocks as the result of errors, poor organization or poor communications, the wrong materials, etc.

Scrap

Defective product, material or stationery that cannot be repaired, used or sold.

Rework or rectification

The correction of defective material or errors to meet the requirements.

Reinspection

The re-examination of products or work that have been rectified.

Downgrading

A product that is usable but does not meet specifications may be downgraded and sold as 'second quality' at a lower price.

Failure analysis

The activity required to establish the causes of internal product or service failure.

External failure costs

These costs occur when products or services fail to reach design quality standards but are not detected until after transfer to the consumer. External failure includes:

Repair and servicing

Either of returned products or those in the field.

Warranty claims

Failed products that are replaced or services re-performed under some form of guarantee.

Complaints

All work and costs associated with handling and servicing of customers' complaints.

Returns

The handling and investigation of rejected or recalled products or materials, including transport costs.

Liability

The result of product or service liability litigation and other claims, which may include a change of contract.

Loss of good will

The impact on reputation and image, which impinges directly on future prospects for sales.

External and internal failure produce the *'costs of getting it wrong'*.

Order re-entry, unnecessary travel and telephone calls, conflict are just a few examples of the wastage or failure costs often excluded. Every organization should be aware of the costs of getting it wrong, and management needs to obtain some idea how much failure is costing each year.

Clearly, this classification of cost elements may be used to interrogate any internal transformation process. Using the internal customer requirements concept as the standard for failure, these cost assessments can be made wherever information, data, materials, service or artefacts are transferred from one person or one department to another. It is the 'internal' costs of lack of quality that lead to the claim that approximately one-third of *all* our efforts are wasted.

The relationship between the quality-related costs of prevention, appraisal, and failure and increasing quality awareness and improvement in the organization is shown in Figure 7.1.

Where the quality awareness is low the total quality-related costs are high, the failure costs predominating. As awareness of the cost to the organization of failure gets off the ground, through initial investment in training, an increase in appraisal costs usually results. As the increased

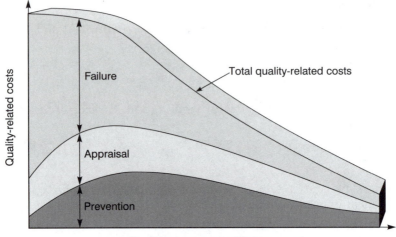

Figure 7.1 Increasing quality awareness and improvement activities. (Source: British Standard BS 6143: 1991)

appraisal leads to investigations and further awareness, further investment in prevention is made to improve design features, processes and systems. As the preventive action takes effect, the failure *and* appraisal costs fall and the total costs reduce.

The first presentations of the P-A-F model suggested that there may be an optimum operating level at which the combined costs are at the minimum. The author, however, has not yet found one organization in which the total costs have risen following investment in prevention.

Case study ■■■

One of BT Retail's strategic objectives has been to reduce the cost of failure in their operations. This delivers three benefits, it is good for customers as they experience fewer things going wrong, it is good for BT's people as they do not have to deal with the hassle of fixing problems, and it saves BT money. In 2001/02 BT Retail saved £47 m (*c*. $70 m) from specific cost of failure reduction programs.

The approach to the reduction of cost of failure is based on clear analysis of where BT spends money on doing things wrong or fixing things that have gone wrong. However, this often simply highlights those operations and processes that are managing failure. Having identified where failure occurred, effort was then put into properly establishing the root cause of the failure. For example, BT engineers sometimes find that they cannot easily get access to customers' premises to carry out work, this can be caused by call centers not taking all the right details when taking the initial customer order.

■ The process model for quality costing

The P-A-F model for quality costing has a number of drawbacks. In the management of quality, prevention of problems, defects, errors, waste, etc., is one of the prime functions, but it can be argued that everything a well-managed organization does is directed at preventing quality problems. This makes separation of *prevention costs* very difficult. There are clearly a range of prevention activities in any manufacturing or service organization that are integral to ensuring quality but may never be included in the schedule of quality-related costs.

It may be impossible and unnecessary to categorize costs into the three categories of P-A-F. For example, a design review may be considered a prevention cost, an appraisal cost, or even a failure cost, depending on how and where it is used in the process. Another criticism of the P-A-F model is that it focuses attention on cost reduction and plays down, or in some cases even ignores, the positive contribution made to price and sales volume by improved quality.

The most serious criticism of the original P-A-F model presented by Feigenbaum and used in, for example, British Standard 6143 (1981) 'Guide to the determination and use of quality related costs', is that it implies an acceptable 'optimum' quality level above which there is a trade-off between investment in prevention and failure costs. Clearly, this is not in tune with the never-ending improvement philosophy. The key focus should be on process improvement, and a cost categorization scheme that does not consider process costs, such as the P-A-F model, has limitations. (BS 6143-2 was republished in 1990 as 'Guide to the economies of quality: prevention, appraisal and failure model'.)

In a quality-related costs system that focuses on processes rather than products or services, the operating costs of generating customer satisfaction will be of prime importance. The so-called 'process cost model', described in the revised BS 6143-1 (1992), sets out a method for applying quality costing to any process or service. It recognizes the importance of process ownership and measurement, and uses process modeling to simplify classification. The categories of the cost of quality (COQ) have been rationalized into the cost of conformance (COC) and the cost of non-conformance (CONC):

$$COQ = COC + CONC$$

The cost of conformance (COC) is the process cost of providing products or services to the required standards, by a given specified process in the most effective manner, i.e. the cost of the ideal process where every activity is carried out according to the requirements first time, every time. The cost of non-conformance (CONC) is the failure cost associated with the process not being operated to the requirements, or the cost due to variability in the process. Part 2 of BS 6143 (1991) still deals with the P-A-F model, but without the 'optimum'/minimum cost theory (see Figure 7.1).

Process cost models can be used for any process within an organization and developed for the process by flowcharting or use of the ICOR methodology (inputs, controls, outputs, resources – see Chapter 10). This will identify the key process steps and the parameters that are monitored in the process. The process cost elements should then be identified and recorded under the four categories. The COC and CONC for each stage of the process will comprise a list of all the parameters monitored.

Steps in process cost modeling

Process cost modeling is a methodology that lends itself to stepwise analysis, and the following are the key stages in building the model:

1 Choose a key process to be analyzed, identify and name it, e.g. Retrieval of Medical Records (Acute Admissions).
2 Define the process and its boundaries.
3 Construct the process diagram:
   (a) identify the outputs and customers (for example, see Figure 7.2);
   (b) identify the inputs and suppliers (for example, see Figure 7.3);
   (c) identify the controls and resources (for example, see Figure 7.4).

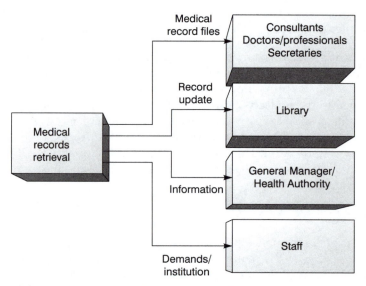

■ **Figure 7.2** Building the model: identify outputs and customers

4 Flow chart the process and identify the process owners (for example, see Figure 7.5). Note, the process owners will form the improvement team.

5 Allocate the activities as COC or CONC (see Table 7.1).

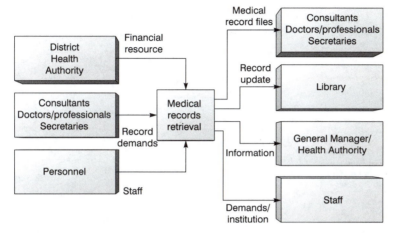

Figure 7.3 Building the model: identify inputs and suppliers

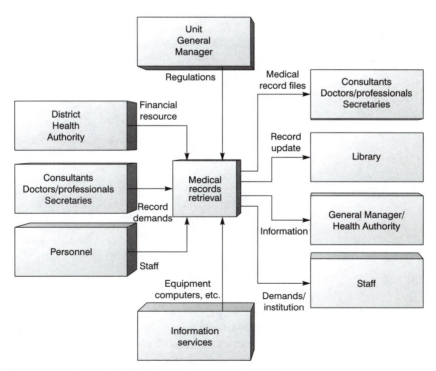

Figure 7.4 Building the model: identify controls and resources

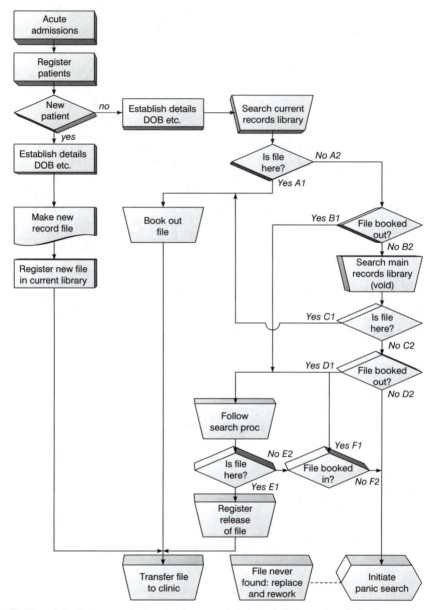

■ **Figure 7.5**   Present practice flowchart for acute admissions medical records retrieval

6 Calculate or estimate the quality costs (COQ) at each stage (COC + CONC). Estimates may be required where the accounting system is unable to generate the necessary information.

7 Construct a process cost report (see Table 7.2). The report summary and results are given in Table 7.3.

Table 7.1 Building the model: allocate activities as COC or CONC

Key activities	COC	CONC
Search for files	Labor cost incurred finding a record while adhering to standard procedure	Labor cost incurred finding a record while unable to adhere to standard procedure
Make up new files	New patient files	Patients whose original files cannot be located
Rework		Cost of labor and materials for all rework files/records never found
Duplication		Cost incurred in duplicating existing files

Table 7.2 Building the model: process cost report

Process cost report
Process: medical records retrieval (acute admissions)
Process owner: various
Time allocation: 4 days (96 hrs)

Process COC	Process CONC	Cost details	Definition	Source	Cost
	Labor cost incurred finding records	# ref. Sample	Cost of time required to find missing records	Medical records	$210
	Cost incurred making up replacement files	#	Labor and material costs multiplied by number of files replaced	Medical records	$108
	Rework	#	Labor and material cost of all rework	Medical records	$80
	Duplication	#		Medical records	$24

There are three further steps carried out by the process owners – the improvement team – which take the process forward into the improvement stage:

8 Prioritize the failure costs and select the process stages for improvement through reduction in costs of non-conformance (CONC).

Table 7.3 Process cost model: report summary

Labor cost
    14 hrs × $12.00/hr = $168
    $168 + overhead and contribution factor 25%
    = $210

Replacement costs
    No. of files unfound 9
    Cost of replace each file $12.00
    Overall cost $108

Rework costs
    2 × Pathology reports to be word processed $80

Duplication costs
    No. of file duplicated 2
    Cost per file $12.00
    Overall cost $24

TOTAL COST $422

RESULTS
Acute admissions operated 24 hrs/day 365 days/year
This project established a cost of non-conformance of approx. $422
This equates to $422 × 365/4 = $38,507.50
Or two personnel fully employed for 12 months

This should indicate any requirements for investment in prevention activities. An excessive cost of conformance (COC) may suggest the need for process redesign.

9 Review the flowchart to identify the scope for reductions in the cost of conformance. Attempts to reduce COC require a thorough process understanding, and a second flowchart of what the new process should be may help (see Chapter 10).

10 Monitor conformance and non-conformance costs on a regular basis, using the model and review for further improvements.

The process cost model approach should be seen as more than a simple tool to measure the financial implications of the gap between the actual and potential performance of a process. The emphasis given to the process, improving the understanding, and seeing in detail where the costs occur, should be an integral part of quality improvement.

A performance measurement framework (PMF)

A performance measurement framework (PMF) is proposed, based on the strategic planning and process management models outlined in

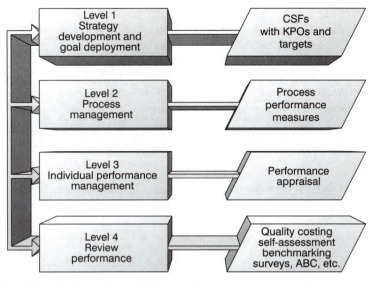

Figure 7.6 Performance measurement framework

Chapters 4 and 10. The framework has four elements related to: strategy development/goal deployment, process management, individual performance management, and review (Figure 7.6). This reflects an amalgamation of the approaches used by a range of organizations in performance measurement.

As we have seen in earlier chapters, the key to strategic planning and goal deployment is the identification of a set of critical success factors (CSFs) and associated key performance outcomes (KPOs). These factors should be derived from the organization's mission, and represent a balanced mix of stakeholders. Action plans over both the short- and long term should be developed, and responsibility clearly assigned for performance. The strategic goals of the organization should then be clearly communicated to all individuals, and translated into measures of performance at the process/functional level.

The key to successful performance measurement at the process level is the identification and translation of customer requirements and strategic objectives into an integrated set of process performance measures or indicators. The documentation and management of processes has been found to be vital in this translation process. Even when a functional organization is retained, it is necessary to treat the measurement of performance between departments as the measurement of customer supplier performance.

Performance measurement at the individual level usually relies on performance appraisal, formal planned performance reviews, and performance management, i.e. day-to-day management of individuals. A major drawback with some performance appraisal systems, of course, is the lack of their integration with other aspects of performance measurement.

Performance review techniques are used by many world class organizations to identify improvement opportunities, and to motivate performance improvement. These companies typically use a wide range of such techniques and are innovative in performance measurement in their drive for continuous improvement.

The links between performance measurement at the four levels of the framework are based on the need for measurement to be part of a systematic process of continuous improvement, rather than for 'control'. The framework provides for the development and use of measurement, rather than prescriptive lists of measures that should be used. It is, therefore, applicable in all types of organization.

The elements of the performance measurement are distinct from the budgetary control process, and also from the informal control systems used within organizations. Having said that, performance measurement should not be treated as a separate isolated system. Instead measurement is documented as and when it is used at the organizational, process and individual levels. In this way it can facilitate the alignment of the goals of all individuals, teams, departments and processes with the strategic aims of the organization and incorporate the voice of the stakeholders in all planning and management activities.

A number of factors have been found to be critical to the success of performance measurement systems. These factors include the level of top management support for non-financial performance measures, the identification of the vital few measures, the involvement of all individuals in the development of performance measurement, the clear communication of strategic objectives, the inclusion of customers and suppliers in the measurement process, and the identification of the key drivers of performance. These factors will need to be taken into account by managers wishing to develop a new performance measurement system, or refine an existing one.

In most organizations there are no separate performance measurement systems. Instead, performance measurement forms part of wider organizational management processes. Although elements of measurement can be identified at many different points within organizations, measurement itself usually forms the 'check' stage of the continuous improvement PDCA cycle. This is important since measurement data that is collected but not acted upon in some way is clearly a waste of resources.

The five elements of the framework in Figure 7.6 are:

Level 1 **Strategy development and goal deployment** leading to mission/vision, critical success factors and **key performance outcomes** (KPOs).

Level 2 Process management and **process performance measurement** through **key performance indicators (KPIs)** (including input, in process and output measures, management of internal and external customer/supplier relationships and the use of management control systems).

Level 3 **Individual** performance management and **performance appraisal.**

Level 4 **Review performance** (including internal and external benchmarking, self-assessment against quality award criteria and quality costing).

Level 1 – Strategy development and goal deployment

The first level of the performance measurement framework is the development of organizational strategy, and the consequent deployment of goals throughout the organization. Steps in the strategy development and goal deployment measurement process are, (see also Chapter 4):

1 Develop a mission statement based on recognizing the needs of all organizational stakeholders, customers, employees, shareholders and society. Based on the mission statement, identify those factors critical to the success of the organization achieving its stated mission. The CSFs should represent all the stakeholder groups, customers, employees, shareholders and society.

2 Define performance measures for each CSF – i.e. key performance outcomes (KPOs). There may be one or several KPOs for each CSF. Definition of KPOs should include:
   (a) title of KPO;
   (b) data used in calculation of KPO;
   (c) method of calculation of KPO;
   (d) sources of data used in calculation;
   (e) proposed measurement frequency;
   (f) responsibility for the measurement process.

3 Set targets for each KPO. If KPOs are new, targets should be based on customer requirements, competitor performance or known organizational criteria. If no such data exists, a target should be set based on best guess criteria. If the latter is used, the target should be updated as soon as enough data is collected to be able to do so.

4 Assign responsibility at the organizational level for achievement of desired performance against KPO targets. Responsibility should rest with directors and very senior managers.

5 Develop plans to achieve the target performance. This includes both action plans for one year, and longer-term strategic plans.

6 Deploy mission, CSFs, KPOs, targets, responsibilities and plans to the core business processes. This includes the communication of goals, objectives, plans, and the assignment of responsibility to appropriate individuals.

7 Measure performance against organizational KPOs, and compare to target performance.

8 Communicate performance and proposed actions throughout the organization.

9 At the end of the planning cycle compare organizational capability to target against all KPOs, and begin again at step 2 above.

10 Reward and recognize superior organizational performance.

Strategy development and goal deployment is clearly the responsibility of senior management within the organization, although there should be as much input to the process as possible by employees to achieve 'buy-in' to the process.

The system outlined above is similar to the policy deployment approach known as Hoshin Kanri, developed in Japan and adapted in the West.

Level 2 – Process management and measurement

The second level of the performance measurement framework is process management and measurement, the steps of which are:

1 If not already completed, identify and map processes. This information should include identification of:
   (a) process customers and suppliers (internal and external);
   (b) customer requirements (internal and external);
   (c) core and non-core activities;
   (d) measurement points and feedback loops.

2 Translate organizational goals, action plans and customer requirements into process performance measures (input, in-process and output) – key performance indicators (KPIs). This includes definition of measures, data collection procedures, and measurement frequency.

3 Define appropriate performance targets, based on known process capability, competitor performance and customer requirements.

4 Assign responsibility and develop plans for achieving process performance targets.

5 Deploy measures, targets, plans and responsibility to all subprocesses.

6 Operate processes.
7 Measure process performance and compare to target performance.
8 Use process performance information to:
   (a) implement continuous improvement activities;
   (b) identify areas for improvement;
   (c) update action plans;
   (d) update performance targets;
   (e) redesign processes, where appropriate;
   (f) manage the performance of teams and individuals (perfor-
        mance management and appraisal) and external suppliers;
   (g) provide leading indicators and explain performance against
        organizational KPOs.
9 At the end of each planning cycle compare process capability to customer requirements against all measures, and begin again at step 2.
10 Reward and recognize superior process performance, including
   subprocesses, and teams.

The same approach can be deployed to subprocesses and to the activity
and task levels.

The above process should be managed by the process owners, with inputs
wherever possible from the owners of subprocesses. The process outlined
should be used whether organization and management are on a process
or functional departmental basis. If functionally organized, the key
task is to identify the customer/supplier relationships between functions,
and for functions to see themselves as part of a customer/ supplier chain.

Key performance outcomes (KPOs) and indicators (KPIs)

The derivation of KPOs and KPIs may follow the 'balanced scorecard'
model, proposed by Kaplan, which divides measures into financial,
customer, internal business and innovation and learning perspectives
(Figure 7.7).

A balanced scorecard derived from the business excellence model
described in Chapters 2 and 8 would include key performance results,
customer results (measured via the use of customer satisfaction surveys
and other measures, including quality), people results (employee development and satisfaction), and society results (including community
perceptions and environmental performance). In the areas of customers,
people, and society there needs to be a clear distinction between
perception measures and other performance measures. Moreover, these
may be used in tandem to ensure improvements in performance or in
perception. Perception can become out of date with actual performance and education of the customer – say about latest delivery
performance – may become necessary.

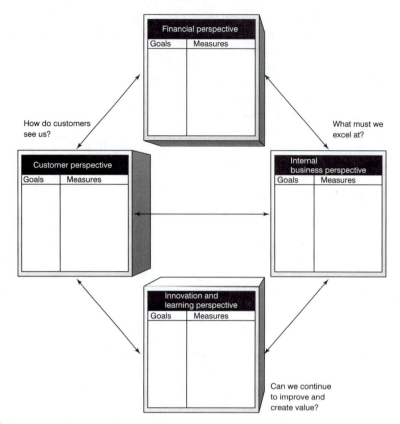

■ **Figure 7.7** The balanced scorecard linking performance measures

Financial performance for external reporting purposes may be seen as a result of performance across the other KPOs, the non-financial KPOs and KPIs assumed to be the leading indicators of performance. The only aspect of financial performance that is cascaded throughout the organization is the budgetary process, which acts as a constraint rather than a performance improvement measure.

In summary then, organizational KPOs and KPIs should be derived from the balancing of internal capabilities against the requirements of identified stakeholder groups. This has implications for both the choice of KPOs/KPIs and the setting of appropriate targets. There is a need to develop appropriate action plans and clearly define responsibility for meeting targets if they are to be taken seriously.

Performance measures used at the process level differ widely between different organizations. Some organizations measure process performance using a balanced scorecard approach, while others monitor performance across different dimensions according to the process. Whichever method is used, measurements should be identified as input (supplier), in-process, and output (or results-customers).

It is usually at the process level that the greatest differences can be observed between the measurement used in manufacturing and services organizations. However, all organizations should measure quality, delivery, customer service/satisfaction, and cost.

Depending on the process, measurement frequency varies from daily, for example in the measurement of delivery performance, to annual, for example in the measurement of employee satisfaction, which has implications for the PDCA cycle time of the particular process(es). Measurement frequency at the process level may, of course, be affected by the use of information technology. Cross-functional process performance measurement is a vital component in the removal of 'functional silos', and the consequent potential for suboptimization and failure to take account of customer requirements. The success of performance measurement at the process level is dependent on the degree of management of processes and on the clarity of the deployment of strategic organizational objectives.

Measuring and managing the whats and the hows

Busy senior management teams find it useful to distil as many things as possible down to one piece of paper or one spreadsheet. The use of key performance outcomes (KPOs), with targets, as measures for CSFs, and the use of key performance indicators (KPIs) for processes may be combined into one matrix which is used by the senior management team to 'run the business'.

Figure 7.8 (also shown in Chapter 4) is an example of such a matrix which is used to show all the useful information and data needed:

▪ the CSFs and their owners – the *whats*;
▪ the KPOs and their targets;
▪ the core business processes and their sponsors – the *hows*;
▪ the process performance measures – KPIs.

It also shows the impacts of the core processes on the CSFs. This may be used in conjunction with a 'business management calendar', which shows when to report/monitor performance in order to identify process areas for improvement. This slick process offers senior teams a way of:

▪ gaining clarity about what is important and how it is measured;
▪ remaining focused on what is important and what the performance is;
▪ knowing where to look if problems occur.

Figure 7.8 CSF/core process reporting matrix

CSFs: We must have	Measures	Year targets	Target CSF owner
Satisfactory financial and non-financial performance	Sales volume. Profit. Costs versus plan. Shareholder return. Associate/employee utilization figures	Turnover $2m. Profit $200k. Return for shareholders. Days/month per person	
A growing base of satisfied customers	Sales/customer. Complaints/recommendations. Customer satisfaction	>$200k = 1 client. $100k-$200k = 5 clients. $50k-$100k = 6 clients. <$50k = 12 clients	
A sufficient number of committed and competent people	No. of employed staff/associates. Gaps in competency matrix. Appraisal results. Perceptions of associates and staff	15 employed staff. 10 associates including 6 new by end of year	
Research projects properly completed and published	Proportion completed on time, in budget with customers satisfied. Number of publications per project.	3 completed on time, in budget with satisfied customers	

Core processes	Satisfactory financial and non-financial performance	A growing base of satisfied customers	A sufficient number of committed and competent people	Research projects properly completed and published	**= Priority for improvement	Process owner	Process performance	Measures and targets
Manage people	X	X	X	X	**			
Develop products		X						
Develop new business	X	X			**			
Manage our accounts	X	X		X	**			
Manage financials	X		X					
Manage int. systems	X	X	X					
Conduct research	X			X				

Level 3 – Individual performance and appraisal management

The third level of the performance measurement framework is the management of individuals. Performance appraisal and management is usually the responsibility of the direct managers of individuals whose performance is to be appraised. At all stages in the process, the individuals concerned must be included to ensure 'buy in'.

Steps in performance and management appraisal are:

1 If not already completed, identify and document job descriptions based on process requirements and personal characteristics. This information should include identification of:
   (a) activities to be undertaken in performing the job;
   (b) requirements of the individual with respect to the identified activities, in terms of experience, skills and training;
   (c) requirements for development of the individual, in terms of personal training and development.
2 Translate process goals and action plans, and personal training and development requirements into personal performance measures.
3 Define appropriate performance targets based on known capability and desired characteristics (or desired characteristics alone if there is no prior knowledge of capability).
4 Develop plans towards achievement of personal performance targets.
5 Document 1 to 4 using appropriate forms, which should include space for the results of performance appraisal.
6 Manage performance. This includes:
   (a) planning tasks on a daily/weekly basis;
   (b) managing performance of the tasks;
   (c) monitoring performance against task objectives using both quantitative and qualitative information on a daily and/or weekly basis;
   (d) giving feedback to individuals of their performance in carrying out tasks;
   (e) giving recognition to individuals for superior performance.
7 Formally appraise performance against range of measures developed, and compare to targets.
8 Use comparison with targets to:
   (a) identify areas for improvement;
   (b) update action plans;
   (c) update performance targets;
   (d) redesign jobs, where appropriate. This impacts step 1 of the process.
9 After a suitable period, ideally more than once a year, compare capability to job requirements and begin again at step 2.
10 Reward and recognize superior performance.

The above activities should be undertaken by the individual whose performance is being managed, together with their immediate superior.

A key to success in the management of individuals lies in the reward of effort as well as achievement and the consequently different measures used, and in the use of information in continuous improvement required to reward and recognize performance, including teamwork. Unlike management by objectives (MBO), where the focus is on measurement of results – which are often beyond the control of the individual whose performance is appraised – good performance management systems attempt to measure a combination of process/task performance (effort and achievement) and personal development.

The frequency of formal performance appraisal is generally defined by the frequency of the appraisal process, usually with a minimum frequency of six months. Between the formal performance appraisal reviews, most organizations rely on the use of other performance management techniques to manage individuals. Measures of team performance, or of participation in teams, should be included in the appraisal systems where possible, to improve team performance. In many organizations, the performance appraisal system is probably the least successfully implemented element of the framework. Appraisal systems are often designed to motivate individuals to achieve process and personal development objectives, but not to perform in teams. One of the limitations of appraisal processes is the frequency of measurement, which could be increased, but few organizations would consider doing so.

Level 4 – Performance review

The fourth level of the performance measurement framework is the use of performance review techniques. Steps in the review are as follows:

1 Identify the need for review, which may come from:
   (a) poor performance at the organizational or process levels against KPO/KPIs;
   (b) identified superior performance of competitors;
   (c) customer inputs;
   (d) the desire to better direct improvement efforts;
   (e) the desire to concentrate attention on the need for performance improvement.
2 Identify the method of performance review to be used. This involves determining whether the review should be carried out internally within the organization, or externally, and the method to be used. Some techniques are mainly internal, e.g. self-assessment, quality costing;

while others, e.g. benchmarking, involve obtaining information from sources external to the organization. The choice should depend on:
  (a) how the need for review was identified (see 1);
  (b) the aim of the review, e.g. if the aim is to improve performance relative to competitors, external benchmarking may be a better option than internally measuring the cost of quality;
  (c) the relative costs and expected benefits of each technique.
3 Carry out the review.
4 Feed results into the planning process at the organizational or process level.
5 Determine whether to repeat the exercise. If it is decided to repeat the exercise, the following points should be considered:
  (a) frequency of review.
  (b) at what levels to carry out future reviews, e.g. organization-wide or process by process.
  (c) decide whether the review technique should be incorporated into regular performance measurement processes, and if so how this will be managed.

Review methods often require the use of a level of resources greater than that normally associated with performance measurement, often due to the need to develop data collection procedures, train people in their use, and the cost of data collection itself. However, review techniques usually give a broader view of performance than most individual measures.

The use of review techniques is most successful when it is based on a clearly identified need, perhaps due to perceived poor performance against existing performance measures or against competitors, and the activity itself is clearly planned and the results used in performance improvement. This is often the difference between the success and failure of quality costing and benchmarking in particular. The use of most of the review techniques has been widely documented, but often without regard to their integration into the wider processes of measurement and management.

Review techniques

Techniques identified for review include:

1 Quality costing, using either prevention-appraisal-failure, or process costing methods.
2 Self-assessment against the Baldrige or EFQM Excellence Model, or internally developed criteria.
3 Benchmarking, internal or external.
4 Customer satisfaction surveys.
5 Activity-based costing (ABC).

Case study ■ ■ ■ ▬▬▬▬▬▬▬▬▬▬▬▬▬▬▬▬▬

■ Best Value in Harrogate Borough Council

'Best Value' is a UK Government initiative which places a duty on all local councils and authorities to deliver the most economic and efficient services possible. Councils must report to their public and the Government each year on their performance, in addition to reviewing all their services to identify and achieve continual improvements. In this way the Government has challenged local councils to look at the way they deliver services and raise their quality at a reasonable cost.

This short case study looks at the way Harrogate Borough Council in North Yorkshire – the author's own district – has addressed the needs and challenges of Best Value through its performance plan.

The Corporate Action Plan sets out the planned actions and targets which deliver the council's corporate objectives and priorities. It enables the authority to look beyond immediate issues and problems and to plan ahead for the longer-term future of the district. The Corporate Action Plan links into both the Best Value Performance Plan and the service and business plans prepared by the council departments to deliver their part of the council's corporate plans and targets.

Some of the actions in the Corporate Action Plan are designed to meet a local need or policy issue while others are to address the council's current performance. All of them are agreed by the council for implementation, following consultation with local communities and partners in the district. The council reviews the Corporate Action Plan twice a year to measure the progress being made in meeting the council's longer-term vision and strategy through the achievement (or not) of service actions and targets each year.

The Corporate Action Plan is divided into action tables – one for each of the council's corporate objectives. The council agrees a number of key priority areas for action to help deliver each of its corporate objectives and these are set out in detail in the plan, together with the actions and targets planned under each priority area, and the links into the relevant service and other council plans. Details of the council's longer-term priorities and targets are set out in a separate 'corporate strategy' document.

The council's budget for the financial year is explained in detail in a separate 'Budget' document and each year, the council allocates funding in its General Fund Revenue Budget to enable it to deliver its annual corporate priorities and targets. Details of the council's funding of corporate priorities is set out in the Best Value Performance Plan.

A Best Value Performance Plan is generated for each coming financial year. This provides a snapshot of the council's performance and achievements for the previous year – what worked/ what did not – and looks forward to what the council needs to do to meet its commitment to provide high-quality, cost-effective services which meet the needs of the people of the Harrogate District.

The objectives and priorities are stated together with the long-term issues facing the district. The council's performance has improved in a number of areas and, where it has not improved, the council has taken action to address this. On the Government's national top 11 indicators for District Councils, Harrogate's performance in the year of the case study preparation was in the top quartile on five indicators, average performance on three indicators and below average performance on three indicators. Over 70 percent of people living in the district were satisfied with the overall service provided and the council met over two-thirds of its performance targets and 'almost met' a further 7 percent.

The implementation of performance measurement systems

It has already been established that a good measurement system will start with the customer and measure the 'right' things. The value of any measure clearly needs to be compared with the cost of producing it. There will be appropriate measures for different parts of the organization, but everywhere they must relate process performance to the needs of the process customer. All critical parts of the process must be measured, and it is often better to start with simple measures and improve them.

There must be a recognition of the need to distinguish between different measures for different purposes. For example, an operator may measure time, various process parameters, and amounts, while at the management level measuring costs and delivery timeliness may be more appropriate.

Participation in the development of measures enhances their understanding and acceptance. Process owners can assist in defining the required performance measures, provided that senior managers have communicated their mission clearly, determined the critical success factors, and identified the core processes.

If all employees participate, and own the measurement processes, there will be lower resistance to the system, and a positive commitment towards future changes will be engaged. This will derive from the 'volunteered accountability', which will in turn make the individual contribution more visible. Involvement in measurement also strengthens the links in the customer/supplier chains and gives quality improvement teams much clearer objectives. This should lead to greater short-term and long-term productivity gains.

There are a number of possible reasons why measurement systems fail:

1 They do not define performance operationally.
2 They do not relate performance to the processes.

3 The boundaries of the processes are not defined.
4 The measures are misunderstood or misused or measure the wrong things.
5 There is no distinction between control and improvement.
6 There is a fear of exposing poor and/or good performance.
7 It is seen as an extra burden in terms of time and reporting.
8 There is a perception of reduced autonomy.
9 Too many measurements are focused internally and too few are focused externally.
10 There is a fear of the introduction of tighter management controls.

These and other problems are frequently due to poor planning at the implementation stage or a failure to assess current systems of measurement. Before the introduction of a total quality-based performance measurement system, an audit of the existing systems should be carried out. Its purpose is to establish the effectiveness of existing measures, their compatibility with the quality drive, their relationship with the processes concerned, and their closeness to the objectives of meeting customer requirements. The audit should also highlight areas where performance has not been measured previously, and indicate the degree of understanding and participation of the employees in the existing systems and the actions that result.

Generic questions that may be asked during the audit include:

- Is there a performance measurement system in use?
- Has it been effectively communicated throughout the organization?
- Is it systematic?
- Is it efficient?
- Is it well understood?
- Is it applied?
- Is it linked to the mission and objectives of the organization?
- Is there a regular review and update?
- Is action taken to improve performance following the measurement?
- Are the people who own the processes engaged in measuring their own performance?
- Have employees been properly trained to conduct the measurement?

Following such an audit, there are 12 basic steps for the introduction of an effective performance measurement. Half of these are planning steps and the other half implementation.

Planning

1 Identify the purpose of conducting measurement, i.e. is it for:
   (a) reporting, e.g. ROI reported to shareholders;

(b) controlling, e.g. using process data on control charts;

(c) improving, e.g. monitoring the results of a quality improvement team project.

2 Choose the right balance between individual measures (activity- or task-related) and group measures (process- and subprocess-related) and make sure they reflect process performance.

3 Plan to measure all the key elements of performance, not just one, e.g. time, cost, and product quality variables may all be important.

4 Ensure that the measures will reflect the voice of the internal/external customers.

5 Carefully select measures that will be used to establish standards of performance.

6 Allow time for the learning process during the introduction of a new measurement system.

Implementation

7 Ensure full participation during the introductory period and allow the system to mold through participation.

8 Carry out cost/benefit analysis on the data generation, and ensure measures that have high 'leverage' are selected.

9 Make the effort to spread the measurement system as widely as possible, since effective decision making will be based on measures from *all* areas of the business operation.

10 Use *surrogate* measures for subjective areas where quantification is difficult, e.g. improvements in morale may be 'measured' by reductions in absenteeism or staff turnover rates.

11 Design the measurement systems to be as flexible as possible, to allow for changes in strategic direction and continual review.

12 Ensure that the measures reflect the quality approach by showing small incremental achievements that match the never-ending improvement approach.

In summary, the measurement system must be designed, planned and implemented to reflect customer requirements, give visibility to the processes and the progress made, communicate the quality effort and engage the never-ending improvement cycle. So it must itself be periodically reviewed.

Chapter highlights

Performance measurement and the improvement cycle

■ Traditional performance measures based on cost-accounting information provide little to support quality management, because they do not map process performance and improvements seen by the customer.

■ Measurement is important in tracking progress, identifying opportunities, and comparing performance internally and externally. Measures, typically non-financial, are used in process control and performance improvement.

■ Some financial indicators, such as ROI, are often inaccurate, irrelevant and too late to be used as measures for performance improvement.

■ The improvement cycle of Plan, Do, Check, Act is a useful design aid for measurement systems, but first four basic questions about measurement should be asked, i.e. why, what, where, and how.

■ In answering the question 'how to measure?' progress is important in five main areas: effectiveness, efficiency, productivity, quality and impact.

■ Activity-based costing (ABC) is based on the activities performed being identified and costs traced to them. ABC uses cost drivers, which reflect the demands placed on activities.

Costs of quality

■ A competitive product or service based on a balance between quality and cost factors is the principal goal of responsible management.

■ The analysis of quality-related costs may provide a method of assessing the effectiveness of the management of quality and of determining problem areas, opportunities, savings, and action priorities.

■ Quality costs may be categorized into prevention, appraisal, internal failure, and external failure costs, the P-A-F model.

■ Prevention costs are associated with doing it right the first time, appraisal costs with checking it is right, and failure costs with getting it wrong.

■ When quality awareness in an organization is low, the total quality-related costs are high, the failure costs predominating. After an initial rise in costs, mainly through the investment in training and appraisal, increasing investment in prevention causes failure, appraisal and total costs to fall.

The process model for quality costing

■ The P-A-F model for quality costing has a number of drawbacks, mainly due to estimating the prevention costs, and its association with an 'optimized' or minimum total cost.

■ An alternative – the process costs model – rationalizes cost of quality (COQ) into the costs of conformance (COC) and the cost of non-conformance (CONC). COQ = COC + CONC at each process stage.

▓ Process cost modeling calls for choice of a process and its definition; construction of a process diagram; identification of outputs and customers, inputs and suppliers, controls and resources; flowcharting the process and identifying owners; allocating activities as COC or CONC; and calculating the costs. A process cost report with summaries and results is produced.

▓ The failure costs of CONC should be prioritized for improvements.

A performance measurement framework

▓ A suitable performance measurement framework (PMF) has four elements related to strategy development and goal deployment, process management, individual performance management, and review.

▓ The key to successful performance measurement at the strategic level is the identification of a set of critical success factors (CSFs) and associated key performance outcomes (KPOs).

▓ The key to success at the process level is the identification and translation of customer requirements and strategic objectives into a process framework, with process performance measures (KPIs).

▓ The key to success at the individual level is performance appraisal and planned formal reviews, through integrated performance management.

▓ The key to success in the review stage is the use of appropriate innovative techniques to identify improvement opportunities followed by good implementation.

▓ A number of factors are critical to the success of performance measurement systems including top management support for non-financial performance measures, the identification of the vital few measures, the involvement of all individuals in the development of performance measurement, the clear communication of strategic objectives, the inclusion of customers and suppliers in the measurement process, and the identification of the key drivers of performance.

The implementation of performance measurement systems

▓ The value of any measure must be compared with the cost of producing it. All critical parts of the process must be measured, but it is often better to start with the simple measures and improve them.

▓ Process owners should take part in defining the performance measures, which must reflect customer requirements.

▓ Prior to introducing performance measurement, an audit of existing systems should be carried out to establish their effectiveness, compatibility, relationship and closeness to the customer.

■ Following the audit, there are 12 basic steps for implementation, six of which are planning steps. The measurement system, then, must be designed, planned and implemented to reflect customer requirements, give visibility to the processes and progress made, communicate the total quality effort and drive continuous improvement. It must also be periodically reviewed.

■
■
■ Self-assessment, audits
■ and reviews
■

■
■ Frameworks for self-assessment

Organizations everywhere are under constant pressure to improve their business performance, measure themselves against world class standards and focus their efforts on the customer. To help in this process, many are turning to total quality models such as the European Foundation for Quality Management's (EFQM) Excellence Model promoted in the UK by the British Quality Foundation (see also Chapter 2).

'Total quality' is the goal of many organizations but it has been difficult until relatively recently to find a universally accepted definition of what this actually means. For some people TQ means statistical process control (SPC) or quality management systems, for others teamwork and involvement of the workforce. More recently, in some organizations, it has been replaced by the terms business excellence, six sigma or lean manufacturing.

Clearly there are many different views on what constitutes the 'excellent' organization and, even with an understanding of a framework, there exists the difficulty of calibrating the performance or progress of any organization towards it.

The so-called excellence models now available recognize that customer satisfaction, business objectives, safety, and environmental considerations are mutually dependent and are applicable in any organization. Clearly the application of these ideas involves investment primarily in people and time, time to implement new concepts, time to train, time for people to recognize the benefits and move forward into new or different organizational cultures. But how will organizations know when they are getting close to excellence or whether they are even on the right road, how will they measure their progress and performance?

There have been many recent developments and there will continue to be many more, in the search for a standard or framework, against which organizations may be assessed or measure themselves, and carry out the so-called 'gap analysis'. To many the ability to judge progress against an accepted set of criteria would be most valuable and informative.

Most quality management approaches strongly emphasize measurement, some insist on the use of cost of quality. The value of a structured discipline using a points system has been well established in quality and safety assurance systems (for example, ISO 9000 and vendor auditing). The extension of this approach to a total quality auditing process has been long established in the Japanese 'Deming Prize' which is perhaps the most demanding and intrusive auditing process and there are other excellence models and standards used throughout the world. Perhaps the most famous and widely used framework for self-assessment is the US Baldrige 'Criteria for Performance Excellence'. Many companies have realized the necessity to assess themselves against the Baldrige and Deming criteria, if not to enter for the awards or prizes, then certainly as a basis for self-audit and review, to highlight areas for priority attention and provide internal and external benchmarking. (See Chapter 2 for details of the Deming Prize and Baldrige Award criteria.)

The European Excellence model for self-assessment

In Europe it has also been recognized that the technique of self-assessment is very useful for any organization wishing to monitor and improve its performance. In 1992 the European Foundation for Quality Management (EFQM) launched a European Quality Award which is now widely used for systematic review and measurement of operations. The EFQM Excellence model recognizes that processes are the means by which a company or organization harnesses and releases the talents of its people to produce results (performance).

Figure 8.1 displays graphically the principle of the full Excellence model.* As described in Chapter 2, customer results, employee results,

* A good resource is 'The Model in Practice – using the EFQM Excellence Model to deliver continuous improvement', Vols 1 and 2, published by the British Quality Foundation (BQF), London, 2002.

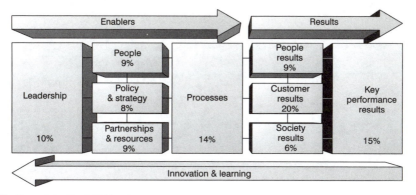

and favorable society results are achieved through leadership driving policy and strategy, people, partnerships, resources, and processes, which lead ultimately to excellence in key performance results – the enablers deliver the results which in turn drive innovation and learning. The EFQM have provided a weighting for each criterion, shown in Figure 8.1, which may be used in scoring self-assessments and making awards. The weightings are not rigid and may be modified to suit specific organizational needs.

The EFQM have thus built a model of criteria and a review framework against which an organization may face and measure itself, to examine any 'gaps'. Such a process is known as self-assessment and organizations such as the EFQM, and in the UK the BQF, publish guidelines for self-assessment, including specific ones directed at public sector and other types of organizations.

Many managers feel the need for a rational basis on which to measure progress in their organization, especially in those companies 'a few years into TQM' which would like the answers to questions such as: 'Where are we now?', 'Where do we need/want to be?', and 'What have we got to do to get there'. These questions need to be answered from internal employees' views, the customers' views, and the views of suppliers.

Self-assessment promotes business excellence by involving a regular and systematic review of processes and results. It highlights strengths and improvement opportunities, and drives continuous improvement.

Enablers

In the Excellence Model, the enabler criteria of leadership, policy and strategy, people, resources and partnerships and processes focus on what is needed to be done to achieve results. The structure of the enabler

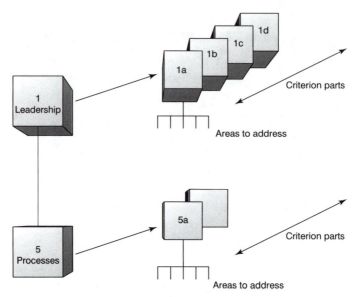

Figure 8.2   Structure of the criteria: enablers

criteria is shown in Figure 8.2. Enablers are assessed on the basis of the combination of two factors (see Figure 8.3):

1   The degree of excellence of the *approach*.
2   The degree of *deployment* of the approach, *assessment* and *review*.

The detailed criterion parts are as follows:

1   **Leadership**
    How leaders develop and facilitate the achievement of the mission and vision, develop values required for long-term success and implement these via appropriate actions and behaviors, and are personally involved in ensuring that the organization's management system is developed and implemented.
    Self-assessment should demonstrate how leaders:
    (a) develop the mission, vision and values and are role models of a culture of excellence;
    (b) are personally involved in ensuring the organization's management system is developed, implemented and continuously improved;
    (c) are involved with customers, partners and representatives of society;
    (d) motivate, support and recognize the organization's people.

2   **Policy and strategy**
    How the organization implements its mission and vision via a clear stakeholder-focused strategy, supported by relevant policies, plans,

Approach	Score	Deployment, assessment and review
Anecdotal or no evidence.	0%	Little effective usage.
Some evidence of soundly based approaches and prevention based processes/systems.	25%	Implemented in about one-quarter of the relevant areas and activities.
Some evidence of integration into normal operations.		Some evidence of assessment and review.
Evidence of soundly based systematic approaches and prevention based processes/systems.	50%	Implemented in about half the relevant areas and activities.
Evidence of integration into normal operations and planning well established.		Evidence of assessment and review.
Clear evidence of soundly based systematic approaches and prevention based processes/systems.	75%	Applied to about three-quarters of the relevant areas and activities.
Clear evidence of integration of approach into normal operations and planning.		Clear evidence of refinement and improved business effectiveness through review cycles.
Comprehensive evidence of soundly based systematic approaches and prevention based processes/systems.	100%	Implemented in all relevant areas and activities.
Approach has become totally integrated into normal working patterns. Could be used as a role model for other organizations.		Comprehensive evidence of refinement and improved business effectiveness through review cycles.

For *Approach, Deployment, Assessment* and *Review* the assessor may choose one of the five levels 0%, 25%, 50%, 75%, or 100% as presented in the chart, or interpolate between these values.

Figure 8.3 Scoring within the self-assessment process: the enablers

objectives, targets and processes. Self-assessment should demonstrate how policy and strategy are:

(a) based on the present and future needs and expectations of stakeholders;
(b) based on information from performance measurement, research, learning and creativity-related activities;
(c) developed, reviewed and updated;
(d) deployed through a framework of key processes;
(e) communicated and implemented.

3 **People**
How the organization manages, develops and releases the knowledge and full potential of its people at an individual, team-based and organization-wide level, and plans these activities in order to support its policy strategy and the effective operation of its processes. Self-assessment should demonstrate how:

(a) resources are planned, managed and improved;
(b) knowledge and competencies are identified, developed and sustained;

(c) people are involved and empowered;

(d) the organization has a dialog;

(e) people are rewarded, recognized and cared for.

4 **Partnerships and resources**

How the organization plans and manages its external partnerships and internal resources in order to support its policy and strategy and the effective operation of its processes. Self-assessment should demonstrate how:

(a) external partnerships are managed;

(b) finances are managed;

(c) buildings, equipment and materials are managed;

(d) technology is managed;

(e) information and knowledge are managed.

5 **Processes**

How the organization designs, manages and improves its processes in order to support its policy strategy and fully satisfy, and generate increasing value, for its customers and stakeholders. Self-assessment should demonstrate how:

(a) processes are systematically designed and managed;

(b) processes are improved, as needed, using innovation in order to fully satisfy and generate increasing value for customers and other stakeholders;

(c) products and services are designed and developed based on customer needs and expectations;

(d) products and services are produced, delivered and serviced;

(e) customer relationships are managed and enhanced.

Assessing the enablers criteria

The criteria are concerned with how an organization or business unit achieves its results. Self-assessment asks the following questions in relation to each criterion:

- What is currently done in this area?
- How is it done? Is the approach systematic and prevention-based?
- How is the approach reviewed and what improvements are undertaken following review?
- How widely used are these practices?

Results

The EFQM Excellence Model's result criteria of customer results, people results, society results, and key performance results focus on what the organization has achieved and is achieving in relation to its:

- external customer;
- people;

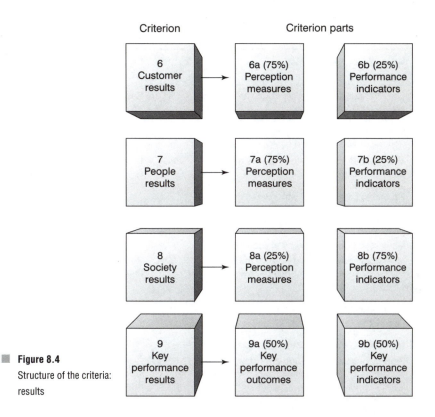

Figure 8.4
Structure of the criteria: results

- local, national, and international society, as appropriate;
- planned performance.

These can be expressed as discrete results, but ideally as trends over a period of years. The structure of the results criteria is shown in Figure 8.4.

'Performance excellence' is assessed relative to the organization's business environment and circumstances, based on information which sets out:

- the organization's actual performance;
- the organization's own targets; and wherever possible:
- the performance of competitors or similar organizations;
- the performance of 'best in class' organizations.

Results are assessed on the basis of the combination of two factors (see Figure 8.5):

- The degree of excellence of the results.
- The scope of the results.

Results	Score	Scope
No results or anecdotal information.	0%	Results address few relevant areas and activies.
Some results show positive trends and/or satisfactory performance. Some favorable comparisons with own targets/external organizations. Some results are caused by approach.	25%	Results address some relevant areas and activities.
Many results show strongly positive trends and/or sustained good performance over the last 3 years. Favorable comparisons with own targets in many areas. Some favorable comparison with external organizations. Many results are caused by approach.	50%	Results address many relevant areas and activities.
Most results show strong positive trends and/or sustained excellent performance over at least 3 years. Favorable comparisons with own targets in most areas. Favorable comparisons with external organizations in many areas. Most results are caused by approach.	75%	Results address most relevant areas and activites.
Stongly positive trends and/or sustained excellent performance in all areas over at least 5 years. Excellent comparisons with own targets and external organizations in most areas. All results are clearly caused by approach. Positive indication that leading position will be maintained.	100%	Results address all relevant areas and facets of the organization.

For both *Results* and *Scope* the assessor may choose one of the five levels 0%, 25%, 50%, 75%, or 100% as presented in the chart, or interpolate between these values.

Figure 8.5   Scoring within the self-assessment process: the results

6 **Customer results**

What the organization is achieving in relation to its external customers. Self-assessment should demonstrate the organization's success in satisfying the needs and expectations of its external customers. Areas to consider:

(a) results achieved for the measurement of customer perception of the organization's products, services and customer relationships;

(b) internal performance indicators relating to the organization's customers.

7 **People results**

What the organization is achieving in relation to its people. Self-assessment should demonstrate the organization's success in satisfying

the needs and expectations of its people. Areas to consider:
(a) results of people's perception of the organization;
(b) internal performance indicators relating to people.

8 **Society results**
What the organization is achieving in relation to local, national and international society as appropriate. Self-assessment should demonstrate the organization's success in satisfying the needs and expectations of the community at large. Areas to consider:
(a) society's perception of the organization;
(b) internal performance indicators relating to the organization and society.

9 **Key performance results**
What the organization is achieving in relation to its planned performance. Areas to consider:
(a) key performance outcomes, including financial and non-financial;
(b) key indicators of the organization's performance which might predict likely key performance outcomes.

Assessing the results criteria

These criteria are concerned with what an organization has achieved and is achieving. Self-assessment address the following issues:

- The measures used to indicate performance.
- The extent to which the measures cover the range of the organization's activities.
- The relative importance of the measures presented.
- The organization's actual performance.
- The organization's performance against targets, and wherever possible.
- Comparisons of performance with similar organizations.
- Comparisons of performance with 'best in class' organizations.

Self-assessment against the Excellence Model may be performed generally using the so-called RADAR system:

- Results
- Approach
- Deployment
- Assessment
- Review

The RADAR 'screen' with the next level of detail is shown in Figure 8.6.

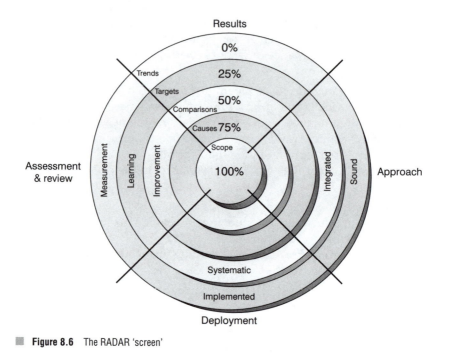

Case study ■■■ ━━━━━━━━━━━━━━━━━━━━━━━━━━

■ Getting started with self-assessment using the
■ Shell SQF

As part of the validation process for the SQF a baseline assessment was carried out across the organization to determine the starting point for performance improvement and to 'prove' the SQF in practice. This yielded valuable data which served both objectives. Some 80 managers and leaders were interviewed and asked to assess where they thought their part of the organization was in comparison to the tiers of practice in level 4 of the SQF. A fundamental finding was confirmation of virtually no performance measurement in many areas – indeed the organization did not rate at all against this component. Other key findings were: good articulation of aspirations and purpose, but limited cascade and execution through the line; very few common processes implemented across the organization; supplier relationships not effectively managed; lots of initiatives activity around knowledge management and virtual team working but little collection of institutional knowledge and intellectual capital; many valuable initiatives in place to improve overall performance but signs of initiative overload and limited capacity to follow-through.

Although a sobering exercise it proved invaluable in demonstrating the need for a systematic approach to improving the business, and in all areas. Figures 8.7 and 8.8 illustrate some of the findings from the baseline activity.

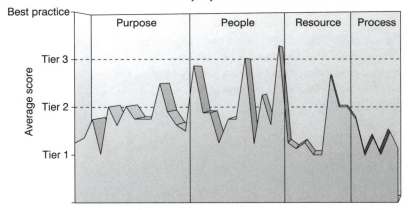

Figure 8.7 Average baseline findings

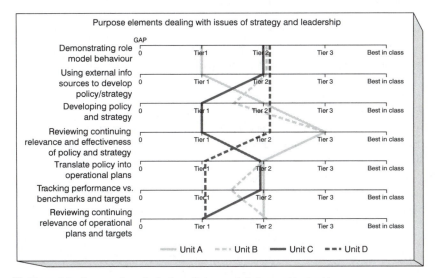

Figure 8.8 Purpose elements dealing with issues of strategy and leadership

Business improvement strategies in the Highways Agency

The Highways Agency is responsible for England's strategic road network, a network that consists of 9400 km (5481 miles) of motorways and trunk roads and carries a third of all road traffic and two-thirds of all freight traffic in the country. This equates to an annual total of around 153 billion kilometers traveled. As an Executive Agency for the Department for Transport the Highways Agency priorities are to:

- Continue to maintain the network in good condition to ensure that it is safe and available for use.

■ Maximize performance from the existing network.
■ Improve the network where necessary.

At the end of the 1990s the UK Cabinet Office introduced a Better Quality Services Review (BQSR) program in response to the Government's white paper 'Modernising Government'. As part of that program the Highways Agency undertook a review of the performance of its activities with a view to considering one of five options:

■ Abolition
■ Market testing
■ Contracting out
■ Privatization or
■ Internal improvement

The Highways Agency Management Board realized the potential of the BQSR proposals, particularly the opportunity to incorporate a long held policy to improve the management and operation of the organization. The first step was to set up a small team to consider how such a program could be delivered and by April 1999 this 'Business Improvement Team' presented a paper to the board detailing a potential improvement strategy, which linked the need to implement a program of better quality service reviews to a structured approach to improving the business.

The proposed BQSR program was authorized by the Highways Agency Management Board and was due for completion in April 2004. Alongside the authorization of the program the board approved the establishment of a Business Improvement Co-ordinator in each of its 23 divisions. This role was to be supported by the Business Improvement Team who were charged with facilitating program delivery.

The approach that was adopted by the Agency was to use a small team of internal consultants supported by external expertise to identify good business practices and assist the functional directorates of the Agency to analyze the services provided using the BQSR criteria, identify areas for improvement and implement any improvements. Each of the services identified would then be subjected to a comparative benchmark with a view to aiding the final BQSR decision-making process. Figure 8.9 is a visual representation of the approach that each directorate used.

The Business Improvement Team worked with the management teams in each area to decide on the service strategy for each of the previously identified directorate key activities. This was effectively the first BQSR analysis designed to identify those services which could be abolished or where there was already consideration of outsourcing. This exercise provided a strategic review helping directorates to clarify their purpose. The end of 1999 saw the completion of this part of the program with a limited number of services put forward as having the potential for outsourcing. Those services that were put forward were subjected to a comparative benchmark and a subsequent improvement program was devised.

The Business Improvement Team then facilitated the delivery of the second phase of the framework – 'Self-assessment using the EFQM Excellence Model'. Several of the divisions

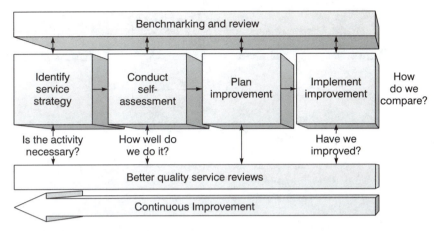

Figure 8.9   BQSR framework

were already using the EFQM Excellence Model to identify and plan improvements, so it was decided early in the program that self-assessment against the Excellence Model was likely to be the most effective way to identify, in a holistic sense, the areas that each functional directorate of the Agency should consider for improvement. The Business Improvement Team assessed a number of methods for self-assessment and eventually settled on two. The first of these would be a simple questionnaire-based self-assessment, designed for use with the smaller directorates, where all staff would participate. The second form of self-assessment, to be utilized by larger directorates, would involve training a small group of staff as EFQM assessors who would then gather evidence of business practice, assess that evidence, identify areas for improvement and plan the implementation.

By mid-2001 over 90 percent of the organization had undergone one of the two forms of self-assessment and a clearer picture of the key areas for improvement was evident. Indeed a number of directorates had already agreed improvement action plans and were well into the delivery of improvement.

Methodologies for self-assessment

The EFQM provides a flow diagram of the general steps involved in undertaking self-assessment. A simplified version of this is shown in Figure 8.10.

There are a number of approaches to carrying out self-assessment including:

- discussion group/workshop methods;
- surveys, questionnaires and interviews (peer involvement);
- pro formas;

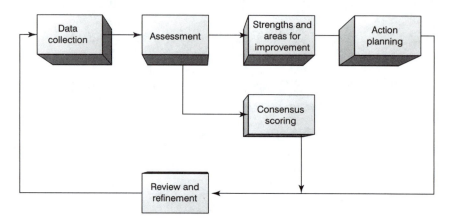

■ **Figure 8.10**   The key steps in self-assessment

- organizational self-analysis matrices (e.g. see Figure 8.11);
- an award simulation;
- activity or process audits;
- hybrid approaches.

Whichever method is used, the emphasis should be on understanding the organization's strengths and areas for improvement, rather than the score. The scoring charts provide a consistent basis for establishing a quantitative measure of performance against the model, and gaining consensus promotes discussion and development of the issues facing the organization. It should also gain the involvement, interest and commitment of the senior management, but the scores should not become an end in themselves. Tito Conti, often called 'the father of self-assessment', following the contribution he made to its establishment and development through the EFQM, when he was head of Fiat, has often expressed concern that organizations can become obsessed with self-assessment scores rather than focusing on the improvement opportunities identified.

Using assessment

There is great overlap between the criteria used by the various awards and it may be necessary for an organization to rationalize them. The main components, however, must be the organization's processes, management systems, people management and results, customer results and key performance results. Self-assessment provides an organization with vital information in monitoring its progress towards its goals and business 'excellence'. The external assessments used in the processes of making awards must be based on these self-assessments that are performed as prerequisites for improvement.

Management jargon is increasingly confused by a vast literature, spiced with acronyms, the generation of which often bends the meaning of words. There is also often in the leadership of large organizations an ego-driven or publicity seeking wish to invent new buzz words. It may be necessary to assess the status of the language to be used before launching a self-assessment process. If recipients are not familiar with certain language, many propositions will be meaningless. A preliminary teach-in or awareness process may be necessary.

Whatever are the main 'motors' for driving an organization towards its vision or mission, they must be linked to the five stakeholders embraced by the values of any organization, namely:

- Customers
- Employees
- Suppliers
- Stakeholders
- Community

In any normal business or organization, measurements are continuously being made, often in retrospect, by the leaders of the organization to reflect the value put on the organization by its five stakeholders. Too often, these continuous readings are made by internal biased agents with short-term priorities, not always in the best long-term interests of the organization or its customers, i.e. narrow firefighting scenarios which can blind the organization's strategic eye. Third party agents, however, can carry out or facilitate periodic assessments from the perspective of one or more of the key stakeholders, with particular emphasis on forward priorities and needs. These reviews will allow realignment of the principal driving motors to focus on the critical success factors and continuous improvement, to maintain a balanced and powerful general thrust which moves the whole organization towards its mission.

The relative importance of the five stakeholders may vary in time but all are important. The first three, customers, employees, and suppliers, which comprise the core value chain, are the *determinant* elements. The application of quality management principles in these areas will provide satisfaction as a *resultant* to the shareholders and the community. Thus, added value will benefit the community and the environment. The ideal is a long way off in most organizations, however, and active attention to the needs of the shareholders and/or community remain a priority for one major reason – they are the 'customers' of most organizational activities and are vital stakeholders.

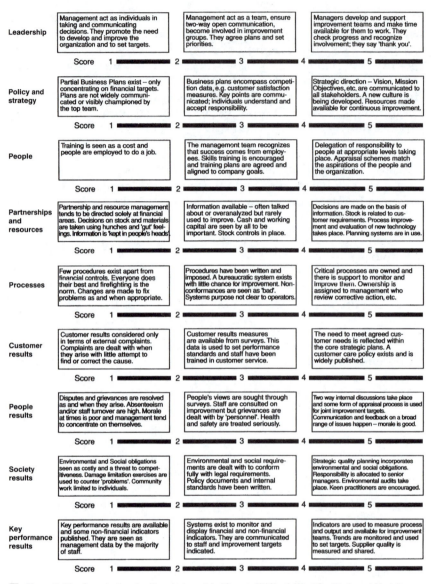

Leadership	Management act as individuals in taking and communicating decisions. They promote the need to develop and improve the organization and to set targets.	Management act as a team, ensure two-way open communication, become involved in improvement groups. They agree plans and set priorities.	Managers develop and support improvement teams and make time available for them to work. They check progress and recognize involvement; they say 'thank you'.

Score 1 ▬ 2 ▬ 3 ▬ 4 ▬ 5 ▬

Policy and strategy	Partial Business Plans exist – only concentrating on financial targets. Plans are not widely communicated or visibly championed by the top team.	Business plans encompass competition data, e.g. customer satisfaction measures. Key points are communicated; individuals understand and accept responsibility.	Strategic direction – Vision, Mission Objectives, etc. are communicated to all stakeholders. A new culture is being developed. Resources made available for continuous improvement.

Score 1 ▬ 2 ▬ 3 ▬ 4 ▬ 5 ▬

People	Training is seen as a cost and people are employed to do a job.	The management team recognizes that success comes from employees. Skills training is encouraged and training plans are agreed and aligned to company goals.	Delegation of responsibility to people at appropriate levels taking place. Appraisal schemes match the aspirations of the people and the organization.

Score 1 ▬ 2 ▬ 3 ▬ 4 ▬ 5 ▬

Partnerships and resources	Partnership and resource management tends to be directed solely at financial areas. Decisions on stock and materials are taken using hunches and 'gut' feelings. Information is 'kept in people's heads'.	Information available – often talked about or overanalyzed but rarely used to set performance. Cash and working capital are seen by all to be important. Stock controls in place.	Decisions are made on the basis of information. Stock is related to customer requirements. Process improvement and evaluation of new technology takes place. Planning systems are in use.

Score 1 ▬ 2 ▬ 3 ▬ 4 ▬ 5 ▬

Processes	Few procedures exist apart from financial controls. Everyone does their best and firefighting is the norm. Changes are made to fix problems as and when appropriate.	Procedures have been written and imposed. A bureaucratic system exists with little chance for improvement. Non-conformances are seen as 'bad'. Systems purpose not clear to operators.	Critical processes are owned and there is support to monitor and improve them. Ownership is assigned to management who review corrective action, etc.

Score 1 ▬ 2 ▬ 3 ▬ 4 ▬ 5 ▬

Customer results	Customer results considered only in terms of external complaints. Complaints are dealt with when they arise with little attempt to find or correct the cause.	Customer results measures are available from surveys. This data is used to set performance standards and staff have been trained in customer service.	The need to meet agreed customer needs is reflected within the core strategic plans. A customer care policy exists and is widely published.

Score 1 ▬ 2 ▬ 3 ▬ 4 ▬ 5 ▬

People results	Disputes and grievances are resolved as and when they arise. Absenteeism and/or staff turnover are high. Morale at times is poor and management tend to concentrate on themselves.	People's views are sought through surveys. Staff are consulted on improvement but grievances are dealt with by 'personnel'. Health and safety are treated seriously.	Two way internal discussions take place and some form of appraisal process is used for joint improvement targets. Communication and feedback on a broad range of issues happen – morale is good.

Score 1 ▬ 2 ▬ 3 ▬ 4 ▬ 5 ▬

Society results	Environmental and Social obligations seen as costly and a threat to competitiveness. Damage limitation exercises are used to counter 'problems'. Community work limited to individuals.	Environmental and social requirements are dealt with to conform fully with legal requirements. Policy documents and internal standards have been written.	Strategic quality planning incorporates environmental and social obligations. Responsibility is allocated to senior managers. Environmental audits take place. Keen practitioners are encouraged.

Score 1 ▬ 2 ▬ 3 ▬ 4 ▬ 5 ▬

Key performance results	Key performance results are available and some non-financial indicators published. They are seen as management data by the majority of staff.	Systems exist to monitor and display financial and non-financial indicators. These are communicated to staff and improvement targets indicated.	Indicators are used to measure process and output and available for improvement teams. Trends are monitored and used to set targets. Supplier quality is measured and shared.

Score 1 ▬ 2 ▬ 3 ▬ 4 ▬ 5 ▬

Figure 8.11 Organizational self-analysis matrix (*Source*: UK North West Quality Award Model)

Case study ■ ■ ■ ▬▬▬▬▬▬

■ Unilever HPCE – the self-assessment journey

This short case study outlines the Unilever business, Home and Personal Care – Europe (HPCE) self-assessment journey from when it was formed in 1996, when it set itself the aim of being 'world class' in the new millennium. For a full version of the case study see reference 1.

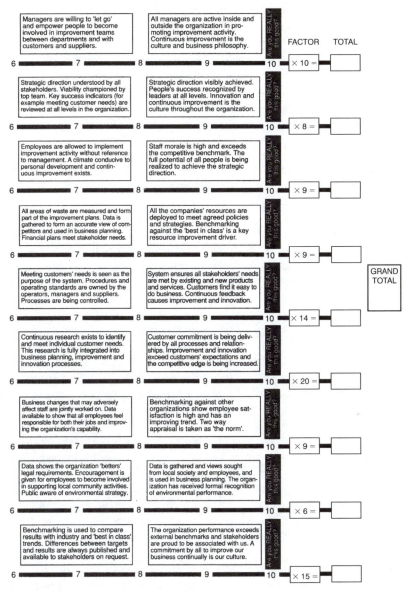

Figure 8.11 (*continued*)

Over 100 years ago, the founder, William Hesketh Lever, stated the following mission for his soap company:

- To make cleanliness commonplace.
- To lessen work for women.
- To foster health and contribute to personal attractiveness, that life may be more enjoyable and rewarding for the people who use our products.

Unilever's mission today is not fundamentally different but better reflects the social and economic climate and the wider scope of its operations.

The HPCE Business Group was established in 1996 as one of the 12 Unilever Business Groups with its headquarters in Brussels, Belgium. At that time HPCE comprised 15 local companies in 18 European countries, although these numbers have increased over recent years. Within Europe there are 18 factories and eight innovation centers. Unilever HPCE has 10 000 employees, of whom 1600 are managers.

HPCE started to use the EFQM Excellence Model from the end of 1995 and carried out the first round of self-assessment in 1996. The intention to conduct self-assessments was announced during the first annual OBJ!* to the top 150 managers.

It was decided to carry out the assessments at country level, category level, factory level as well as at HPCE board level. Each unit nominated a facilitator and a first 'Business Excellence Conference' was held attended by the excellence manager from Texas Instruments, award winner in 1995. The conference was organized with the support of the Birds Eye Wall's business excellence manager, who was a highly experienced award assessor. During the conference, syndicate groups were used to select the key results at the country, category and factory levels.

It was decided to use the simplified Unilever checklist developed three years before by representatives from three UK companies, Birds Eye Wall's, Van Den Bergh Foods and Elida Fabergé. The simplified checklist provided a way for areas to commence self-assessment that was easy for everyone to relate to, which was especially important for the top team in order to gain their commitment. Much of the management understanding of the longer-term self-assessment process was gained at this stage.

The HPCE board assessment was conducted by consolidating the output from the unit assessments and agreeing a score, plus strengths and AFIs. The first opportunities for improvement were quite major and included both enabler and results issues.

On the enabler side the need for a policy deployment approach was identified and so 'Strategy into Action' was introduced. This was based on an approach used in Unilever Australia and the transfer of the approach to HPCE is an early example of 'external' learning.

In 1996 Unilever already had many soundly based practices in place. Examples included the 'Integrated Approach', which combines objective setting, development planning and remuneration, and the approach to product development, which has been benchmarked on several occasions. However, many of the major strategic improvements for HPCE stemmed from the first self-assessment. Improvements were identified in the areas of communication, process management and goal deployment, which led to the 'Strategy into Action' approach being deployed. Self-assessment also identified gaps in the results areas such as employee satisfaction, with limited measurement systems and, with the exception of the financial area, there were few targets and external comparisons.

*OBJ! is a Unilever wide annual communication event. OBJ! stands for 'Oh be Joyful!'

The value of the simple questionnaire route into self-assessment should not be understated. Without such an approach an organization increases the difficulty to take the decision to experiment with or start the process. People then get the impression that very soon in the process they will have to put in place a complex support system and the self-assessment becomes too heavy a process. This need not be the case.

As HPCE moved into the third round of self-assessment there was the need to review the first two years' experience and improve the self-assessment approach. Hence, the central team established a strategy for Business Excellence in HPCE. The mission was:

■ To provide the philosophy, the structure and the means for continuous improvement so that the organization can realize its goals.

There was also a set of aims:

■ Create an understanding of the process of business improvement.
■ Create cross-functional working based around key business processes.
■ Establish self-assessment as the performance measure within the integrating business frameworks.

It was also recognized that the various units of HPCE were at potentially different levels of maturity. As feedback from the business units indicated that there was the requirement to maintain the way that the self-assessments were conducted, only four changes were made to the 1998 self-assessment approach. The first change, which was small but extremely significant, was a change to the way that the enabler questions were presented. In the simple checklist most of the questions were phrased 'What do you ...?' and this was changed to 'How do you ...?'. Answering 'How' questions tends to be much more probing than answering 'What' questions.

The second change was to the checklist format and the inclusion of a question relating to review and improvement for each enabler. This made the checklist more rigorous and closer to an external assessment approach.

The third change related to scoring was that some facilitators were exposed to the external scoring based on assessing a number of factors. This approach had not been used in the past as the view was held that training was required before the technique could be used properly. However, with limited training use of the approach led to more valuable feedback.

The final change was the introduction of improved support materials to aid planning and improvement activity.

A key aspect of the development of self-assessment was to open a discussion on the use of the output of the self-assessment. For example, it was stated that a unit in the 'Entry' phase would:

■ Be using self-assessment to increase awareness.
■ Be selling the need to change.
■ Focus on high-level areas for improvement.

Most units agreed that they meet all these requirements but acknowledged that they did not meet all the 'User' phase requirements, which were defined as all the above plus:

■ Have an improvement plan detailing specific actions to be taken.
■ Will know that the improvement activity is linked to business need.
■ Will be actively seeking best practice from external organizations.

To help with improvement planning two options were tabled. These were either to have a separate improvement plan or to integrate the output of the self-assessment with business planning activities. It was suggested that with a separate improvement plan approach there was a need to have:

■ SMART objectives.
■ Actionable steps breaking the improvement action down.
■ Responsibilities assigned to specific people.
■ Achievement date set.

For units that chose to integrate their improvement planning with their business planning, advice was given to 'screen' each of their areas of improvement with a set of questions.

The ongoing project concluded that the approach to self-assessment needed to evolve to meet current unit needs. The evolution of the EFQM Excellence Model in 1999 presented an opportunity to make this change.

Under the cover of 'the new model', pro forma-based assessment was introduced, together with the more rigorous RADAR assessment scoring approach. At the annual conference facilitators were trained in the new approach in preparation for the 2000 self-assessment round. In the most recent self-assessment cycles, units have been aligning their improvement activities with the organization's strategic thrusts.

■ Securing prevention by audit and review of the ■ management systems

Error or defect prevention is the process of removing or controlling error/defect causes in the management systems. There are two major elements of this:

■ Checking the systems.
■ Error/defect investigation and follow-up.

These have the same objectives – to find, record and report *possible* causes of error, and to recommend future preventive or corrective action.

Checking the systems _____

There are six methods in general use:

1 *Quality audits and reviews*, which subject each area of an organization's activity to a systematic critical examination. Every component of the total system is included, i.e. policy, attitudes, training, processes, decision features, operating procedures, documentation. Audits and reviews, as in the field of accountancy, aim to disclose the strengths and the main areas of vulnerability or risk – the areas for improvement.
2 *Quality survey*, a detailed, in-depth examination of a narrower field of activity, i.e. major key areas revealed by system audits, individual sites/plants, procedures or specific problems common to an organization as a whole.
3 *Quality inspection*, which takes the form of a routine scheduled inspection of a unit or department. The inspection should check standards, employee involvement and working practices, and that work is carried out in accordance with the agreed processes and procedures.
4 *Quality tour*, which is an unscheduled examination of a work area to ensure that, for example, the standards of operation are acceptable, obvious causes of defects or errors are removed, and in general quality standards are maintained.
5 *Quality sampling*, which measures by random sampling, similar to activity sampling, the error/defect potential. Trained observers perform short tours of specific locations by prescribed routes and record the number of potential errors or defects seen. The results may be used to portray trends in the general quality situation.
6 *Quality scrutinies*, which are the application of a formal, critical examination of the process and technological intentions for new or existing facilities, or to assess the potential for maloperation or malfunction of equipment and the consequential effects on quality. There are similarities between quality scrutinies and FMECA studies (see Chapter 6).

The design of an error prevention program, combining all these elements, is represented in Figure 8.12.

Error or defect investigations and follow-up _____

The investigation of errors and defects can provide valuable error prevention information. The general method is based on:

- *Collecting* data and information relating to the error or defect.
- *Checking* the validity of the evidence.

■ **Figure 8.12** An error prevention program combining various elements of 'checking' the quality system

■ *Selecting* the evidence without making assumptions or jumping to conclusions.

The results of the analysis are then used to:

■ *Decide* the most likely cause(s) of the errors or defect.
■ *Notify* immediately the person(s) able to take corrective action.
■ *Record* the findings and outcomes.
■ *Record* the findings and outcomes to everyone concerned, to prevent recurrence.

The investigation should not become an inquisition to apportion blame, but focus on the positive preventive aspects. The types of follow-up to errors and their effects are shown in Table 8.1.

It is hoped that errors or defects are not normally investigated so frequently that the required skills are developed by experience, nor are these skills easily learned in a classroom. One suggested way to overcome this problem is the development of a programmed sequence of questions to form the skeleton of an error or defect investigation questionnaire. This can be set out with the following structure:

■ *People* – duties, information, supervision, instruction, training, attitudes, etc.
■ *Systems* – procedures, instructions, monitoring, control methods, etc.
■ *Plant/equipment* – description, condition, controls, maintenance, suitability, etc.
■ *Environment* – climatic, space, humidity, noise, etc.

Table 8.1 Following up errors

System type	Aim	General effects
Investigation	To prevent a similar error or defect	*Positive:* identification notification correction
Inquisition	To identify responsibility	*Negative:* blame claims defence

Internal and external quality management system audits and reviews

A good quality management system will not function without adequate audits and reviews. The system reviews, which need to be carried out periodically and systematically, are conducted to ensure that the system achieves the required effect, while audits are carried out to make sure that actual methods are adhering to the documented procedures. The reviews should use the findings of the audits, for failure to operate according to the plan often signifies difficulties in doing so. A re-examination of the processes and procedures actually being used may lead to system improvements unobtainable by other means.

A schedule for carrying out the *audits* should be drawn up, different activities perhaps requiring different frequencies. All procedures and systems should be audited at least once during a specified cycle, but not necessarily all at the same audit. For example, every three months a selected random sample of the processes could be audited, with the selection designed so that each process is audited at least once per year. There must be, however, a facility to adjust this on the basis of the audit results.

A quality management system *review* should be instituted, perhaps every 12 months, with the aims of:

- ensuring that the system is achieving the desired results;
- revealing defects or irregularities in the system;
- indicating any necessary improvements and/or corrective actions to eliminate waste or loss;
- checking on all levels of management;

■ uncovering potential danger areas;
■ verifying that improvements or corrective action procedures are effective.

Clearly, the procedures for carrying out the audits and reviews and the results from them should be documented, and themselves be subject to review. Useful guidance on quality management system audits is given in the international standard ISO 10 011.

The assessment of a quality management system against a particular standard or set of requirements by internal audit and review is known as a *first party* assessment or approval scheme. If an *external* customer makes the assessment of a supplier against either its own or a national or international standard, a *second party* scheme is in operation. The assessment by an independent organization, not connected with any contract between customer and supplier, but acceptable to them both, is known as an *independent third party* assessment scheme. The latter often results in some form of certification or registration by the assessment body.

One advantage of the third party schemes is that they obviate the need for customers to make their own detailed checks, potentially saving both suppliers' and customers' time and money, and avoiding issues of commercial confidentiality. Just one knowledgeable organization has to be satisfied, rather than a multitude with varying levels of competence. This method can be used to certify suppliers for contracts without further checking, but good customer/supplier relations often include second party extensions to the third party requirements and audits.

Each certification body usually has its own recognized mark, which may be used by registered organizations of assessed capability in their literature, letter headings, and marketing activities. There are also publications containing lists of organizations whose quality management systems and/or products and services have been assessed. To be of value, the certification body must itself be recognized and, usually, assessed and registered with a national or international accreditation scheme.

Many organizations have found that the effort of designing and implementing a quality management system, good enough to stand up to external independent third party assessment, has been extremely rewarding in:

■ involving staff and improving morale;
■ better process control and improvement;
■ reduced wastage and costs;
■ reduced customer-service costs.

This is also true of those organizations that have obtained third party registrations and supply companies which still insist on their own second party assessment. The reason for this is that most of the standards on quality management systems, whether national, international, or company specific, are now very similar indeed. A system that meets the requirements of the ISO 9000:2000 standard, for example, should meet the requirements of most other standards, with only the slight modifications and small emphases here and there required for specific customers. It is the author's experience, and that of his colleagues, that an assessment carried out by one of the good independent certified assessment bodies is a rigorous and delving process.

Internal system audits and reviews should be positive and conducted as part of the preventive strategy and not as a matter of expediency resulting from problems. They should not be carried out only prior to external audits, nor should they be left to the external auditor – whether second or third party. An external auditor, discovering discrepancies between actual and documented systems, will be inclined to ask why the internal review methods did not discover and correct them.

Any management team needs to be fully committed to operating an effective quality management system for all the people within the organization, not just the staff in the 'quality department'. The system must be planned to be effective and achieve its objectives in an uncomplicated way. Having established and documented the processes it is necessary to ensure that they are working and that everyone is operating in accordance with them. The system once established is not static, it should be flexible to enable the constant seeking of improvements or streamlining.

Quality auditing standard

The growing use of standards internationally emphasizes the importance of auditing as a management tool for this purpose. There are available several guides to management systems auditing (e.g. ISO 10011*) and the guidance provided in these can be applied equally to any one of the three specific and yet different auditing activities:

1 **First party or internal audits**, carried out by an organization on its own systems, either by staff who are independent of the systems being audited, or by an outside agency.
2 **Second party audits**, carried out by one organization (a purchaser or its outside agent) on another with which it has either contracts to purchase goods or services or intends to do so.

* ISO 9001 'Guidelines on quality and/or environmental management systems auditing' (revision of ISO 10011).

3 **Third party audits**, carried out by independent agencies, to provide assurance to existing and prospective customers for the product or service.

Audit objectives and responsibilities, including the roles of auditors and their independence, and those of the 'client' or auditee should be understood. The generic steps involved are as follows:

- **initiation**, including the audit scope and frequency;
- **preparation**, including review of documentation, the program, and working documents;
- **execution**, including the opening meeting, examination and evaluation, collecting evidence, observations, and closing the meeting with the auditee;
- **report**, including its preparation, content and distribution;
- **completion**, including report submission and retention.

Attention should be given at the end of the audit to corrective action and follow-up and the improvement process should be continued by the auditee after the publication of the audit report. This may include a call by the client for a verification audit of the implementation of any corrective actions specified.

Any instrument which is developed for assessment, audit or review may be used at several stages in an organization's history:

- before starting an improvement program to identify 'strengths' and 'areas for improvement', and focus attention (at this stage a parallel cost of quality exercise may be a powerful way to overcome skepticism and get 'buy-in');
- as part of a program launch, especially using a 'survey' instrument;
- every one or two years after the launch to steer and benchmark.

The systematic measurement and review of operations is one of the most important management activities of any organization. Self-assessment, audit and review should lead to clearly discerned strengths and areas for improvement by focusing on the relationship between the people, processes, and performance. Within any successful organization these will be regular activities.

Reference

1. John S. Oakland, *TQM: text & cases* 3rd ed. Butterworth-Heinemann, Oxford, 2003.

Chapter highlights

Frameworks for self-assessment

- Many organizations are turning to total quality models to measure and improve performance. These frameworks include the Japanese Deming Prize, the US Baldrige Award and in Europe the EFQM Excellence Model.
- The nine components of the Excellence Model are: leadership, policy and strategy, people, partnerships and resources and processes (ENABLERS), people results, customer results, society results, and key performance results (RESULTS).
- The various award criteria provide rational bases against which to measure progress towards TQM in organizations. Self-assessment against, for example, the EFQM Excellence model should be a regular activity, as it identifies opportunities for improvement in performance through processes and people.

Methodologies for self-assessment

- Self-assessment against the Excellence Model may be performed using RADAR: results, approach, deployment, assessment and review.
- There are a number of approaches for self-assessment, including groups/workshops, surveys, pro formas, matrices, award simulations, activity/process audits or hybrid approaches.

Securing prevention by audit and review of the management system

- There are two major elements of error or defect prevention: checking the system, and error/defect investigations and follow-up. Six methods of checking the quality management systems are in general use: audits and reviews, surveys, inspections, tours, sampling, and scrutinies.
- Investigations proceed by collecting, checking and selecting data, and analyzing it by deciding causes, notifying people, recording and reporting findings and outcomes.

Internal and external quality management system audits and reviews

- A good quality management system will not function without adequate audits and reviews. Audits make sure the actual methods are adhering to documented procedures. Reviews ensure the system achieves the desired effect.
- System assessment by internal audit and review is known as first party, by external customer as second party, and by an independent organization as third party certification. For the latter to be of real value the certification body must itself be recognized.

Chapter 9

Benchmarking

The why and what of benchmarking

Product, service and process improvements can take place only in relation to established standards, with the improvements then being incorporated into new standards. *Benchmarking*, one of the most transferable aspects of Rank Xerox's approach to quality management, and thought to have originated in Japan, measures an organization's operations, products and services against those of its competitors in a ruthless fashion. It is a means by which targets, priorities and operations that will lead to competitive advantage can be established.

There are many drivers for benchmarking including the external ones:

■ customers continually demand better quality, lower prices, shorter lead times, etc;
■ competitors are constantly trying to get ahead and steal markets;
■ legislation – changes in our laws place ever greater demands for improvement.

Internal drivers include:

■ targets which require improvements on our 'best ever' performance;
■ technology – a fundamental change in processes is often required to benefit fully from introducing new technologies;
■ self-assessment results, which provide opportunities to learn from adapting best practices.

The word 'benchmark' is a reference or measurement standard used for comparison, and benchmarking is the continuous process of identifying, understanding and adapting best practice and processes that will lead to superior performance.

Benchmarking is *not*:

- a panacea to cure the organization's problems, but simply a practical tool to drive up process performance;
- primarily a cost reduction exercise, although many benchmarking studies will result in improved financial performance;
- industrial tourism – study tours have their place, but proper benchmarking goes beyond 'tourism' – to really understanding the enablers to outstanding results;
- spying – use of a benchmarking code of conduct ensures the work is done with the agreement and openness of all parties;
- catching up with the best – the aim is to reach out and extend the current best practice (by the time we have caught up, the benchmark will have moved anyway).

There may be many reasons for carrying out benchmarking. Some of them are set against various objectives in Table 9.1. The links between benchmarking and quality management are clear – establishing

Table 9.1 Reasons for benchmarking

Objectives	Without benchmarking	With benchmarking
Becoming competitive	• Internally focused • Evolutionary change	• Understanding of competitiveness • Ideas from proven practices
Industry best practices	• Few solutions • Frantic catch-up activity	• Many options • Superior performance
Defining customer requirements	• Based on history or gut feeling • Perception	• Market reality • Superior performance
Establishing effective goals and objectives	• Lacking external focus • Reactive	• Credible, unarguable • Proactive
Developing true measures of productivity	• Pursuing pet projects • Strength and weaknesses not understood • Route of least resistance	• Solving real problems • Understanding outputs • Based on industry best practices

objectives based on industry best practice should directly contribute to better meeting of the internal and external customer requirements.

The benefits of benchmarking can be numerous but include:

- creating a better understanding of the current position;
- heightening sensitivity to changing customer needs;
- encouraging innovation;
- developing realistic stretch goals;
- establishing realistic action plans.

Data from the American Productivity and Quality Center's International Benchmarking Clearinghouse suggests that an average benchmarking study takes six months to complete, occupies more than a quarter of the team member's time, and costs $50 000–100 000. The same source identified that the average return was **five times** the cost of the study, in terms of reduced costs, increased sales, greater customer retention and enhanced market share.

There are four basic categories of benchmarking:

- **Internal** – the search for best practice of internal operations by comparison, e.g. multi-site comparison of polymerization processes and performance.
- **Functional** – seeking functional best practice outside an industry, e.g. mining company benchmarking preventive maintenance of pneumatic/hydraulic equipment with Disney.
- **Generic** – comparison of outstanding processes irrespective of industry or function, e.g. restaurant chain benchmarking kitchen design with US nuclear submarine fleet to improve restaurant to kitchen space ratios.
- **Competitive** – specific competitor to competitor comparisons for a product, service, or function of interest, e.g. retail outlets comparing price performance and efficiency of internet ordering systems.

■ The purpose and practice of benchmarking

The evolution of benchmarking in an organization is likely to progress through four focuses. Initially attention may be concentrated on competitive products or services, including, for example, design, development and operational features. This should develop into a focus on industry best practices and may include, for example, aspects of distribution or service. The real breakthroughs are when organizations focus on all aspects of the total business performance, across all functions and aspects, and address current *and projected* performance gaps. This should lead to the focus on processes and true continuous improvement.

At its simplest, competitive benchmarking, the most common form, requires every department to examine itself against the counterpart in the best competing companies. This includes a scrutiny of all aspects of their activities. Benchmarks which may be important for *customer satisfaction*, for example, might include:

- Product or service consistency.
- Correct and on-time delivery.
- Speed of response or new product development.
- Correct billing.

For internal *impact* the benchmarks may include:

- Waste, rejects or errors.
- Inventory levels/work-in-progress.
- Costs of operation.
- Staff turnover.

The task is to work out what has to be done to improve on the competition's performance in each of the chosen areas.

Benchmarking is very important in the administration areas, since it continuously measures services and practices against the equivalent operation in the toughest direct competitors or organizations renowned as leaders in the areas, even if they are in the same organization. An example of quantitative benchmarks in absenteeism is given in Table 9.2.

■ **Table 9.2** Quantitative benchmarking in absenteeism

Organization's absence level (%)	Productivity opportunity
Under 3	This level matches an aggressive benchmark that has been achieved in 'excellent' organizations.
3–4	This level may be viewed within the organization as a good performance – representing a moderate productivity opportunity improvement.
5–8	This level is tolerated by many organizations but represents a major improvement opportunity.
9–10	This level indicates that a serious absenteeism problem exists.
Over 10	This level of absenteeism is extremely high and requires immediate senior management attention.

Technologies and conditions vary between different industries and markets, but the basic concepts of measurement and benchmarking are of general validity. The objective should be to produce products and services that conform to the requirements of the customer in a never-ending improvement environment. The way to accomplish this is to use the continuous improvement cycle in all the operating departments – nobody should be exempt. Benchmarking is not a separate science or unique theory of management, but rather another strategic approach to getting the best out of people and processes, to deliver improved performance.

The purpose of benchmarking then is predominantly to:

■ **change** the perspectives of executives and managers;
■ **compare** business practices with those of world class organizations;
■ **challenge** current practices and processes;
■ **create** improved goals and practices for the organization.

As a managed process for change, benchmarking uses a disciplined structured approach to identify **what** needs to change, **how** it can be changed, and the **benefits** of the change. It also creates the desire for change in the first place. Any process or practice that can be defined can be benchmarked but the focus should be on those which impact on customer satisfaction and/or business results – financial or non-financial.

For organizations which have not carried out benchmarking before, it may be useful initially to carry out a simple self-assessment of their readiness in terms of:

■ how well processes are understood;
■ how much customers are listened to;
■ how committed the senior team is.

Table 9.3 provides a simple pro forma for this purpose. The score derived gives a crude guide to the readiness of the organization for benchmarking:

32–48   Ready for benchmarking
16–31   Some further preparation required before the benefits of benchmarking can be fully derived
  0–15   Some help is required to establish the foundations and a suitable platform for benchmarking.

The benchmarking process has five main stages which are all focused on trying to measure comparisons and identify areas for action and change (Figure 9.1). The detail is as follows.

Table 9.3 Is the organization ready for benchmarking?

After studying the statements below tick one box for each to reflect the level to which the statement is true for the organization.

	Most	Some	Few	None
Processes have been documented with measures to understand performance	☐	☐	☐	☐
Employees understand the processes that are related to their own work	☐	☐	☐	☐
Direct customer interactions, feedback or studies about customers influence decisions about products and services	☐	☐	☐	☐
Problems are solved by teams	☐	☐	☐	☐
Employees demonstrate by words and deeds that they understand the organization's mission, vision and values	☐	☐	☐	☐
Senior executives sponsor and actively support process improvement projects	☐	☐	☐	☐
The organization demonstrates by words and by deeds that continuous improvement is part of the culture	☐	☐	☐	☐
Commitment to change is articulated in the organization's strategic plan				
Add the columns:	☐	☐	☐	☐
	× 6 =	× 4 =	× 2 =	× 0 =
Multiply by the factor	☐	☐	☐	☐
Obtain the grand total?				

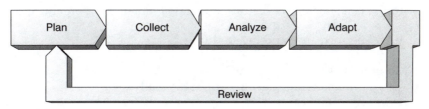

Figure 9.1 The five main stages of benchmarking

PLAN the study:

- Select processes for benchmarking.
- Bring together the appropriate team to be involved and establish roles and responsibilities.
- Identify benchmarks and measures for data collection.
- Identify best competitors or operators of the process(es), perhaps using customer feedback or industry observers.
- Document the current process(es).

COLLECT data and information:

- Decide information and data collection methodology, including desk research.
- Record current performance levels.
- Identify benchmarking partners.
- Conduct a preliminary investigation.
- Prepare for any site visits and interact with target organizations.
- Use data collection methodology.
- Carry out site visits.

ANALYZE the data and information:

- Normalize the performance data, as appropriate.
- Construct a matrix to compare current performance with benchmarking competitors'/partners' performance.
- Identify outstanding practices.
- Isolate and understand the process enablers, as well as the performance measures.

ADAPT the approaches:

- Catalogue the information and create a 'competency profile' of the organization.
- Develop new performance level objectives/targets/standards.
- Vision alternative process(es) incorporating best practice enablers.
- Identify and minimize barriers to change.
- Develop action plans to adapt and implement best practices, make process changes, and achieve goals.
- Implement specific actions and integrate them into the organization.

REVIEW performance and the study:

- Monitor the results/improvements.
- Assess outcomes and learnings from the study.
- Review benchmarks.
- Share experiences and best practice learnings from implementation.

- Review relationships with target/partner organizations.
- Identify further opportunities for improving and sustaining performance.

In a typical benchmarking study involving several organizations, the study will commence with the **Plan** phase. Participants will be invited to a 'kick-off' meeting where they will share their aspirations and objectives for the study and establish roles and responsibilities. Participants will analyze their own organization to understand the strengths and areas for improvement. They will then agree appropriate measures for the study.

In the **Collect** phase, participants will collect data on their current performance, based on the agreed measures. The benchmarking partners will be identified, using a suitable screening process, and the key learning points will be shared. The site visits will then be planned and conducted, with appropriate training. Five to seven site visits might take place in each study.

Data collected from the site visits will be **Analyzed** in the next phase to identify best practices and the enablers which deliver outstanding performance. The reports from this phase will capture the learning and key outcomes from the site visits and present them as the main process enablers, linked to major performance outcomes.

In the **Adapt** phase, the participants will attend a feedback session where the conclusions from the study will be shared, and they will be assisted in adapting them to their own organization. Reports to partners should be issued after this session. (A 'subject expert' is often useful in benchmarking studies, to ensure good learning and adaptation at this stage.)

The final phase of the study will be a post-completion **Review.**

This will give all the participants and partners valuable feedback and establish, above all else, what actions are required to sustain improved performance. Best practice databases may be created to enable further sharing and improvement among participants and other members of the organization.

The role of benchmarking in change

One aspect of benchmarking is to enable organizations to gauge how well they are performing against others who undertake similar tasks and activities. But a more important aspect of best practice benchmarking is gaining an understanding of *how* other organizations achieve superior performance. A good benchmarking study, for example, in customer satisfaction and retention, will provide its participants with data and ideas on how excellent organizations undertake their activities and demonstrate best practices that may be adopted, adapted and used.

This new knowledge will result in the benchmarking team being able to judge the gap between leading and less good performance, as well as planning considered actions to bring about changes to bridge that gap. These changes may be things that can be undertaken quickly, with little adaptation and at a minimum of cost and disruption. Such changes, often brought about by the affected operational team, may be called 'quick wins'. This type of change is incremental and carries low levels of risk but usually lower levels of benefit.

Quick wins will often give temporary or partial relief from the problems associated with poor performance and tend to address symptoms not the underlying 'diseases'. They can have a disproportionately favorable psychological impact upon the organizations. Used well, quick wins should provide a platform from which longer lasting changes may be made, having created a feeling of movement and success. All too often however, once quick wins are implemented there is a tendency to move on to other areas, without either fully measuring the impact of the change or getting to the root cause of a performance issue.

Quick wins are clearly an important weapon in effecting change but must be followed up properly to deliver sustainable business improve-ment through the adoption of best practice. The changes needed to do this will usually be of a more fundamental nature and require invest-ment in effort and money to implement. Such changes will need to be carefully planned and systematically implemented as a discrete change project or program of projects. They carry substantial risk if not sys-tematically managed and controlled, but they have the potential for sig-nificant improvement in performance. These types of change projects are sometimes referred to as 'step change' or 'breakthrough' projects/ programs.

Whatever type of change is involved, a key ingredient of success is tak-ing the people along. A first class communication strategy is required throughout and beyond any change activity, as well as the linked activity of stakeholder management. The benchmarking efforts need to fit into the change model deployed – such a framework is proposed in Figure 9.2. Many change models exist in diagrammatic form and are often, in both intent and structure, quite similar. Such a model may be considered as a 'footprint' that will lead to the chosen destination, in this case the desired performance improvements through adoption of best practice. The foot-print in Figure 9.2 demonstrates where benchmarking activities link into the general flow of change activity leading to better results.

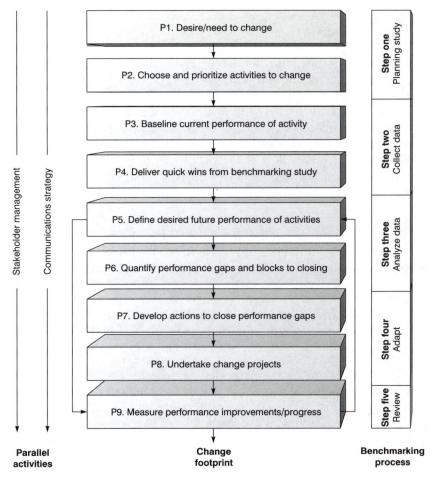

Figure 9.2 The benchmarking-change footprint

The success and benefits derived from any benchmarking and change-related activity are directly related to the excellence of the preparation. It is necessary to consider both the 'hard' and 'soft' aspects represented in Figure 9.2 and to systematically plan to meet and overcome any difficulties and challenges identified.

Communicating, managing stakeholders and lowering barriers

The importance of first class communication during any change can never be overemphasized. A vital element of excellent communication is targeting the right audience with the right message in the right way at the right time. A scattergun approach to communication rarely has the intended impact.

In any benchmarking study it is a wise and well-founded investment in time and effort to define and understand the key stakeholders. The rise of the term 'stakeholder' in business language is relatively recent – used to describe any group or individual that has some, however small, vested interest or influence in the business. Stakeholders are frequently referred to by generic groupings and may be either internal or external to an organization or business. The importance of forming, managing and maintaining good working relations with these groups is widely acknowledged and accepted.

The reality is that this activity is frequently not performed well in benchmarking. A disgruntled or ignored stakeholder with high direct organizational power or influence can easily derail the intent and hard work of others. Stakeholders with less direct power or influence can, at best, provide an unwelcome and costly distraction from the main objectives of a benchmarking study. The art of stakeholder management is to proactively head off any major confrontations. This means really understanding the stakeholders' needs and their potential to do both good and ill.

The burden of effective stakeholder management rests with the benchmarking team charged with stimulating change. They may need the ongoing patronage and support of people outside their direct control. In any good benchmarking study early thought will be given to who the stakeholders are and this will be valuable input to developing a robust stakeholder management strategy.

The elements of successful stakeholder management should include:

1 Defining and mapping the stakeholder groupings.
2 Analyzing and prioritizing these groupings.
3 Researching the key players in the most important groupings.
4 Developing a management strategy.
5 Deploying the strategy by tactical actions.
6 Reviewing effectiveness of the strategy and improving the future approach.

Objective measurement is also key to targeting change activity wisely. Benchmarking project budgets are often limited and it is good practice to target such discretionary spend at changes and improvements that will deliver the best return for their investment. Systematic measurement will provide a reliable baseline for making such decisions. By relating current performance against desired performance it should be possible to define both the gaps and appreciate the scale of improvements required to achieve the desired change.

Benchmarking studies add an extra dimension by understanding the levels of performance that best practices and leading organizations achieve. This allows realistic and sometimes uncomfortable comparisons with what an organization is currently able to achieve and what is possible. This is especially useful when setting stretch but realistic targets for future performance.

Baselining performance will allow teams to monitor and understand how successful they have been in delivering beneficial change. Used with care, as part of an overall communications strategy, successes on the road to achieving superior performance through change is a powerful motivator and useful influencing tool. Many organizations have clearly defined sets of performance measures, some self-imposed and some statute-based. These should be used, if in existence. If the interest is in customer satisfaction and retention, for example, a generic but good starting point might be:

- internal measures (the lead/predictor measures) – production cycle times, unit costs, defect rate found – quality and complaints resolved;
- external measures (the lag/reality measures) – customer satisfaction (perception), customer retention and complaints received.

The benchmarking activity may provide teams with ideas on how they might change the way goods or services are produced and delivered. They will need to prioritize this opportunity, however, to deliver best value for time and money invested and to ensure the organization does not become paralyzed by initiative overload – while making improvements the day job has to continue!

The benchmarking data collected will give a clear steer to the areas that require the most urgent attention but decisions will still have to be made. Measurement and benchmarking are tools not substitutes for management and leadership – the data on its own cannot make the decisions.

Choosing benchmarking-driven change activities wisely

As we have seen, benchmarking studies should fuel the desire to undertake change activities, but the excitement generated can allow the desire for change to take on a life of its own and irrational and impractical decisions can follow. These may result in full or partial failure to deliver the desired changes and waste of the valuable financial and people resource spent on the benchmarking itself.

Organizations should resist the temptation to start yet another series of improvement initiatives, without any consideration of their impact upon existing initiatives and the 'business as usual' activities. It is important to target the change wisely and a number of key questions need to be answered including:

- Do we fully understand the scale of the change?
- Do we have the financial resources to support the change?
- Do we have the people resources to undertake the change?
- Do we have the right skills available to undertake the change?
- Do we fully understand the operational impact during the change?
- Can beneficial changes be made without major disruption to the business?
- Will the delivered change support achievement of our business goals?
- What will the new changes do to existing change initiatives?
- Is the organization culturally ready for change?

Table 9.4 shows a simple decision-making tool to help consider the opportunities that are presented. The process may be viewed as a series of filters – it is assumed that the organization has defined business goals.

Case study ■■■

■ Benchmarking the Business Management System ■ in QinetiQ (DERA)

In a published case study[1] benchmarking was used to examine the contribution that business management systems (BMS) made to the achievement of organizational objectives. The Defence Evaluation and Research Agency (DERA), with the help of Oakland Consulting, conducted the study, which was based on the approach set out in this chapter. This included the following 12 key steps:

1. A benchmarking team was formed and educated.
2. Background research was conducted.
3. A decision was made on precisely what to benchmark (including metrics).
4. A questionnaire was produced and sent out to prospective companies.
5. The returns were analyzed.
6. Partners were selected.
7. Partners were site visited.
8. Data collected during the site visit was analyzed.
9. Good practice was distilled from the data.
10. A final report was produced.
11. Recommendations were made.
12. The project was reviewed.

The benchmarking project was conducted on the basis of sharing best practice to the benefit of both DERA and its external partners. The project team adhered to 'The European Benchmarking Code of Conduct' (1999).

The details of the study and its findings will not be repeated here, but what is relevant to this chapter are the actions resulting from the study. The benchmarking project and its recommendations were key in the development of revisions to the DERA BMS. It was not the sole input but, as a result of the work, an improvement project was initiated with the aim of making the BMS more process-based than it had previously been. A top-level process model was derived, in parallel with the development of the future strategy for DERA by the senior management. This included, of course, the part-privatization of DERA to QinetiQ. From this model, key processes were further developed and better use of web-based technology stimulated.

Table 9.4 Simple decision tool for choosing change activities

No.	Filter test	Yes	No
1	Does the benchmarking-driven proposed change support the achievement of one or more of the defined business goals?	Allow the opportunity to move forward for consideration.	Decline the opportunity or defer taking forward and schedule a review.
2	Does the change require financial and people resources above those agreed for the current budget round?	Prepare a business case within a project definition for consideration by senior management.	Pass the opportunity to local operational management to undertake the changes as a 'quick win' initiative.
3	Will current improvement activity be adversely impacted by the envisaged new changes?	Consider the relative merits and benefits of new and existing change initiates and amalgamate or amend or cancel existing initiates.	Allow change projects to proceed and add to the controlled list of overall change projects.
4	Is the required additional financial and people resource needed to undertake new change projects available?	Senior management agree and sign off project definition and project begins.	Senior management prioritize change activity agreeing necessary slippage or deferment or cancellation of some change projects.

Benchmarking studies in the BBC have provided insight on, for example, the potential for new technology to radically change the existing program making processes. Benchmarking should be an integral part of each process re-engineering project undertaken. The external perspective provided by the benchmarking studies helped BBC employees to see how things could be different (thinking outside the box), and provided valuable input to the steps required to implement new processes (see also Chapters 10 and 11).

The drivers of change are everywhere but properly conducted systematic benchmarking studies can define clearer objectives and help their effective deployment through well-executed change management. Best practice benchmarking and change management clearly are bedfellows. If well understood and integrated they can deliver lasting improvements in performance, which satisfy all stakeholder needs. Benchmarking is an efficient way to promote effective change by learning from the successful experiences of others and putting that learning to good effect.

■ Acknowledgement

The author is grateful for the contribution made in the preparation of this chapter by his colleagues Dr Steve Tanner and Robin Walker.

■ Reference

1. Morling, P. and Tanner, S. J. 'Benchmarking a public service business management system', *Total Quality Management*, Vol. 11, p. 417, 2000.

Chapter highlights

The why and what of benchmarking

- Benchmarking measures an organization's products, services and processes to establish targets, priorities and improvements, leading in turn to competitive advantage and/or cost reductions.
- Benefits of benchmarking can be numerous and include creating a better understanding of the current position, heightening sensitivity to changing customer needs, encouraging innovation, developing stretch goals, and establishing realistic action plans.
- Data from APQC suggests an average benchmarking study takes six months to complete, occupies a quarter of the team members' time and costs $50 000–100 000. The average return was five times the costs.

- The four basic types of benchmarking are internal, functional, generic and competitive, although the evolution of benchmarking in an organization is likely to progress through focus on continuous improvement.

The purpose and practice of benchmarking

- The evolution of benchmarking is likely to progress through four focuses: competitive products/services; industry best practices; all aspects of the business in terms of performance gaps; and focus on processes and true continuous improvement.
- The purpose of benchmarking is predominantly to change perspective, compare business practices, challenge current practices and processes, and to create improved goals and practices, with the focus on customer satisfaction and business results.
- A simple scoring proforma may help an organization to assess whether it is ready for benchmarking, if it has not engaged in it before. Help may be required to establish the right platforms if low scores are obtained.
- The benchmarking process has five main stages: plan, collect, analyze, adapt and review. These are focused on trying to measure comparisons and identify areas for action and change.

The role of benchmarking in change

- An important aspect of benchmarking is gaining an understanding of how other organizations achieve superior performance. Some of this knowledge will result in 'quick wins', with low risk but relatively low levels of benefit.
- Step changes are of a more fundamental nature, usually require further investment in time and money, will need to be carefully planned and systematically implemented, and typically carry a higher risk.
- A change model or 'footprint' should lead to the chosen destination – improved performance through the adoption of best practice – and show the role of benchmarking.

Communicating, managing stakeholders and lowering barriers

- Communication is vital during change, and a vital element is targeting the right audience, with the right message, in the right way at the right time.
- Defining and understanding the key stakeholders is a wise investment of time. This should be followed by building and managing good relationships. This falls on the benchmarking team.
- Elements of successful stakeholder group management include: defining and mapping; analyzing and prioritizing; researching key

players/groups; developing and deploying a strategy; and review-ing effectiveness.

■ Objective measurement is key to targeting change wisely and provides a reliable baseline for decisions. Baselining performance allows teams to monitor and understand success in delivering beneficial change.

Choosing benchmarking-driven change activities wisely

■ Organizations should start benchmarking-driven improvement activities with consideration of their impact on existing initia-tives. Questions to be asked include those related to the scale of the change, the financial and people resources (including skills) required, the impact and disruption aspects, the degree of sup-port to the business goals, and the cultural implications.

■ Benchmarking may be used to drive revisions in business man-agement systems, facilitate the application of new technologies, and generally help people to see how processes might be dif-ferent.

■ Properly conducted systematic benchmarking studies can aid the definition of clearer objectives and help their deployment through well-executed change management.

Part 4

Processes

I must Create A System, or be enslav'd
by another Man's.

William Blake, 1757–1827,
from 'Jerusalem'

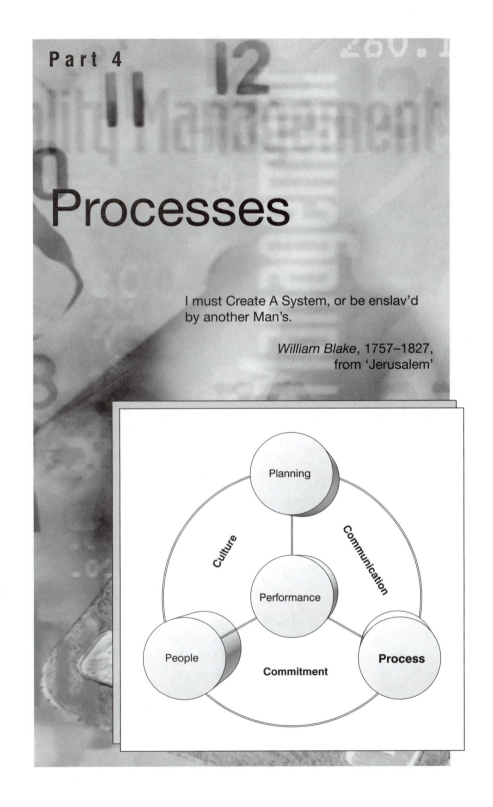

Process management

■ Process management vision

Organizations create value by delivering their products and/or services to customers. Everything they do in that whole chain of events is a process. So to perform well in the eyes of the customers and the stakeholders all organizations need very good process management – underperformance is primarily caused by poor processes.

This is recognized, of course, in the EFQM Excellence Model, in which the processes criterion is the central 'anchor' box linking the other enablers and the results together. Performance can be improved often by improving or changing processes but the devil is in the detail and successful exponents of process management understand all the dimensions related to:

Process strategy – particularly deployment
Operationalizing processes – including definition/design/systems
Process performance – measurement and improvement
People and leadership roles – values, beliefs, responsibilities, accountabilities, authorities and rewards
Information and knowledge – capturing and leveraging throughout the supply chains

Where process management is established and working, executives no longer see their organizations as sets of discrete vertical functions with

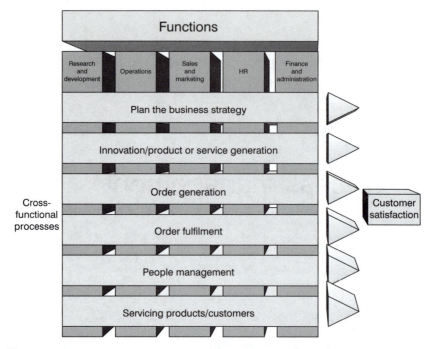

■ **Figure 10.1**   Cross-functional approach to managing core business processes

silo-type boundaries. Instead they visualize things from the customer perspective – as a series of interconnected work and information flows that cut horizontally across the business. Effectively the customers are pictured as 'taking a walk' through some or all of these 'end-to-end' processes and interfacing with the company or service organization, experiencing how it generates demand for products and services, how it fulfills orders, services products, etc. (Figure 10.1). All these processes need managing – planning, measuring and improving – sometimes discontinuously.

In monitoring process performance, measurement will inevitably identify necessary improvement actions. In many process managed companies, management has shifted the focus of the measurement systems from functional to process goals and even based remuneration and career advancement on process performance.

Operationalizing process management

Top management in many organizations now base their approach to business on the effective management of 'key or core business processes'. These are well-defined and developed sequences of steps with clear rationale, which add value by producing required outputs from a variety of inputs. Moreover these management teams have aligned the core processes with their strategy, combining related activities and cutting out ones that do not add value. This has led in some cases to a fundamental change in the way the place is managed.

The changes required have caused these organizations to emerge as true 'process enterprises'. There are many such organizations, including Texas Instruments, ST Microelectronics, Philips, IBM, Celestica, BBC, Highways Agency and QinetiQ, that are familiar to the author and his colleagues.

Companies comprising a number of different business units, such as outsourcing companies, face an early and important strategic decision when introducing process management – should all the business units follow the same process framework and standardization, or should they tailor processes to their own particular and diverse needs? Each organization must consider this question carefully and there can be no one correct approach.

Deployment of a common high-level process framework throughout the organization gives many benefits, including presenting 'one company' to the customers and suppliers, lower costs, and increased flexibility, particularly in terms of resource allocation.

In research on award-winning companies[1,2] the author and his colleagues identified process management best practices as:

- Identifying the key business processes:
  - prioritizing on the basis of the value chain, customer needs and strategic significance, and using process models and definitions.
- Managing processes systematically:
  - giving process ownership to the most appropriate individual or group and resolving process interface issues through meetings or ownership models.
- Reviewing processes and setting improvement targets:
  - empowering process-owners to set targets and collect data from internal and external customers.
- Using innovation and creativity to improve processes:
  - adopting self-managed teams, business process improvement and idea schemes.
- Changing processes and evaluating the benefits:
  - through process improvement or re-engineering teams, project management and involving customers, and suppliers.

Too many businesses are still not process oriented, however; they focus instead on tasks, jobs, the people who do them, and on structures.

■ Process Classification Framework and process
■ modeling

In establishing a core process framework, many organizations have found inspiration in the Process Classification Framework developed and copyrighted by the American Productivity and Quality Center (APQC) International Benchmarking Clearinghouse with the assistance of several major international corporations. The intent was to create a high-level generic enterprise model that encourages businesses and other organizations to see their activities from a cross-industry, process viewpoint rather than from a narrow functional viewpoint.

The Process Classification Framework supplies a generic view of business processes often found in multiple industries and sectors – manufacturing and service companies, health care, government, education, and others. As we saw earlier, many organizations now seek to understand their inner workings from a horizontal, process viewpoint, rather than from a vertical, functional viewpoint.

The Process Classification Framework seeks to represent major processes and subprocesses, not functions, through its structure (Figure 10.2) and vocabulary (Table 10.1). The framework does not list all processes within any specific organization. Likewise, not every process listed in the framework is present in every organization.

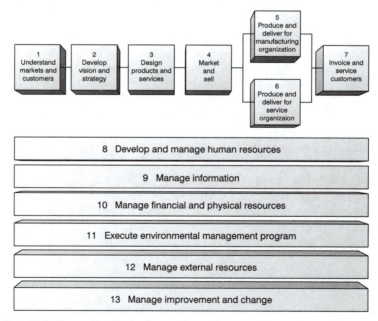

■ **Figure 10.2**   Process Classification Framework: overview

Table 10.1 APQC Process Classification Framework vocabulary

OPERATING PROCESSES

1   UNDERSTAND MARKETS AND CUSTOMERS
1.1   Determine customer needs and wants
    1.1.1 Conduct qualitative assessments
        1.1.1.1 Conduct customer interviews
        1.1.1.2 Conduct focus groups
    1.1.2 Conduct quantitative assessments
        1.1.2.1 Develop and implement surveys
    1.1.3 Predict customer purchasing behavior
1.2   Measure customer satisfaction
    1.2.1 Monitor satisfaction with products and services
    1.2.2 Monitor satisfaction with complaint resolution
    1.2.3 Monitor satisfaction with communication
1.3   Monitor changes in market or customer expectations
    1.3.1 Determine weaknesses of product/service offerings
    1.3.2 Identify innovations that meet customer needs
    1.3.3 Determine customer reactions to competitive offerings

2   DEVELOP VISION AND STRATEGY
2.1   Monitor the external environment
    2.1.1 Analyze and understand competition
    2.1.2 Identify economic trends
    2.1.3 Identify political and regulatory issues
    2.1.4 Assess new technology innovations
    2.1.5 Understand demographics
    2.1.6 Identify social and cultural changes
    2.1.7 Understand ecological concerns
2.2   Define the business concept and organizational strategy
    2.2.1 Select relevant markets
    2.2.2 Develop long-term vision
    2.2.3 Formulate business unit strategy
    2.2.4 Develop overall mission statement
2.3   Design the organizational structure and relationships between organizational units
2.4   Develop and set organizational goals

3   DESIGN PRODUCTS AND SERVICES
3.1   Develop new product/service concept and plans
    3.1.1 Translate customer wants and needs into products and/or service
        requirements
    3.1.2 Plan and deploy quality targets
    3.1.3 Plan and deploy cost targets

(continued)

Table 10.1 (*continued*)

OPERATING PROCESSES

     3.1.4 Develop product life cycle and development timing targets
     3.1.5 Develop and integrate leading technology into product/service concept
3.2   Design, build, and evaluate prototype products and services
     3.2.1 Develop product/service specifications
     3.2.2 Conduct concurrent engineering
     3.2.3 Implement value engineering
     3.2.4 Document design specifications
     3.2.5 Develop prototypes
     3.2.6 Apply for patents
3.3   Refine existing products/services
     3.3.1 Develop product/service enhancements
     3.3.2 Eliminate quality/reliability problems
     3.3.3 Eliminate outdated products/services
3.4   Test effectiveness of new or revised products or services
3.5   Prepare for production
     3.5.1 Develop and test prototype production process
     3.5.2 Design and obtain necessary materials and equipment
     3.5.3 Install and verify process or methodology
3.6   Manage the product/service development process

4   **MARKET AND SELL**
4.1   Market products or services to relevant customer segments
     4.1.1 Develop pricing strategy
     4.1.2 Develop advertising strategy
     4.1.3 Develop marketing messages to communicate benefits
     4.1.4 Estimate advertising resource and capital requirements
     4.1.5 Identify specific target customers and their needs
     4.1.6 Develop sales forecast
     4.1.7 Sell products and services
     4.1.8 Negotiate terms
4.2   Process customer orders
     4.2.1 Accept orders from customers
     4.2.2 Enter orders into production and delivery process

5   **PRODUCE AND DELIVER FOR MANUFACTURING-ORIENTED ORGANIZATIONS**
5.1   Plan for and acquire necessary resources
     5.1.1 Select and certify suppliers
     5.1.2 Purchase capital goods
     5.1.3 Purchase materials and supplies
     5.1.4 Acquire appropriate technology
5.2   Convert resources or inputs into products
     5.2.1 Develop and adjust production delivery process (for existing process)

Table 10.1 (*continued*)

OPERATING PROCESSES

    5.2.2 Schedule production

    5.2.3 Move materials and resources

    5.2.4 Make product

    5.2.5 Package product

    5.2.6 Warehouse or store product

    5.2.7 Stage products for delivery

5.3 Deliver products

    5.3.1 Arrange product shipment

    5.3.2 Deliver products to customers

    5.3.3 Install product

    5.3.4 Confirm specific service requirements for individual customers

    5.3.5 Identify and schedule resources to meet service requirements

    5.3.6 Provide the service to specific customers

5.4 Manage production and delivery process

    5.4.1 Document and monitor order status

    5.4.2 Manage inventories

    5.4.3 Ensure product quality

    5.4.4 Schedule and perform maintenance

    5.4.5 Monitor environmental constraints

6   **PRODUCE AND DELIVER FOR SERVICE-ORIENTED ORGANIZATIONS**

6.1 Plan for and acquire necessary resources

    6.1.1 Select and certify suppliers

    6.1.2 Purchase materials and supplies

    6.1.3 Acquire appropriate technology

6.2 Develop human resource skills

    6.2.1 Define skill requirements

    6.2.2 Identify and implement training

    6.2.3 Monitor and manage skill development

6.3 Deliver service to the customer

    6.3.1 Confirm specific service requirements for individual customer

    6.3.2 Identify and schedule resources to meet service requirements

    6.3.3 Provide the service to specific customers

6.4 Ensure quality of service

7   **INVOICE AND SERVICE CUSTOMERS**

7.1 Bill the customer

    7.1.1 Develop, deliver, and maintain customer billing

    7.1.2 Invoice the customer

    7.1.3 Respond to billing inquiries

(*continued*)

■ **Table 10.1** (*continued*)

OPERATING PROCESSES

7.2   Provide after-sales service
    7.2.1 Provide post-sales service
    7.2.2 Handle warranties and claims
7.3   Respond to customer inquiries
    7.3.1 Respond to information requests
    7.3.2 Manage customer complaints

MANAGEMENT & SUPPORT PROCESSES

---

8   DEVELOP AND MANAGE HUMAN RESOURCES
8.1   Create and manage human resource strategies
    8.1.1 Identify organizational strategic demands
    8.1.2 Determine human resource costs
    8.1.3 Define human resource requirements
    8.1.4 Define human resource's organizational role
8.2   Cascade strategy to work level
    8.2.1 Analyze, design, or redesign work
    8.2.2 Define and align work outputs and metrics
    8.2.3 Define work competencies
8.3   Manage deployment of personnel
    8.3.1 Plan and forecast workforce requirements
    8.3.2 Develop succession and career plans
    8.3.3 Recruit, select and hire employees
    8.3.4 Create and deploy teams
    8.3.5 Relocate employees
    8.3.6 Restructure and rightsize workforce
    8.3.7 Manage employee retirement
    8.3.8 Provide outplacement support
8.4   Develop and train employees
    8.4.1 Align employee and organization development needs
    8.4.2 Develop and manage training programs
    8.4.3 Develop and manage employee orientation programs
    8.4.4 Develop functional/process competencies
    8.4.5 Develop management/leadership competencies
    8.4.6 Develop team competencies
8.5   Manage employee performance, reward and recognition
    8.5.1 Define performance measures
    8.5.2 Develop performance management approaches/feedback
    8.5.3 Manage team performance
    8.5.4 Evaluate work for market value and internal equity

Table 10.1 (*continued*)

MANAGEMENT & SUPPORT PROCESSES

     8.5.5 Develop and manage base and variable compensation
     8.5.6 Manage reward and recognition programs
8.6   Ensure employee well-being and satisfaction
     8.6.1 Manage employee satisfaction
     8.6.2 Develop work and family support systems
     8.6.3 Manage and administer employee benefits
     8.6.4 Manage workplace health and safety
     8.6.5 Manage internal communications
     8.6.6 Manage and support workforce diversity
8.7   Ensure employee involvement
8.8   Manage labor-management relationships
     8.8.1 Manage collective bargaining process
     8.8.2 Manage labor-management partnerships
8.9   Develop Human Resource Information Systems (HRIS)

9   MANAGE INFORMATION RESOURCES
9.1   Plan for information resource management
     9.1.1 Derive requirements from business strategies
     9.1.2 Define enterprise system architectures
     9.1.3 Plan and forecast information technologies/methodologies
     9.1.4 Establish enterprise data standards
     9.1.5 Establish quality standards and controls
9.2   Develop and deploy enterprise support systems
     9.2.1 Conduct specific needs assessments
     9.2.2 Select information technologies
     9.2.3 Define data life cycles
     9.2.4 Develop enterprise support systems
     9.2.5 Test, evaluate, and deploy enterprise support systems
9.3   Implement systems security and controls
     9.3.1 Establish systems security strategies and levels
     9.3.2 Test, evaluate, and deploy systems security and controls
9.4   Manage information storage and retrieval
     9.4.1 Establish information repositories (data bases)
     9.4.2 Acquire and collect information
     9.4.3 Store information
     9.4.4 Modify and update information
     9.4.5 Enable retrieval of information
     9.4.6 Delete information
9.5   Manage facilities and network operations
     9.5.1 Manage centralized facilities

(*continued*)

■ **Table 10.1** (*continued*)

MANAGEMENT & SUPPORT PROCESSES

Table 10.1 (*continued*)

MANAGEMENT & SUPPORT PROCESSES

---

11   EXECUTE ENVIRONMENTAL MANAGEMENT PROGRAM
11.1   Formulate environmental management strategy
11.2   Ensure compliance with regulations
11.3   Train and educate employees
11.4   Implement pollution prevention program
11.5   Manage remediation efforts
11.6   Implement emergency response programs
11.7   Manage government agency and public relations
11.8   Manage acquisition/divestiture environmental issues
11.9   Develop and manage environmental information system
11.10   Monitor environmental management program

12   MANAGE EXTERNAL RELATIONSHIPS
12.1   Communicate with shareholders
12.2   Manage government relationships
12.3   Build lender relationships
12.4   Develop public relations program
12.5   Interface with board of directors
12.6   Develop community relations
12.7   Manage legal and ethical issues

13   MANAGE IMPROVEMENT AND CHANGE
13.1   Measure organizational performance
        13.1.1 Create measurement systems
        13.1.2 Measure product and service equality
        13.1.3 Measure cost of quality
        13.1.4 Measure costs
        13.1.5 Measure cycle time
        13.1.6 Measure productivity
13.2   Conduct quality assessments
        13.2.1 Conduct quality assessments based on external criteria
        13.2.2 Conduct quality assessments based on internal criteria
13.3   Benchmark performance
        13.3.1 Develop benchmarking capabilities
        13.3.2 Conduct process benchmarking
        13.3.3 Conduct competitive benchmarking
13.4   Improve processes and systems
        13.4.1 Create commitment for improvement
        13.4.2 Implement continuous process improvement
        13.4.3 Reengineer business processes and systems
        13.4.4 Manage transition to change

(*continued*)

Table 10.1 (*continued*)

MANAGEMENT & SUPPORT PROCESSES

13.5   Implement TQM
    13.5.1   Create commitment for TQM
    13.5.2   Design and implement TQM systems
    13.5.3   Manage TQM life cycle

About the framework

The Process Classification Framework was originally envisioned as a 'taxonomy' of business processes during the design of the American Productivity and Quality Center's International Benchmarking Clearinghouse. That design process involved more than 80 organizations with a strong interest in advancing the use of benchmarking in the USA and around the world. The Process Classification Framework can be a useful tool in understanding and mapping business processes. In particular, a number of organizations have used the framework to classify both internal and external information for the purpose of cross-functional and cross-divisional communication.

The Process Classification Framework is an evolving document and the Center will continue to enhance and improve it on a regular basis. To that end, the Center welcomes your comments, suggestions for improvement, and any insights you gain from applying it within your organization. The Center would like to see the Process Classification Framework receive wide distribution, discussion, and use. Therefore, it grants permission for copying the framework, as long as acknowledgement is made to the American Productivity and Quality Center.*

Process modeling

More than 25 years ago, the US Airforce adopted 'Integration Definition Function Modeling' (IDEF-0), as part of its Integrated Computer-Aided Manufacturing (ICAM) architecture. The IDEF-0 modeling language, now described in a Federal Information Processing Standards Publication (FIPS PUBS), provides a useful structured graphical framework for

---

* Please direct your comments, suggestions, and questions, to: APQC International Benchmarking Clearinghouse Information Services Department, 123 North Post Oak Lane, 3rd Floor, Houston, Texas 77024–7797, tel: 713–681–4020, 713–681–8578 fax: Internet: apqcinfo@apqc.org. *For updates, visit the website at http: www.apqc.org.*

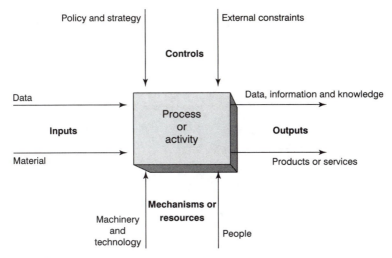

describing and improving business processes. The associated 'Integration
Definition for Information Modeling' (IDEF-1X) language allows the
development of a logical model of data associated with processes, such
as measurement.

These techniques are widely used in business process re-engineering
(BPR) and business process improvement (BPI) projects, and to integrate process information. A range of specialist software (including
Windows/PC-based) is also available to support the applications. IDEF-0
may be used to model a wide variety of new and existing processes,
define the requirements, and design an implementation to meet the
requirements.

An IDEF-0 model consists of a hierarchical series of diagrams, text and
glossary, cross-referenced to each other through boxes (process components) and arrows (data and objects). The method is expressive and
comprehensive and is capable of representing a wide variety of business, service and manufacturing processes. The relatively simple language allows coherent, rigorous and precise process expression, and
promotes consistency. Figure 10.3 shows the basis of the approach.

For a full description of the IDEF-0 methodology, it is necessary to consult the FIPS PUBS standard (Federal Information Processing Standard
Publication 183 (December 1993), National Institute of Standards and
Technology (NIST)). It should be possible, however, from the simple
description given here, to begin process modeling (or mapping) using
the technique.

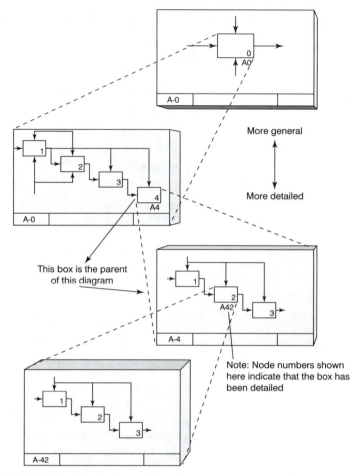

More general

More detailed

This box is the parent
of this diagram

Note: Node numbers shown
here indicate that the box has
been detailed

■ **Figure 10.4**   Decomposition structure – subprocesses

Processes can be any combination of things, including people, information, software, equipment, systems, products or materials. The IDEF-0 model describes what a process does, what controls it, what things it works on, what means it uses to perform its functions, and what it produces. The combined graphics and text are comprised of:

■ **Boxes** – which provide a description of what happens in the form of an active verb or verb phrase.
■ **Arrows** – which convey data or objects related to the processes to be performed (they do not represent flow or sequence as in the traditional process flow model).

Each side of the process box has a standard meaning in terms of box/arrow relationships. Arrows on the left side of the box are **inputs**,

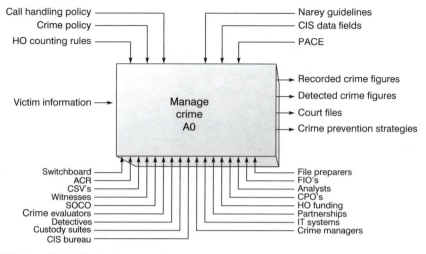

Figure 10.5   A0 crime management

which are transformed or consumed by the process to produce **output** arrows on the right side. Arrows entering the top of the box are **controls** which specify the conditions required for the process to generate the correct outputs. Arrows connected to the bottom of the box represent '**mechanisms**' or **resources**. The abbreviation ICOR (inputs, controls, outputs, resources) is sometimes used.

Using these relationships, process diagrams are broken down or decomposed into more detailed diagrams, the top-level diagram providing a description of the highest-level process. This is followed by a series of child diagrams providing details of the subprocesses (see Figure 10.4).

Each process model has a top-level diagram on which the process is represented by a single box with its surrounding arrows (e.g. Figure 10.5). Each subprocess is modeled individually by a box, with parent boxes detailed by child diagrams at the next lower level. An example of the application of IDEF or ICOR modeling in Crime Management and Reporting in West Yorkshire Police is given in Figures 10.6 to 10.8.

Text and glossary

An IDEF-0 diagram may have associated structured text to give an overview of the process model. This may also be used to highlight features, flows, and interbox connections and to clarify significant patterns. A glossary may be used to define acronyms, key words and phrases used in the diagrams.

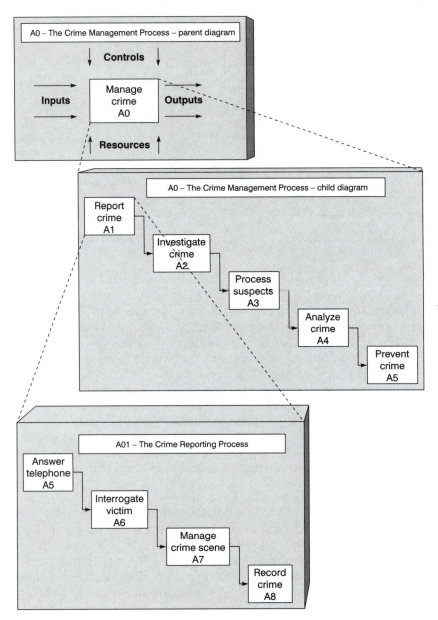

Figure 10.6 IDEF-0 decomposition structure – subprocesses – for crime management

Arrows

Arrows on high-level IDEF-0 diagrams represent data or objects as constraints. Only at low levels of detail can arrows represent flow or sequence. These high-level arrows may usefully be thought of as

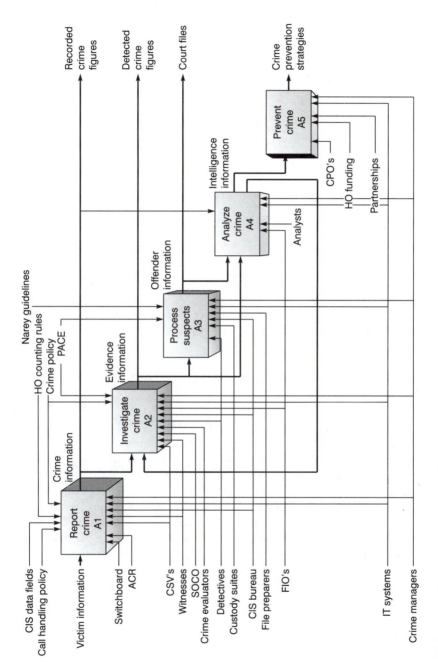

Figure 10.7 A0 crime management child diagram

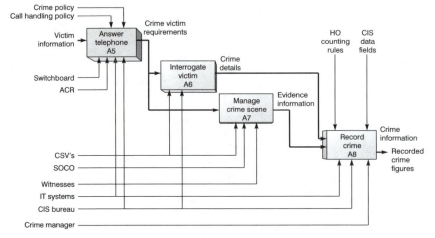

pipelines or conduits with general labels. An arrow may branch, fork, or join, indicating that the same kind of data or object may be needed or produced by more than one process or subprocess.

IDEF-0 process modeling, improvement and teamwork

The IDEF-0 methodology includes procedures for developing and critiquing process models by a group or team of people. The creation of an IDEF-0 process model provides a disciplined teamwork procedure for process understanding and improvement. As the group works on the process following the discipline, the diagrams are changed to reflect corrections and improvements. More detail can be added by creating more diagrams, which in turn can be reviewed and altered. The final model represents an agreement on the process for a given purpose and from a given viewpoint, and can be the basis of new process or system improvement projects.

In the West Yorkshire Police example (Figures 10.5–10.8), teams involved in constructing these models gained appreciation of the role everyone plays in the overall process. This encouraged ownership of the process and acted as a spur to making improvements.

Initially the overall crime management process was considered at the macro level, identifying its various input, output, control and resource factors through a combination of brainstorming and consultation with managers. This allowed a high-level model to be compiled, as shown at Figure 10.8. Although the model reveals little in terms of the process interactions, it is helpful in illustrating the importance of crime management to the police force.

The outputs of the crime management process arguably represent the most important outputs of any police force. They include the levels of recorded crime, as well as the proportion of these offences that have been successfully detected. The publication of this information is of great interest to the public, the force and its political stakeholders, forming perceptions of safety and assessment of organizational performance. Also produced by the process are prosecution files for the courts and crime prevention strategies, often part of community safety partnerships. These outputs, though often less publicized, are measures of activities vital for the medium- and long-term reduction of crime.

The range of resources applied by West Yorkshire Police to its crime management process covers a large proportion of its functional areas. It includes uniform patrol, CID, Intelligence, Communication and Case Preparation. It gives an indication of the complexity, but also the appropriateness, of using a process approach to manage improvement across the departmental structures of the force.

Having established the process at its macro level, the process was broken down into the main subprocesses of which it was composed. It was not possible to be prescriptive as to how many subprocesses this entailed. However, they needed to be sufficient to cover the range of the crime management process, and ensure that all the original input, output, control and resource factors could be reassigned on the child diagram.

Crime reporting was the initial process providing the interface with the victim and included the telephone crime reporting system. Crime investigation was the next process involving the greatest concentration of organizational resources, particularly those of an operational nature. The processing of suspects similarly attracted a heavy concentration of resources, reflecting the sometimes onerous and even bureaucratic demands of its main output – court files. The crime analysis process would not have featured in a similar process model years ago, but is now an indispensable process in tackling crime, made possible through advances in IT. Finally, the crime prevention process has risen in prominence through community safety partnerships.

The IDEF-0/ICOR methodology allows processes to be broken down to as many levels as are required by the process under investigation. In this study, one further level was required in order to examine the crime reporting process in sufficient detail. This involved taking the crime reporting subprocess A1 from Figure 10.7 and decomposing its own constituent subprocesses.

This final level was composed of four subprocesses: telephone answering where the victim's call was handled between the switchboard, ACR and the CIS Bureau; victim interrogation where information pertaining to the crime was obtained directly through dialog with the victim, either in person or by phone; crime scene management, which is the obtaining of forensic evidence or identifying potential witnesses at the crime scene; crime recording, where a detailed bundle of crime information is entered onto the CIS database as a recorded crime. The detailed subprocess model is shown in Figure 10.8.

In using such techniques in process management, there can be a propensity for maps to assume disproportionate importance. This can result in participants becoming distracted in pursuit of accuracy or even the overall purpose of process improvement being supplanted by the modeling process itself.

IDEF-1X

This is used to produce structural graphical information models for processes, which may support the management of data, the integration of information systems, and the building of computer databases. It is described in detail in the FIPS PUB 184 (December 1993, NIST). Its use is facilitated by the introduction of IDEF-0 modeling for process understanding and improvement.

A number of commercial software packages, which support IDEF-0 and IDEF-1X implementation, are available.

■ Process flowcharting

In the systematic planning or detailed examination of any process, whether that be an administrative service delivery, manufacturing, or managerial activity, it is necessary to record the series of events and activities, stages and decisions, in a form that can be easily understood and communicated to all. If improvements are to be made, the facts relating to the existing method must be recorded first. The statements defining the process should lead to its understanding and will provide the basis of any critical examination necessary for the development of improvements. It is essential therefore that the descriptions of processes are accurate, clear and concise.

The usual method of recording facts is to write them down, but this is not suitable for recording the complicated processes that exist in any

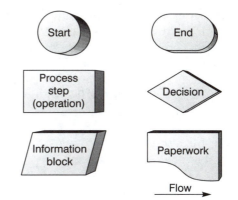

Figure 10.9
Flowcharting symbols

organization, particularly when an exact record is required of a long process, and its written description would cover several pages requiring careful study to elicit every detail. To overcome this difficulty, certain methods of recording have been developed, and the most powerful of these is flowcharting. This method of describing a process owes much to computer programming, where the technique is used to arrange the sequence of steps required for the operation of the program. It has a much wider application, however, than computing.

Certain standard symbols are used on the chart, and these are shown in Figure 10.9. The starting point of the process is indicated by a circle. Each processing step, indicated by a rectangle, contains a description of the relevant operation, and where the process ends is indicated by an oval. A point where the process branches because of a decision is shown by a diamond. A parallelogram relates to process information but is not a processing step. The arrowed lines are used to connect symbols and to indicate direction of flow. For a complete description of the process, all operation steps (rectangles) and decisions (diamonds) should be connected by pathways to the start circle and end oval. If the flowchart cannot be drawn in this way, the process is not fully understood.

It is a salutary experience for most people to sit down and try to draw the flowchart for a process in which they take part every working day. It is often found that:

- The process flow is not fully understood.
- A single person is unable to complete the flowchart without help from others.

The very act of flowcharting will improve knowledge of the process, and will begin to develop the teamwork necessary to find improvements.

In many cases, the convoluted flow and octopus-like appearance of the chart will highlight unnecessary movements of people and materials and lead to common-sense suggestions for waste elimination.

Example of flowcharting in use – improving a travel procedure

We start by describing the original process for a male employee, though clearly it applies equally to females.

The process starts with the employee explaining his travel plans to his secretary. The secretary then calls the travel agent to enquire about the possibilities and gives feedback to the employee. The employee decides if the travel arrangements, e.g. flight numbers and dates, are acceptable and informs his secretary, who calls the agent to make the necessary bookings or examine alternatives. The administrative procedure, which starts as soon as the bookings have been made, is as follows:

1 The employee's secretary prepares the travel request (which is in four parts, A, B, C and D), and gives it to the employee. The request is then sent to the employee's manager, who approves it. The manager's secretary sends it back to the employee's secretary.
2 The employee's secretary sends copies A, B and C to the agent and gives copy D to the employee. The travel agent delivers the ticket to the employee's secretary, together with copy B of the travel request. The secretary endorses copy B for receipt of the ticket, sends it to Accounting, and gives ticket to the employee.
3 The travel agent bills the credit card company, and sends Accounting a pro-forma invoice with copy C of the travel request. Accounting matches copies B and C, and charges the employee's 181 account.
4 Accounting receives the monthly bill from the credit card company, matches it against the travel request, and then pays the credit card company.
5 The employee reports the travel request on his expense statement. Accounting matches and balances the employee's 181 account.

The total time taken for the administrative procedure, excluding the correction of errors and the preparation of overview reports is 23 minutes per travel request.

The flowchart for the process is drawn in Figure 10.10. A quality-improvement team was set up to analyze the process and make recommendations for improvement, using brainstorming and questioning techniques. They made the following proposal to change the procedure. The preparation for the trip remained the same but

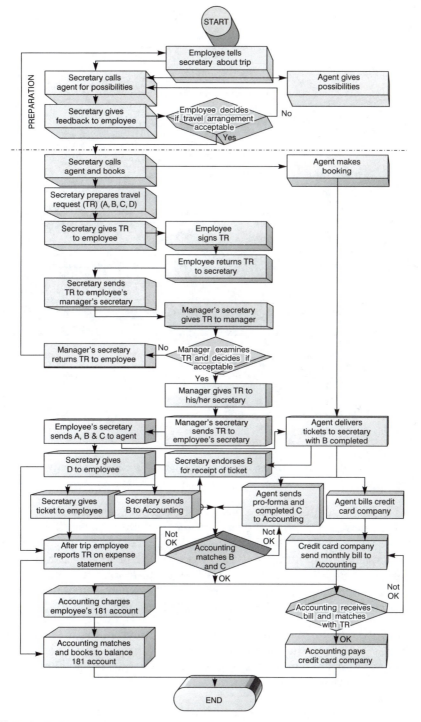

Figure 10.10 Original process for travel procedure

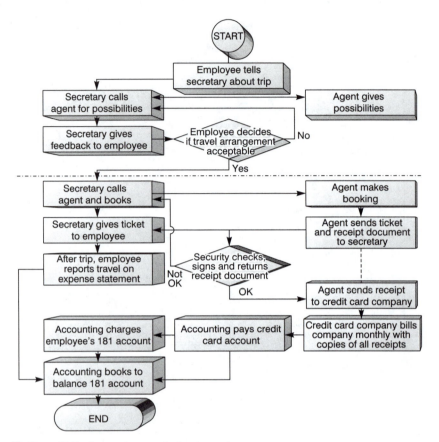

Figure 10.11   Improved process for travel procedure

the administrative steps, following the bookings being made, became:

1 The travel agent sends the ticket to the secretary, along with a receipt document, which is returned to the agent with the secretary's signature.
2 The agent sends the receipt to the credit card company, which bills the company on a monthly basis with a copy of all the receipts. Accounting pays the credit card company and charges the employee's 181 account.
3 The employee reports the travel on his expense statement, and Accounting balances the employee's 181 account.

The flowchart for the improved process is shown in Figure 10.11. The proposal reduced the total administrative effort per travel request (or per travel arrangement, because the travel request was eliminated) from 23 minutes to 5 minutes.

The details that appear on a flowchart for an existing process must be obtained from direct observation of the process, not by imagining what

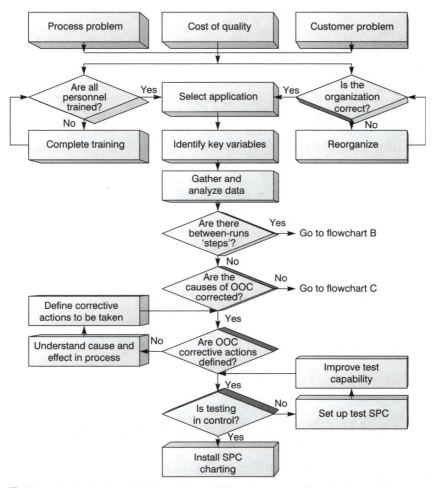

■ **Figure 10.12** Flowchart (A) for installation of SPC charting systems (The author is grateful to Exxon Chemical International for permission to use and modify this chart)

is done or what should be done. The latter may be useful, however, in the planning phase, or for outlining the stages in the introduction of a new concept. Such an application is illustrated in Figure 10.12 for the installation of statistical process control charting systems (see Chapter 13). Similar charts may be used in the planning of quality management systems.

It is surprisingly difficult to draw flowcharts for even the simplest processes, particularly managerial ones, and following the first attempt it is useful to ask whether:

■ The facts have been correctly recorded.
■ Any oversimplifying assumptions have been made.
■ All the factors concerning the process have been recorded.

The author has seen too many process flowcharts that are so incomplete as to be grossly inaccurate.

Summarizing, then, a flowchart is a picture of the steps used in performing an activity or task. Lines connect the steps to show the flow of the various tasks or steps. Flowcharts provide excellent documentation and are useful troubleshooting tools to determine how each step is related to the others. By reviewing the flowchart, it is often possible to discover inconsistencies and determine potential sources of variation and problems. For this reason, flowcharts are very useful in process improvement when examining an existing process to highlight the problem areas. A group of people, with the knowledge about the process, should take the following simple steps:

1 Draw a flowchart of existing process.
2 Draw a second chart of the flow the process could or should follow.
3 Compare the two to highlight the changes necessary.

A number of commercial software packages which support process flowcharting are available.

Case study ■■■ ▬▬▬▬▬▬▬▬▬▬▬▬▬▬▬

■ Process improvement in the Highways Agency

The improvement strategy has had an impact on the way the Highways Agency manages and operates its day-to-day business. For example, work completed in the financial payments division has led to improvements in key performance, particularly in terms of the handling time of invoices for payment (down from an average of over 15 minutes to less than four minutes per invoice). This has led to an improvement in the prompt payment initiative targets, from less than 75 percent being paid within the mandated 28 days to greater than 95 percent being paid on time, with fewer staff employed in the process.

The planned improvement is led by the customer facing and program delivery directorates as they are responsible for the Agency's key or core delivery activity, working to a directorate improvement framework (see Figure 10.13).

From their clarified purpose (what they are there to achieve), both directorates have identified their key activities (what they are there to do), and their key processes (how they will operate). This was developed using ICOR (inputs, controls, outputs and resources techniques and has

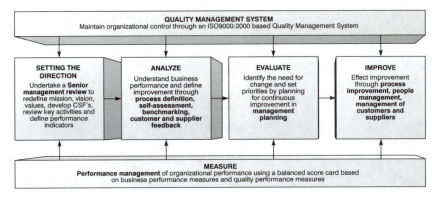

Figure 10.13   Directorate improvement frameworks

enabled the production of agreed consistent practices, recorded as flowcharts and working procedures in the form of process tables (see Figure 10.14). These enable anyone in these directorates to understand their contribution to the overall work of the Highways Agency and provide clear practical working guidance. The delivery processes, held in an electronic web format, are continuously improved and developed alongside their support processes for customer management, people management, supplier partnerships and management planning, following ISO 9000:2000 principles.

This delivery process development work has enabled the picture of the Agency's overall delivery process to be developed, as shown in Figure 10.15. This is enabling the wider process picture and to be addressed supporting the Directorates in clarifying their contributions. Measures from these key delivery processes feed forward into the Agency's balanced scorecard.

Other areas include the identification of potential improvements to project delivery areas that, once realized, should improve control over delivery processes, reducing wasted effort through failures, etc.

Work in developing customer satisfaction and management systems allow the line managers to identify key areas for improvement, based on the needs and expectations of the customers, as well as clearly measuring, in both lead and lag terms, how the organization is performing in meeting those needs and expectations.

There has also been a number of improvements identified as the result of detailed benchmarking studies conducted with a range of public and private sector partners and it would appear that the impact of these has been mainly positive.

A notable achievement following the adoption of the Highways Agency business improvement strategy is the greater awareness and desire by all staff to embrace business improvement as a way of resolving problems and delivering improved services to customers.

Item Ref	Task/Gateway	Approach	Forms/Ref Doc's Used	Records Kept	Performed by	Performance standard
2.2.1	Is Procurement Division's involvement required?	Determine type of value of project/service required. Refer to HA Procedures Manual – Procurement Section	HA Procedures Manual – Procurement Section	Record of decision kept on file	Project Sponsor in liaison with Procurement Division	
2.2.1a	Is request for procurement of Utility Services?	Determine whether utility equipment is involved	New Roads and Streets Works Act 1991 / Code of Practice	Record of decision kept on file	Project Sponsor	
2.2.1b	Perform tender process for services	Procurement Manual?				
2.2.2	Determine members of project team (inc. virtual team and/or DBFO Transaction Team)	Contact potential team members and arrange meeting	Commissioning forms for SSR	Record of Commissioning form	Project Sponsor	
2.2.2a	Undertake Tender Process for Non-Roads Services	Procurement Manual?				
2.2.3	Review options with Procurement	Arrange to meet with Procurement Division to regarding options	HAPM4	HAPM4	Project Sponsor in liaison with Procurement Division	
2.2.4	Are Framework Contracts applicable?	Consider advice from Procurement Division to determine whether service can be procured from Agreements inc. Framework Contracts	Minute/Email from Procurement	Minute/Email from Procurement	Project Sponsor in liaison with Procurement Division	

Figure 10.14 Example of a process table

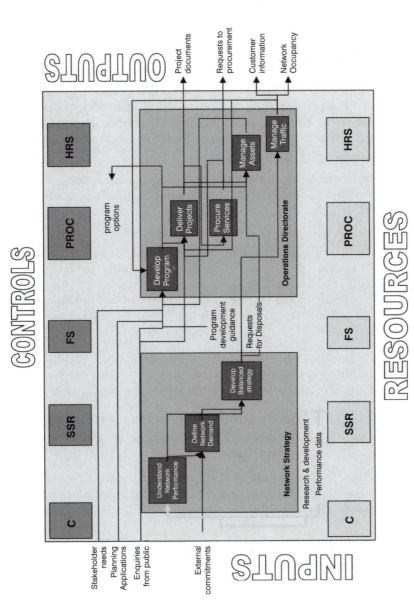

Figure 10.15 End-to-end Highways Agency process ICOR

Leadership, people and implementation aspects of process management

There are many top executives who have famously used process management to great effect, including Richard J. Leo, President and General Manager of Xerox, who suggested that process orientation is one of the key factors that delivered the remarkable turnaround in the performance of that company:

> Business processes are designed to be customer driven, cross-functional and value based. They create knowledge, eliminate waste and abandon unproductive work yielding world-class productivity and higher perceived service levels for customers.

Alan Jones, Group Managing Director of multiple quality award winning TNT Express, operates a process-driven company and he believes this approach helps provide a logical framework for people in the organization to satisfy customers. The processes provide awareness for each person of his or her role in the business and this leads to superior performance:

> Publishing the process on the wall helps people understand their place in the big picture ... the continuously improving profits earned by TNT Express are a consequence of superlative performance that is derived from well thought out processes, ongoing measurement of a few carefully selected key indicators, good communications and the full involvement everywhere of all people working in the company.

The approach manifests itself in complete understanding by everyone of the end-to-end TNT Express process and this is described in the 'perfect transaction' (Figure 10.16).

Perhaps the most visible difference between a process management enterprise and a more traditional, functionally based one is the existence of process owners – top management with end-to-end responsibilities for individual processes. They have real responsibility for and authority over the process design, operation and measurement of performance. This requires working, at least initially, through functional heads, which usually leads to a major cultural challenge for the organization. Process owners cannot simply instruct workers in the process to do as they say, so style and the ability to influence is at least as important to process management as structure.

Managing the people who work in the processes demands attention to:

- designing, developing and delivering training programs;
- setting performance targets;

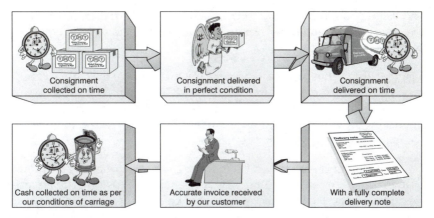

Figure 10.16   TNT Express Delivery Services: the perfect transaction process

- regular communication, preferably face to face:
  - keeping them informed of changing customer needs,
  - listening to concerns and ideas,
- negotiation and collaboration.

Case study ■■■

■ Process management at Celestica

Celestica (previously D2D*) realized in the early 1980s that to be a cost-effective, competitive and indeed world class organization, it must ensure that all processes are understood, measured and in control. Everyone is trained in process management and improvement and shown how they are part of a supplier–process–customer chain. The training reinforces that these chains are interdependent and that the processes all support the delivery of products or services to customers.

Operators of every process are properly trained, have necessary work instructions available, and have the appropriate tools (such as statistical process control), facilities and resources to perform the process to its optimum capability. This applies to all processes throughout the organization, whatever the outputs, including those in finance and human resources.

* Celestica acquired Design to Distribution (D2D) Limited in 1997.

In many process-managed organizations process management approach has changed the way they assign and train employees, emphasizing the whole process rather than narrowly focused tasks. It has made fundamental changes to cultures, stressing process-based teamwork and customers rather than functionally driven command and control. Creativity and innovation in process improvement are recognized as core competencies, and the annual performance reviews and personal development plans are linked to these.

The first thing that top management must recognize is that moving to process management requires much more than redrawing the organizational chart or structure. The changes needed are fundamental, leading to new ways of working and managing, and they will challenge any company or public service organization.

Many organizations today are facing a large number of changes and initiatives, often driven by public/private transitions, customer or government demands, technology, and so on. Before implementing process management, therefore, a senior management team needs to examine closely all its current change initiatives to prune those that are not relevant to a process-managed business and combining/rationalizing those that are.

The introduction of process management is often driven or directly connected to a strategic initiative, such as reducing cycle times, increasing customer satisfaction, reducing working capital (perhaps tied up in work-in-progress), an enterprise resource planning (ERP) implementation, changes in technology, or introduction of e-business. The application of new enabling technologies is an ideal time to review the design and configuration of key processes. Failure to do so could lead to missed opportunities to extract maximum benefit from the technology.

Implementing process management like many change initiatives cannot be a quick fix and it will not happen overnight. Top management need the resolution and commitment for major changes in the way things are structured, carried out and measured. As with all such implementation, it needs careful planning and an understanding of what needs to be done first. Things high on the list will be the establishment of a core process framework, aligned to the needs of the business, and the appointment of key process owners. A process-based performance measurement framework[2] then needs to track progress. As with all change initiatives, delivering some tangible measurable benefits early on will help overcome the inevitable resistance. In one pharmaceutical company, for example, the success of the work on the product development and product promotion processes helped significantly the cause of process management and the company extended its approach into the supply-chain management and other processes.

As companies and public service organizations move inexorably towards the wider introduction of e-commerce to do business, this will place a premium on rapid and fault-free execution of business processes. Putting a website in front of an inefficient, ineffectual or even broken process will soon bring it to its knees, together with everyone working in and around it. This will also bring 'back-office' mistakes to the attention of the market place. Some of these processes will, of course, need to be redesigned – from customer order fulfillment to procurement. They will need to change 'shape' as demands, technology and markets change. Without good process management in place this is going to be very difficult for functionally driven organizations.

■ Acknowledgement

The author is grateful to Mark Milsom of West Yorkshire Police for permission to use the material related to crime reporting in this chapter.

■ References

1. 'The X-factor, winning performance through business excellence', European Centre for Business Excellence/British Quality Foundation, London 1999.
2. 'The Model in Practice' and 'The Model in Practice 2', British Quality Foundation, London 2000 and 2002 (prepared by European Centre for Business Excellence).

Chapter highlights

Process management vision

- Everything organizations do to create value for customers of their products or services is a process. Process management is key to improving performance.
- Process-managed organizations see things from a customer perspective – as a series of interconnected work and information flows that cut horizontally across the business functions.
- The key or core business processes are well-defined and developed sequences of steps with clear rationale, which add value by producing required outputs from a variety of inputs.
- Deployment of a common high-level process framework throughout the organization gives many benefits, including reduced costs and increased flexibility.
- Process management best practices include: identifying the key business processes, managing processes systematically, reviewing processes and setting improvement targets, using innovation and creativity to improve processes, changing processes, and evaluating the benefits.

Process Classification Framework and process modeling

■ The APQC's Process Classification Framework creates a high-level generic, cross-functional process view of an enterprise – a taxonomy of business processes.

■ The IDEF (Integrated Definition Function Modeling) language provides a useful structured graphical framework for describing and improving business processes. It consists of a hierarchical series of diagrams and text, cross-referenced to each other through boxes. The processes are described in terms of inputs, controls, outputs and resources (ICOR).

Process flowcharting

■ Flowcharting is a method of describing a process in pictures, using symbols – rectangles for operation steps, diamonds for decisions, parallelograms for information, and circles/ovals for the start/end points. Arrow lines connect the symbols to show the 'flow'.

■ Flowcharting improves knowledge of the process and helps to develop the team of people involved.

■ Flowcharts document processes and are useful as troubleshooting tools and in process improvement. An improvement team would flowchart the existing process and the improved or desired process, comparing the two to highlight the changes necessary.

Leadership, people and implementation aspects of process management

■ Top management who have used process management to great effect recognize its contribution in creating knowledge and eliminating waste, yet they understand the importance of involving people, measurement and good communications.

■ Process owners are key to effective process management. They have responsibility for and authority over process design, operation and measurement of performance.

■ Managing the people who work in the processes requires attention to training programs, performance targets, communicating changing customer needs, negotiation and collaboration.

■ Moving to process management requires some challenging and fundamental changes, leading to new ways of working and managing. Current initiatives should be carefully examined to ensure good planning and an understanding of what needs to be done first.

■ As with all change initiatives, delivering some tangible measurable benefits early on will help overcome the inevitable resistance.

■ With the wider introduction of e-commerce systems, there will be greater pressure to run rapid, fault-free business processes. Some of the processes will need to change 'shape' as demands, technologies and markets change.

Chapter 11

Process redesign/engineering

■ Re-engineering the organization?

When it has been recognized that a major business process requires radical reassessment, business process re-engineering or redesign (BPR) methods are appropriate. In their book *Re-Engineering the Corporation*, Hammer and Champy talked about reinventing the nature of work, 'starting again – reinventing our corporations from top to bottom'. BPR was launched on a wave of organizations needing to completely rethink how and why they do what they do in order to cope with the ever-changing world, particularly the development of technology-based solutions.

The reality is, of course, that many processes in many organizations are very good and do not need re-engineering, redesigning or reinventing, not for a while anyway. These processes should be subjected to a regime of continuous improvement (Chapter 13) at least until we have dealt with the very poorly performing processes that clearly do need radical review.

Some businesses and industries more than others have been through some pretty hefty changes – technological, political, financial and/or cultural. Customers of these organizations may be changing and demanding certain new relationships. Companies are finding leaner competitors encroaching into their market place, increased competition from other countries where costs are lower, and start-up competitors which do not share the same high bureaucracy and formal structures.

Enabling an organization, whether in the public or private sector, to be capable of meeting these changes is not a case of working harder but working differently. There have been many publicized BPR success stories and, equally, there have been some abject failures. In some cases, radical changes to major business processes have brought corresponding radical improvements in productivity. However, knowing how to reap such benefit, or indeed knowing if and how to apply BPR, has proved difficult for some organizations.

The concept of BPR was introduced to the world via two articles that described the radical changes to business processes being performed by a handful of Western business. These were also among the first to embark on quality management initiatives in the 1980s and included Xerox, Ford, AT&T, Baxter Healthcare, and Hewlett-Packard.

Many companies adopted quality management initiatives in the 1980s hoping to win back business lost to Japanese competition. When Ford benchmarked Mazda's accounts payable department, however, they discovered a business process being run by five people, compared to Ford's 500. Even with the difference in scale of the two companies, this still demonstrated the relative inefficiency of Ford's accounts payable process. At Xerox, taking a customer's perspective of the company identified the need to develop systems rather than stand-alone products, which highlighted Xerox's own inefficient office systems.

Both Ford and Xerox realized that incremental improvement alone was not enough. They had developed high infrastructure costs and bureaucracies that made them relatively unresponsive in customer service. Focusing on internal customer/supplier interfaces improved quality, but preserved the current process structure and they could not hope to achieve in a few years what had taken the Japanese 30 years. To achieve the necessary improvements required a radical rethink and redesign of processes.

What was being applied by organizations such as Ford and Xerox was **discontinuous improvement**. In order to respond to the competitive threats of Canon and Honda, Xerox and Ford needed quality management to catch up, but to get ahead they felt they required radical breakthroughs in performance. Central to these breakthrough improvements was information technology (IT).

Information technology as a driver for BPR_____

BPR is often based on new possibilities for breakthrough performance provided by the emergence of new enabling technologies. The most important of these, the one that is the nominal ingredient in many BPR recipes, is IT.

Explosive advances in IT have enabled the dissemination, analysis, and use of information from and to customers and suppliers and within enterprises, in new ways and in time frames that impact processes, organization designs and strategic competencies. Computer networks, open systems, client/server architecture, groupware, and electronic data interchange have opened up the possibilities for the integrated automation of business processes. Neural networks, enterprise analyzer approaches, computer-assisted software engineering, and object-oriented programming now facilitate systems design around office processes.

The pace of change has, of course, been enormous and IT systems unavailable just 10 to 15 years ago have enabled sweeping changes in business process improvement, particularly in office systems. Just as statistical process control (SPC) enabled manufacturing processes to be improved by controlling variation and improving efficiency, so IT is enabling non-manufacturing processes to be fundamentally restructured.

IT in itself, however, did not offer all the answers, automation frequently being claimed not to produce the gains expected. Many companies putting in major new computer systems have achieved only the automation of existing processes. Frequently, different functions within the same organization have systems that are incompatible with each other. Locked into traditional functional structures, managers have spent large amounts on IT systems that have not been used cross-functionally. Yet it is in this cross-functional area that the big improvement gains through IT are to be made. Once a process view is taken to designing and installing an IT system, it becomes possible to automate cross-functional, cross-divisional, even cross-company processes.

What is BPR and what does it do?

There are almost as many definitions of BPR as there are of quality management! However, most of them boil down to the same substance – the fundamental rethink and radical redesign of a business process, its structure and associated management systems, to deliver major or step improvements in performance (which may be in process, customer, or business performance terms).

Of course, BPR and quality management are complementary under the umbrella of process management. The continuous and step change improvements must live side by side – when does continuous change become a step change anyway? There has been over the years much debate, including some involving the author, about this issue. Whether it gets resolved is not usually the concern of the organization facing today's uncertainties with the realization that 'business as usual' will not do and some major changes in the ways things are done are required.

Put into a strategic context, BPR is a means of aligning work processes with customer requirements in a dynamic, flexible way, in order to achieve long-term corporate objectives. This requires the involvement of customers and suppliers and thinking about future requirements. Indeed the secrets to redesigning a process successfully lie in thinking about how to reshape it for the future.

BPR then challenges managers to rethink their traditional methods of doing work and to commit to customer-focused processes. Many outstanding organizations have achieved and/or maintained their leadership through process re-engineering, especially where they found processes which were not customer focused. Companies using these techniques have reported significantly improved results, including better customer relations, reductions in cycle time to market, increased productivity, fewer defect/errors and increased profitability. BPR uses recognized methods for improving business results and questions the effectiveness of the traditional organizational structure. Defining, measuring, analyzing and re-engineering work processes to improve customer satisfaction can pay off in many different ways.

For example, Motorola had set stretch goals of tenfold improvement in defects and twofold improvement in cycle time within five years. The time period was subsequently revised to three years and the now famous six sigma goal of 3.4 defects per million became a slogan for the company and probably one of the real drivers (see also Chapter 13). These stretch goals represent a focus on discontinuous improvement and there are many examples of other companies that have made dramatic improvements following major organizational and process redesign as part of quality management initiatives, including approaches such as the 'clean sheet' design of a 'green field' plant around work cells and self-managed teams.

Most organizations have vertical functions: experts of similar backgrounds grouped together in a pool of knowledge and skills capable of completing any task in that discipline. This focus, however, can foster a vertical view and limit the organization's ability to operate effectively. Barriers to customer satisfaction may evolve, resulting in unnecessary work, restricted sharing of resources, limited synergy between functions, delayed development time and no clear understanding of how one department's activities affect the total process of attaining customer satisfaction. Managers who remain tied to managing singular functions with rewards and incentives for their narrow missions, can inhibit a shared external customer perspective.

BPR breaks down these internal barriers and encourages the organization to work in cross-functional teams with a shared horizontal view of the business. As we have seen in earlier chapters, this requires shifting the work focus from managing functions to managing processes. Process owners, accountable for the success of major cross-functional processes, are charged with ensuring that employees understand how their individual work processes affect customer satisfaction. The interdependence between one group's work and the next becomes quickly apparent when everyone understands who the customer is and the value they add to the entire process of satisfying that customer.

Processes for redesign

IT provided the means to achieve the breakthrough in process performance in some organizations. The inspiration, however, came from understanding both the current and potential processes. This required a more holistic view than that taken in traditional quality programs, involving wholesale redesigns of the processes concerned.

Ford estimated a 20 percent reduction in head count if it automated the existing processes in accounts payable. Taking an overall process perspective, Ford achieved a 75 percent reduction in one department. Xerox took an organizational view and concentrated on the cross-functional processes to be re-engineered, radically changing the relationship between supplier and external customer.

Clearly, the larger the scope of the process, the greater and further reaching are the consequences of the redesign. At a macro level, turning raw materials into a product used by a delighted customer is a process made up of subsets of smaller processes. The aim of the overall process is to add value to the raw materials. Taking a holistic view of the process makes it possible to identify non-value-adding elements and remove them. It enables people to question why things are done, and to determine what should be done.

Some of the early re-engineering literature advised starting with a blank sheet of paper and redesigning the process anew. The problems inherent in this approach are:

- the danger of designing another inefficient system; and
- not appreciating the scope of the problem.

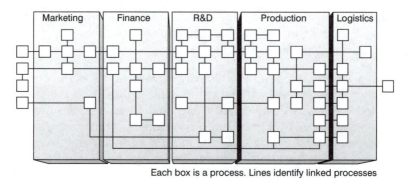

Each box is a process. Lines identify linked processes

■ **Figure 11.1**   Simplified process map

Therefore, the author and his colleagues recommend a thorough understanding of current processes before embarking on a re-engineering project.

Current processes can be understood and documented by process mapping and flowcharting. As processes are documented, their interrelationships become clear and a map of the organization emerges. Figure 11.1 shows a much simplified process map. As the aim of BPR is to make discontinuous, major improvements, this invariably means organizational change, the extent of which depends on the scope of the process re-engineered.

Taking the organization depicted in Figure 11.1 as an example, if the decision is made to redesign the processes in finance, the effect may be that in Figure 11.2a: eight individual processes have become three. There has been no organizational effect on the processes in the other functions, but finance has been completely restructured. In Figure 11.2b, a chain of processes, crossing all the functions, has been re-engineered. The effect has been the loss of redundant processes and possibly many heads but much of the organization has been unaffected. Figure 11.2c shows the organization after a thorough re-engineering of all its processes. Some elements may remain the same, but the effect is organization-wide.

Whatever the scope of the redesign, head count is not the only change. When work processes are altered, the way people work alters. Figures 11.1 and 11.2 show an organization's functional departments with process running through them. These are the handful of core processes that make up what an organization does (see Figure 11.3) and in many organizations these would benefit from re-engineering to improve added value output and efficiency.

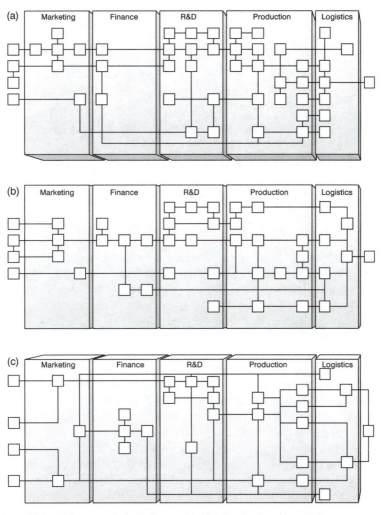

Figure 11.2 (a) Process redesign in finance, (b) cross-functional process redesign, (c) organizational process redesign

Focus on results

BPR is not intended to preserve the status quo, but to fundamentally and radically change what is done; it is *dynamic*. Therefore, it is essential for a BPR effort to focus on required customers which will determine the scope of the BPR exercise. A simple requirement may be a 30 percent reduction in costs or a reduction in delivery time of two days. These would imply projects with relatively narrow scope, which are essentially inwardly focused and probably involve only one department; for example, the finance department in Figure 11.2a.

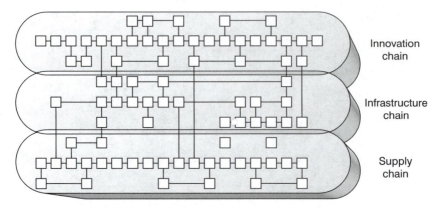

Figure 11.3   Process organization

When Wal-Mart focused on satisfying customer needs as an outcome it started a redesign that not only totally changed the way it replenished inventory, but also made this the centerpiece of its competitive strategy. The system put in place was radical, and required tremendous vision. In ten years, Wal-Mart grew from being a small niche retailer to the largest and most profitable retailer in the world.

Focusing on results rather than just activities can make the difference between success and failure in change projects. The measures used, however, are crucial. At every level of redesign and re-engineering, a focus on results gives direction and measurability; whether it be cost reduction, head count reduction, increase in efficiency, customer focus, identification of core processes and non-value-adding components, or strategic alignment of business processes. Benchmarking is a powerful tool for BPR and is the trigger for many BPR projects, as in Ford's accounts payable process. As shown in Chapter 9, the value of benchmarking does not lie in what can be copied, but in its ability to identify goals. If used well, benchmarking can shape strategy and identify potential competitive advantage.

The redesign process

Central to BPR is an objective overview of the processes to be redesigned. Whereas information needs to be obtained from the people directly involved in those processes it is never initiated by them. Even at its lowest level, BPR has a top-down approach and most BPR efforts, therefore, take the form of a major project. There are numerous methodologies proposed, but all share common elements. Typically, the project takes the form of seven phases, shown in Figure 11.4.

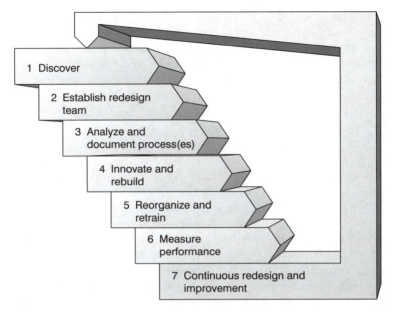

1 Discover

2 Establish redesign team

3 Analyze and document process(es)

4 Innovate and rebuild

5 Reorganize and retrain

6 Measure performance

7 Continuous redesign and improvement

Figure 11.4 The seven phases of BPR

1 **Discover**

This involves first identifying a problem or unacceptable outcome, followed by determining the desired outcome. This usually requires an assessment of the business need and will certainly include determining the processes involved, including the scope, identifying process customers and their requirements, and establishing effectiveness measurements.

2 **Establish redesign team**

Any organization, even a small company, is a complex system. There are customers, suppliers, employees, functions, processes, resources, partnerships, finances, etc. and many large organizations are incomprehensible – no one person can easily get a clear picture of all the separate components. Critical to the success of the redesign is the makeup of a redesign team.

The team should comprise as a minimum the following:

- senior manager as sponsor;
- steering committee of senior managers to oversee overall re-engineering strategy;
- process owner;
- team leader;
- redesign team members.

It is generally recommended that the redesign team have between five and ten people; represent the scope of the process (that is, if the process to be re-engineered is cross-functional, so is the team); only work on one redesign at a time; and is made up of both insiders and outsiders. Insiders are people currently working within the process concerned who help gain credibility with co-workers. Outsiders are people from outside the organization who bring objectivity and can ask the searching questions necessary for the creative aspects of the redesign. Many companies use consultants for this purpose.

3 **Analyze and document process(es)**

Making visible the invisible, documenting the process(es) through mapping and/or flowcharting is the first crucial step that helps an organization see the way work really is done and not the way one thinks or believes it is done. Seeing the process as it is provides a baseline from which to measure, analyze, test and improve.

Collecting supporting process data, including benchmarking information and IT possibilities, allows people to weigh the value each task adds to the total process, to rank and select areas for the greatest improvement, and to spot unnecessary work and points of unclear responsibility. Clarifying the root causes of problems, particularly those that cross department lines, safeguards against quick-fix remedies and assures proper corrective action, including the establishment of the right control systems.

4 **Innovate and rebuild**

In this phase the team rethink and redesign the new process, using the same process mapping techniques, in an iterative approach involving all the stakeholders, including senior management. A powerful method for challenging existing practices and generating breakthrough ideas is 'assumption busting' – see later section.

5 **Reorganize and retrain**

This phase includes piloting the changes and validating their effectiveness. The new process structure and operation/system will probably lead to some reorganization, which may be necessary for reinforcement of the process strategy and to achieve the new levels of performance.

Training and/or retraining for the new technology and roles play a vital part in successful implementation. People need to be equipped to assess, re-engineer, and support – with the appropriate technology – the key processes that contribute to customer satisfaction and corporate objectives. Therefore, BPR efforts can involve substantial investment in training but they also require considerable top management support and commitment.

6 **Measure performance**

It is necessary to develop appropriate metrics for measuring the performance of the new process(es), subprocesses, activities, and tasks. These must be meaningful in terms of the inputs and outputs of the processes, and in terms of the customers of and suppliers to the process(es) (see Chapter 7).

7 **Continuous redesign and improvement**
  The project approach to BPR suggests a one-off approach. When the project is over, the team is disbanded, and business returns to normal, albeit a radically different normal. It is generally recommended that an organization does not attempt to re-engineer more than one major process at a time, because of the disruption and stress caused. Therefore, in major re-engineering efforts of more than one process, as one team is disbanded, another is formed to redesign yet another process. Considering that Ford took five years to redesign its accounts payable process, BPR on a large scale is clearly a long-term commitment.

In a rapidly changing, ever more competitive business environment, it is becoming more likely that companies will re-engineer one process after another. Once a process has been redesigned, continuous improvement of the new process by the team of people working in the process should become the norm.

Assumption busting

Within BPR is a powerful method for challenging existing practices and generating breakthrough ideas for improvement. 'Assumption busting', as it was named by Hammer and Champy, aims to identify the **rules** that govern the way we do business and then uncover the real underlying **assumptions** behind the adoption of these rules. Business processes are governed by a number of rules that determine the way the process is designed, how it interfaces with other activities within the organization, and how it is operated. These rules can exist in the form of explicit policies and guidelines or, what is more often the case, in the mind of the people who operate the process. These unwritten rules are the product of assumptions about the process environment that have been developed over a number of years and often emerge from uncertainties surrounding trading relationships, capabilities, resources, authorities, etc. Once these underlying assumptions are uncovered they can be challenged for relevance and, in many cases, can be found to be false. This opens up new opportunities for process redesign and, as a consequence, the creation of new value and improved performance. The eight steps of the technique are outlined in Table 11.1.

Case study ▨▩■ ▬▬▬▬▬▬▬▬▬▬

■ BPR in BBC Resources

BBC Resources Ltd, a supplier of TV and radio studio and outside broadcast resource services, faced the requirement to improve business performance. The business was losing money and

Table 11.1 Assumption busting – a proven technique

Step 1 Identify the core value that must be delivered to the customer, the business and the key stakeholders.
Step 2 Map the process or processes to be improved at a high level only, identifying key problems.
Step 3 Select a particular problem to resolve (e.g. process cost efficiency, process quality, process speed) and collect supporting performance data.
Step 4 Brainstorm the rules that have an effect on the problem being resolved. Test the rule statements for validity and prioritize them for further analysis.
Step 5 Undertake a rigorous review of each rule, uncovering the underlying assumption behind each.
Step 6 Identify the modified assumptions and, in turn, a modified set of process rules.
Step 7 Identify the impact of these rules on the process and construct a new set of process design principles.
Step 8 Develop a revised process design and test for validity.

faced competition from independent providers. They needed to improve the efficiency of their processes while retaining their core capability that created competitive advantage.

A team was commissioned to review the core value-adding processes, setting challenging targets for improvements in performance in order to stimulate breakthrough thinking. The team decided to take a more radical approach by using assumption busting and prepared a six-week program of work. Within that timeframe they used an established eight-step method (shown in Table 11.1) to redesign the core end-to-end service delivery processes for two major business units – studios and outside broadcasts. The work involved identifying the key areas of cost consumption, challenging the rules and assumptions that governed the existing process and generating a set of improvement opportunities. When they had evaluated their findings, the team presented ideas to deliver an improvement in excess of 15 percent in process efficiency.

One of the process rules concerned the use of a highly technically qualified member of staff for the planning and delivery of all of the programs supported. The core underlying assumption was that all of the programs were complex in nature. When this assumption was challenged the team in BBC Resources realized that, as not all programs were so complex in nature, less technically qualified members of staff could be utilized at lower cost to the business.

Assumption busting in Grattan

The mail order company Grattan was presented with a different challenge. With the catalog shopping market facing intense competition from other retailing methods, the Grattan team realized the need to look at their value chain processes to find ways of improving the service they provide to their agent community. One target for service improvement was the reduction in the value of query calls coming into the service centers.

Once the relevant processes had been defined, a cross-functional team decided to employ the assumption busting method to identify areas for improvement. One rule that quickly came to light concerned the issue of order acknowledgements, which, in the form of account statements, are posted out to agents on the same day that the order is processed. However, the current rule stated that statements should be sent via third class mail. This rule was created out of an assumption that this was the most cost-efficient way of communicating order details to the agent. Third class mail can take up to ten days, which is beyond the current expected delivery time. So, when consignments did not arrive as expected, the agent called to progress the order. As a consequence of changing the process, statements are now being sent out within the expected delivery time leading to a reduction in query calls and improved service.

In practice the author's colleagues have found this technique to be of greatest value when applied by a cross-functional group of process operators and supervisors who are given a specific problem to fix. In using the technique, care must be exercised in the use of terms such as rule and assumption. They often cause initial confusion and there can be real difficulty in uncovering the core underlying assumptions. Rules should be clearly stated and tested for validity before proceeding down what eventually could become a blind alley. Furthermore, a rigorous approach to the identification of the core assumption is vital to uncovering the real opportunities for improvement. An assumption by definition 'is a statement/belief that is accepted or supposed to be true without proof or demonstration'. In some cases, rules are created from specific knowledge about the business and its environs and not based on assumptions. Assumptions spring from our beliefs about the environment and not our specific knowledge.

Assumption busting is of particular benefit when applied by partners within a supply chain. The trading relationships and practices that exist in a modern supply chain, such as a supermarket and its multiple tiers of suppliers, are the product of a number of assumptions made by the supply chain partners about what is possible. Once teams from each of the partnering businesses work collaboratively to uncover the rules and assumptions that govern their trading relationships the door is unlocked to new methods and economies.

The method can also be immensely powerful when companies are introducing new technology. Breakthrough technologies can lead to breakthrough performance as new technologies make possible what is considered impossible today. Hence, a number of current rules and assumptions are there to be challenged as processes are redesigned to take advantage of the new technology. We are often just as constrained by our lack of imagination regarding the possibilities of tomorrow as we are by our knowledge of what is possible today. One example of this was in the BBC World Service where the introduction of digital technology to replace analog was accompanied by assumption busting led process redesign to take advantage of the new technological capability.

While assumption busting has been primarily applied to the generation of new process designs, it exists in its own right as a method for developing more 'lateral' solutions to problems. In the early 1970s, Dr Edward de Bono introduced the concept of lateral thinking as an alternative method of generating ideas to that of the more traditional logical or 'vertical' thinking. Dr de Bono argued that our thinking is constrained by patterns that form in our minds over time and channel our future thoughts. Assumption busting helps people break out of this 'channeled thinking' to develop creative ideas. Managers could benefit from applying assumption busting to a number of problems or opportunities in their businesses – assumptions constrain us everywhere, not just within our business processes.

Whether it is in response to specific customer requirements, new technology, or in the quest for competitive advantage assumption busting provides a simple but effective method for breaking into new areas of adding value. World class performance will not be achieved by effort alone; creativity and innovation are cornerstones of future success and innovative ways of delivering new value will be rewarded. Assumption busting provides a powerful method for generating new ideas from looking at today and tomorrow in a different way.

■ BPR – the people and the leaders

For an organization to focus on its core processes almost certainly requires an understanding of its core competencies. Moreover, core process redesign can channel an organization's competencies into an outcome that gives it strategic competitive advantage and the key element is visioning that outcome. Visioning the outcome may not be enough, however, since many companies' 'vision' desires results without simultaneously 'visioning' the systems that are required to generate them. Without a clear vision of the systems, processes, methods, and approaches that will allow achievement of the desired results, dramatic improvement is frequently not obtained as the organization fails to align around a common tactical strategy. Such an 'operational' vision is lacking in many organizations.

The fallout from BPR has profound impacts on the employees in any enterprise at every level – from executives to operators. In order for

BPR to be successful, therefore, significant changes in organization design and enterprise culture are also often required. Unless the leaders of the enterprise are committed to undertake these changes, the BPR initiative will flounder. The point is, of course, that organization design and culture change are much more difficult than modifying processes to take advantage of new IT.

While the enabling IT is often necessary and is clearly going to play a role in many BPR exercises, it is by no means sufficient, nor is it the most difficult hurdle on the path to success. Thanks to IT we can radically change the processes an organization operates and, hopefully, achieve dramatic improvements in performance. However, in any BPR project there will be considerable risk attached to building the information system that will support the new, redesigned processes. Information systems should be but rarely are described so that they are easy for people to understand.

While BPR may be a distinct, short-term activity for a specific business function, the record indicates that BPR activities are most successful when they occur within the framework of a long-term thrust for excellence. A BPR effort is more likely to find the process focus supportive workforce, organization design, and mindset changes needed for its success where there is a solid foundation of good quality management.

Process improvement is sometimes positioned as a bottom-up activity, but quality management involves setting longer-term goals at the top and modifying the business as necessary to achieve the goals. Often, the modifications to the business required to achieve the goals are extensive and groundbreaking. The history of successful quality management in award-winning companies in Europe and the United States is replete with new organization designs, with flattened structures, and with empowered employees in the service of end customers. In many successful organizations, BPR has been an integral part of the culture – a process-driven change dedicated to the ideals and concepts of quality. That change must create something that did not exist before, namely a 'learning organization' capable of adapting to a changing competitive environment. When process, or even the whole business, need to be re-engineered, the radical change may not be readily accepted.

■ Acknowledgement

The author is grateful to the contribution made by his colleagues Ken Gadd and Mike Turner in the preparation of this chapter.

Chapter highlights

■ ■ ■

Re-engineering the organization?

- When a major business process requires radical reassessment, perhaps through the introduction of new technology, discontinuous methods of business process re-engineering or redesign (BPR) are appropriate.
- Drives for process change include information technology (IT), political, financial, cultural and competitive aspects. These often require a change of thinking about the ways processes are and could be operated.
- IT often creates opportunities for breakthrough performance but BPR is needed to deliver it. Successful practitioners of BPR have made striking improvements in customer satisfaction and productivity in short periods of time.

What is BPR and what does it do?

- There are many definitions of BPR but the basic elements involve a fundamental rethink and radical redesign of a business process, its structure and associated management systems to deliver step improvements in performance.
- BPR and quality management are complementary under the umbrella of process management – the continuous and discontinuous improvements living side by side. Both require the involvement of customers and suppliers and their future requirements.
- BPR challenges managers to rethink their traditional methods of doing work and to commit to customer-focused processes. This breaks down, organizational barriers and encourages cross-functional teams.

Processes for redesign

- Much larger savings and head count reductions are possible through properly applied BPR than simply automating existing processes. The larger the scope of the process the greater the further reaching the consequences of the redesign.
- A thorough understanding of the current process is needed before embarking on a re-engineering project. Documentation of processes through mapping and flowcharting allows interrelationships to be clarified.
- Focusing on results rather than activities can make the difference between success and failure in BPR and other change projects, but the measures used are critical. Benchmarking is a powerful tool for BPR and often the trigger for many projects.

The redesign process

- BPR has a top-down approach and needs an objective overview of the process to be redesigned to drive the project.
- Typically a BPR project will have seven phases: discover – identifying the problem or unacceptable outcome; establish redesign team; analyze and document processes; innovate and rebuild; reorganize and retrain; measure performance; continuous redesign and improvement.
- Assumption busting is a useful eight-step BPR method which aims to identify and challenge the 'rules' and assumptions that govern and underlie the way business is done. A team is formed to: identify the core value to be delivered to customer and stakeholders; map the process at high level; select problems to resolve and collect performance data; brainstorm and test the rules; rigorously review each rule to uncover underlying assumptions; identify modified assumptions and process rules; identify impact and construct a new set of process principles; develop revised process and test validity.

BPR – the people and the leaders

- For an organization to focus and its core processes requires an understanding of its core competencies, and the channeling of these into outcomes that deliver strategic competitive advantage.
- BPR has profound impacts on employees from the top to the bottom of an organization. In order to be successful significant changes in organization design and enterprise culture are also often required. This requires commitment from the leaders to undertake these changes.
- Quality management ideals and concepts provide a perfect platform for BPR projects and the creation of a 'learning organization' capable of adapting to a radically changing environment.

Chapter 12

■ Quality management
■ systems

■ Why a quality management system?

In earlier chapters we have seen how the keystone of quality management is the concept of customer and supplier working together for their mutual advantage. For any particular organization this becomes 'total' quality management if the supplier/customer interfaces extend beyond the immediate customers, back inside the organization, and beyond the immediate suppliers. In order to achieve this, a company must organize itself in such a way that the human, administrative and technical factors affecting quality will be under control. This leads to the requirement for the development and implementation of a quality management system that enables the objectives set out in the quality policy to be accomplished. Clearly, for maximum effectiveness and to meet individual customer requirements, the management system in use must be appropriate to the type of activity and product or service being offered.

It may be useful to reflect on why such a device is necessary to achieve control of processes. The author still remembers being at a table in a restaurant with eight people who all ordered the 'Chef's Special Individual Soufflé'. All eight soufflés arrived together at the table, magnificent in their appearance and consistency, each one exhibiting an almost identical size and shape – a truly remarkable demonstration of culinary skill. How had this been achieved? The chef had *managed* such consistency by making sure that, for each soufflé, he used the same ingredients (materials), the same equipment (plant), the same method

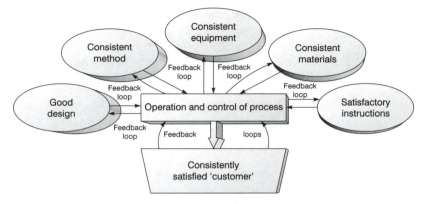

Figure 12.1   The systematic approach to process management

(procedure) in exactly the same way every time. The process was under control. This is the aim of a good quality management system, to provide the 'operator' of the process with consistency and satisfaction in terms of methods, materials, equipment, etc. (Figure 12.1). Two feedback loops are also required: the 'voice' of the customer (marketing activities) and the 'voice' of the process (measurement activities).

The chef's soufflés – they were neither British Standard, NIST Standard, Australian Standard, nor ISO Standard soufflés – were the 'chef's special soufflés'. It is not conceivable that the chef sat down with a blank piece of paper to invent a soufflé recipe. Why reinvent wheels? He probably used a standard formula and changed it slightly to make it his own. This is exactly the way in which successful organizations use the international standards on quality management systems that are available. The 'wheel' has been invented but it must be built in a way that meets the specific organizational and product or service requirements. The international family of standards ISO 9000:2000 'Quality Management Systems' specifies systems which can be implemented in an organization to ensure that all the product/service performance requirements and needs of the customer are fully met.

Let us return to the chef in the restaurant and propose that his success leads to a desire to open eight restaurants in which are served his special soufflés. Clearly he cannot rush from each one of these establishments to another every evening making soufflés. The only course open to him to ensure consistency of output in all eight restaurants is for him to write down in some detail the system he uses, and then make sure that it is used on all sites, every time a soufflé is produced. Moreover, he must periodically visit the different sites to ensure that:

1 The people involved are operating according to the designed system (a system audit).
2 The soufflé system still meets the requirements (a system review).

If, in his system audits and reviews, he discovers that an even better product or less waste can be achieved by changing the method or one of the materials, then he may wish to effect a change. To maintain consistency, he must ensure that the appropriate changes are made to the management system, *and* that everyone concerned is issued with the revision and begins to operate accordingly.

A good quality management system will ensure that two important requirements are met:

■ *The customer's requirements* – for confidence in the ability of the organization to deliver the desired product or service consistently.
■ *The organization's requirements* – both internally and externally including regulatory, and at an optimum cost, with efficient utilization of the resources available – material, human, technological, and information.

The requirements can be truly met only if objective evidence is provided, in the form of information and data, which supports the system activities, from the ultimate suppliers through to the ultimate customers.

A *quality management system* may be defined, then, as an assembly of components, such as the management structure, responsibilities, processes and resources for implementing quality management. These components interact and are affected by being in the system, so the isolation and study of each one in detail will not necessarily lead to an understanding of the system as a whole. Often the interactions between the components – such as materials and processes, people and responsibilities – are just as important as the components themselves, and problems can arise from these interactions as much as from the components. Clearly, if one of the components is removed from the system, the whole thing will change.

The adoption of a quality management system is, of course, a strategic decision and its design should be influenced by the organization's objectives, structure and size, the products or services offered, and its processes.

■ Quality management system design

The quality management system should apply to and interact with all processes in the organization. It begins with the identification of the customer requirements and ends with their satisfaction, at every transaction interface. The activities may be classified in several ways – generally as processing, communicating and controlling, but more usefully and specifically as shown in the quality management process

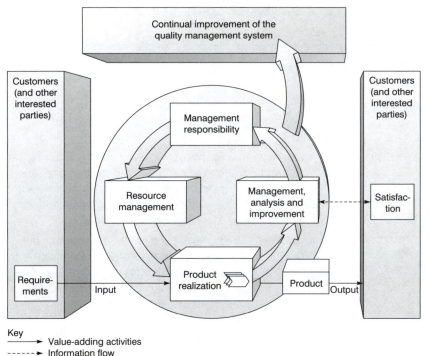

Key
———▶ Value-adding activities
------▶ Information flow

Figure 12.2   Model of a process-based quality management system

model described in ISO 9001:2000, Figure 12.2. This reflects graphically the integration of four major areas:

- Management responsibility.
- Resource management.
- Product realization.
- Measurement, analysis and improvement.

The management system requirements under these headings are specified in the international standard. Table 12.1 lists the ISO 9000:2000 family which together form a coherent set of quality management system standards to hopefully facilitate mutual understanding across national and international trade.

ISO 9000:2000 – the fundamentals and vocabulary – sets down the principles behind quality management which formed the basis for the quality management system standards in the ISO 9000 family.

Eight principles were identified to be used by top management as they lead their organizations and improve performance.

- Customer focus – see Chapter 1.
- Leadership – see Chapter 3.

Table 12.1 The ISO 9000:2000 family of standards on quality management systems

BS EN ISO	Name	Purpose
9000:2000	Quality management systems – Fundamentals and vocabulary	Describes the fundamentals and specifies the terminology for a QMS.
9001:2000	Quality management systems – Requirements	Specifies the requirement for a QMS where an organization needs to demonstrate its ability to provide products that fulfill customer and applicable regulatory requirements and aims to enhance customer satisfaction.
9004:2000	Quality management systems – Guidelines for performance improvements	Provides guidelines that consider both the effectiveness and efficiency of the QMS, with the aim of improving the performance of the organization and satisfaction of customers and other interested parties.

Note: ISO 10011 provides guidance on auditing quality and environmental management systems – see Chapter 8.

- Involvement of people – see Chapters 14–16.
- Process approach – see Chapter 10.
- System approach to management – see Chapters 4 and 10.
- Continual improvement – see Chapter 13.
- Factual approach to decision making – see Chapters 7 and 13.
- Mutually beneficial supplier relationships – see Chapter 5.

BS EN ISO 9001:2000 'Quality Management Systems – Requirements' is the British Standard official English language version of EN ISO 9001:2000. It is identical with ISO 9001:2000 and supersedes BS EN ISO 9001:1994. BS EN ISO 9002:1994 and BS EN ISO 9003:1994 are now obsolescent. A new ISO 9004:2000 provides a set of guidelines for continuous improvement. Detailed information on the ISO 9000:2000 family of standards may be found on the following websites: www.iso.ch and www.bsi.org.uk. The European (EN) standard exists in three official versions (English, French, German) but a version in any other language translated under the responsibility of a member of the European Committee for Standardization CEN (Comité European de Normalisation; Europaisches Komitee für Normung) into its own language and notified to the CEN Management Centre in Brussels has the same status as the official versions. CEN members are the national standards bodies of Austria, Belgium, Czech Republic, Finland, France, Germany,

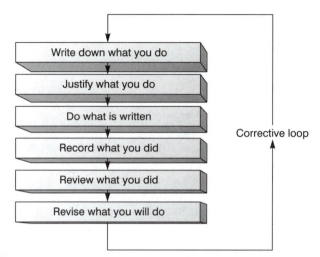

Write down what you do

Justify what you do

Do what is written

Record what you did   Corrective loop

Review what you did

Revise what you will do

Figure 12.3   The quality system and never-ending improvement

Greece, Iceland, Ireland, Italy, Luxembourg, Netherlands, Norway, Portugal, Spain, Sweden, Switzerland and the United Kingdom.

It is interesting to bring together Deming's concept of a cycle of continuous improvement – Plan, Do, Check, Act – and quality management systems. A simplification of what a good management system is trying to do is given in Figure 12.3, which follows the improvement cycle.

In many organizations established methods of working already exist around identified processes, and all that is required is the *documenting of what is currently done*. In some instances companies may not have procedures to satisfy the requirements of a good standard, and they may have to begin to devise them. Alternatively, it may be found that two people, supposedly performing the same task, are working in different ways, and there is a need to standardize procedures. Some organizations use the effective slogan 'If it isn't written down, it doesn't exist'. This can be a useful discipline, provided it doesn't lead to bureaucracy.

Justify that the *system* as it is designed *meets the requirements of a good international standard*, such as ISO 9001:2000. There are other excellent standards that are used, and these provide similar checklists of things to consider in the establishment of the quality system. For example, EN 9100 provides an 'Aerospace series' of quality management system requirements, based on ISO 9001.

One person alone cannot document a quality management system; the task is the job of all personnel who have responsibility for any part of it. The quality system must be a *practical working one* – that way it ensures that consistency of operation is maintained and it may be used as a training aid.

In the operation of any process, a useful guide is:

- No process without data collection (measurement).
- No data collection without analysis.
- No analysis without decisions.
- No decisions without actions (improvement) – which can include doing nothing.

This excellent discipline is built into any good quality management system, primarily through the audit and review mechanism. The requirement to *audit or 'check'* that the system is functioning according to plan, and to *review* possible system improvements, utilizing audit results, should ensure that the *improvement* cycle is engaged through the *corrective action* procedures. The overriding requirement is that the systems must reflect the established practices of the organization, improved where necessary to bring them into line with current and future requirements.

Case study ■ ■ ■

Underpinning all of BT's operations is BT's management system. First registered to ISO 9001 in 1994, this is one of the largest single corporate-wide registrations in the world. The management system was later refined and improved to take account of environmental and people management standards and BT is also registered to ISO 14001 and accredited as an Investor in People (IiP).

Achievement of ISO 9001 registration was not seen as an end in itself and after considering the Malcolm Baldrige National Quality Award framework, BT adopted the EFQM Business Excellence Model as a driver of organizational improvement. Since 1995 many BT business units have used the Business Excellence Model to identify strengths and areas for improvement as input to their business planning process. The extensive use of self-assessment against the Business Excellence Model has ensured that BT has a rigorous and structured approach to organizational improvement. That this approach was effective is demonstrated by the success of BT business units in national and international quality awards in the late 1990s. BT's Yellow Pages, National Business Communications and Northern Ireland units all won the British Quality Award. BT Northern Ireland won European quality prizes in 1998 and 1999 with Yellow Pages winning the European Quality Award in 1999. Following Yellow Pages' success BT ceased entering external quality awards; however, business excellence principles remain in everyday use, particularly for the periodic comprehensive reviews of business unit performance known in BT as health checks.

Quality management system requirements

The quality management system that needs to be documented and implemented will be determined by the nature of the process carried out to ensure that the product or service conforms to customer requirement. Certain fundamental principles are applicable, however, throughout industry, commerce, and the services. These fall into generally well-defined categories which are detailed in ISO 9001:2000.

1 Management responsibility

Customer needs/requirements (see Chapter 1)

The organization must focus on customer needs and specify them as defined requirements for the organization. The aim of this is to achieve customer confidence in the products and/or services provided. It is also necessary to ensure that the defined requirements are understood and fully met.

Quality policy (see Chapter 3)

The organization should define and publish its quality policy, which forms one element of the corporate policy. Full commitment is required from the most senior management to ensure that the policy is communicated, understood, implemented and maintained at all levels in the organization. It should, therefore, be authorized by top management and signed by the chief executive, or equivalent, who must also ensure that it:

- is suitable for the needs/requirements of the customers and the purpose of the organization;
- includes commitment to meeting requirements and continual improvement for all levels of the organization;
- provides a framework for establishing and reviewing quality objectives;
- is regularly reviewed for its suitability and objectiveness.

Quality objectives and planning

Organizations should establish written quality objectives and define the responsibilities of each function and level in the organization.

One manager reporting to top management, with the necessary authority, resources, support, and ability should be given the responsibility to co-ordinate, implement, and maintain the quality management system, resolve any problems and ensure prompt and effective corrective action. This includes responsibility for ensuring proper handling of the system and reporting on needs for improvement. Those who control sales, service operations, warehousing, delivery, and reworking of non-conforming product or service processes should also be identified.

Management review

Management reviews of the system must be carried out, by top management at defined intervals, with records to indicate the actions decided upon. The effectiveness of these actions should be considered during subsequent reviews. Reviews typically include data on the internal quality audits, customer feedback, product conformance analysis, process performance, and the status of preventive, corrective and improvement actions.

Quality manual

The organization should prepare a 'quality manual' that is appropriate. It should include but not necessarily be limited to:

(a) the quality policy;
(b) definition of the quality management system – scope, exclusions, etc.;
(c) description of the interaction between the processes of the quality management system;
(d) documented procedures required by the QMS, or reference to them.

In the quality manual for a large organization it may be convenient to indicate simply the existence and contents of other manuals, those containing the details of procedures and practices in operation in specific areas of the system.

Before an organization can agree to supply to a specification, it must ensure that:

(a) the processes and equipment (including any that are subcontracted) are capable of meeting the requirements;
(b) the operators have the necessary skills and training;
(c) the operating procedures are documented and not simply passed on verbally;
(d) the plant and equipment instrumentation is capable (e.g. measuring the process variables with the appropriate accuracy and precision);
(e) the quality-control procedures and any inspection, check, or test methods available provide results to the required accuracy and precision, and are documented;
(f) any subjective phrases in the specification, such as 'finely ground', 'low moisture content', 'in good time', are understood, and procedures to establish the exact customer requirement exist.

Control of documents

The organization needs to establish procedures for controlling the new and revised documents required for the operation of the quality management system. Documents of external origin must also be controlled. These procedures should be designed to ensure that:

(a) documents are approved;
(b) documents are periodically reviewed, and revised as necessary;
(c) the current versions of relevant documents are available at all locations where activities essential to the effective functioning of the processes are performed;
(d) obsolete documents are promptly removed from all points of issue and use, or otherwise controlled to prevent unplanned use;
(e) any obsolete documents retained for legal or knowledge-preservation purposes are suitably identified.

Documentation needs to be legible, revision controlled, readily identifiable and maintained in an orderly manner. Of course, the documentation may be in any form or any type of media.

Control of quality records

Quality records are needed to demonstrate conformance to requirements and effective operation of the quality management system. Quality records from suppliers also need to be controlled. This aspect includes record identification, collection, indexing, access, filing, storage and disposition. In addition, the retention times of quality records needs to be established.

2 Resource management

The organization should determine and provide the necessary resources to establish and improve the quality management system, including all processes and projects. The general infrastructure needed to achieve conformity to the product or service requirements needs to be provided and maintained, including buildings, equipment, any supporting services, and the work environment.

Human resources

The organization needs to select and assign people who are competent, on the basis of applicable education, training, skills and experience, to those activities which impact the conformity of product and/or service.

The organization also needs to:

(a) determine the training needed to achieve conformity of product and/or service;
(b) provide the necessary training to address these needs;
(c) evaluate the effectiveness of the training on a continual basis.

Individuals clearly need to be educated and trained to qualify them for the activities they perform. Competence, including qualification levels achieved, needs to be demonstrated and documented.

Information is ever increasingly a vital resource and any organization needs to define and maintain the current information and the infrastructure necessary to achieve conformity of products and/or services. The management of information, including access and protection of information to ensure integrity and availability, needs also to be considered.

3 Product realization

As we have seen in Figure 12.2, any organization needs to determine the processes required to convert customer requirements into customer satisfaction, by providing the required product and/or service. In determining such processes the organization needs to consider the outputs from the quality planning process.

The sequence and interaction of these processes need to be determined, planned and controlled to ensure they operate effectively, and there is a need to assign responsibilities for the operation and monitoring of the product/service generating processes.

These processes clearly need to be operated under controlled conditions and produce outputs which are consistent with the organization's quality policy and objective and it is necessary to:

(a) determine how each process influences the ability to meet product and/or service requirements;
(b) establish methods and practices relevant to process activities, to the extent necessary to achieve consistent operation of the process;
(c) verify processes can be operated to achieve product and/or service conformity;
(d) determine and implement the criteria and methods to control processes related to the achievement of product and/or service conformity;
(e) determine and implement arrangements for measurement, monitoring and follow-up actions, to ensure processes operate effectively and the resultant product/service meets the requirements;

(f) ensure availability of process documentation and records which provide operating criteria and information, to support the effective operation and monitoring of the processes. (This documentation needs to be in a format to suit the operating practices, including written quality plans);

(g) provide the necessary resources for the effective operation of the processes.

Customer-related processes

One of the first processes to be established is the one for identifying customer requirements. This needs to consider the:

(a) extent to which customers have specified the product/service requirements;
(b) requirements not specified by the customer but necessary for fitness for purpose;
(c) obligations related to the product/service, including regulatory and legal requirements;
(d) other customer requirements, e.g. for availability, delivery and support of product and/or service.

The identified customer requirements need also to be reviewed before a commitment to supply a product/service is given to the customer (e.g. submission of a tender, acceptance of a contract or order). This should determine that:

(a) identified customer requirements are clearly defined for the product and/or service;
(b) the order requirements are confirmed before acceptance, particularly where the customer provides no written statement of requirements;
(c) the contract or order requirements differing from those in any tender or quotation are resolved.

This should also apply to amended customer contracted orders. Moreover, each commitment to supply a product/service, including amendment to a contract or order, needs to be reviewed to ensure the organization will have the ability to meet the requirements.

Any successful organization needs to implement effective communication and liaison with customers, particularly regarding:

(a) product and/or service information;
(b) enquiry and order handling, including amendments;
(c) customer complaints and other reports relating to non-conformities;
(d) recall processes, where applicable;
(e) customer responses relating to conformity of product/service.

Where an organization is supervising or using customer property, care needs to be exercised to ensure verification, storage and maintenance. Any customer product or property that is lost, damaged or otherwise found to be unsuitable for use should, of course, be recorded and reported to the customer. Customer property may, of course, include intellectual property, e.g. information provided in confidence.

Design and development

The organization needs to plan and control design and development of products and/or services, including:

(a) stages of the design and development process;
(b) required review, verification and validation activities;
(c) responsibilities for design and development activities.

Interfaces between different groups involved in design and development need to be managed to ensure effective communication and clarity of responsibilities, and any plans and associated documentation should be:

(a) made available to personnel that need them to perform their work;
(b) reviewed and updated as design and development evolves.

The requirements to be met by the product/service need to be defined and recorded, including identified customer or market requirements, applicable regulatory and legal requirements, requirements derived from previous similar designs, and any other requirements essential for design and development. Incomplete, ambiguous or conflicting requirements must be resolved.

The outputs of the design and development process need to be recorded in a format that allows verification against the input requirements. So, the design and development output should:

(a) meet the design and development input requirements;
(b) contain or make reference to design and development acceptance criteria;
(c) determine characteristics of the design essential to safe and proper use, and application of the product or service;
(d) output documents to be reviewed and approved before release.

Validation needs to be performed to confirm that resultant product/ service is capable of meeting the needs of the customers or users under the planned conditions. Wherever possible, validation should be defined, planned and completed prior to the delivery or implementation of the product or service. Partial validation of the design or development output may be necessary at various stages to provide confidence in their correctness, using such methods as:

(a) reviews involving other interested parties;
(b) modeling and simulation studies;
(c) pilot production, construction or delivery trials of key aspects of the product and/or service.

Design and development changes or modifications need to be determined as early as possible, recorded, reviewed and approved, before implementation. At this stage, the effect of changes on compatibility requirements and the usability of the product or service throughout its planned life need to be considered.

Purchasing

Purchasing processes need to be controlled to ensure purchased products/services conform to the organization's requirements. The type and extent of methods to do this are dependent on the effect of the purchased product/service upon the final product/service. Clearly suppliers need to be evaluated and selected on their ability to supply the product or service in accordance with the organization's requirements. Supplier evaluations, supplier audit records and evidence of previously demonstrated ability should be considered when selecting suppliers and when determining the type and extent of supervision applicable to the purchased materials/services.

The purchasing documentation should contain information clearly describing the product/service ordered, including:

(a) requirements for approval or qualification of product and/or service, procedures, processes, equipment and personnel;
(b) any management system requirements.

Review and approval of purchasing documents, for adequacy of the specification of requirements prior to release, are also necessary.

Any purchased products/services need some form of verification. Where this is to be carried out at the supplier's premises, the organization needs to specify the arrangements and methods for product/service release in the purchasing documentation.

Production and service delivery processes

The organization needs to control production and service delivery processes through:

(a) information describing the product/service characteristics;
(b) clearly understandable work standards or instructions;

(c) suitable production, installation and service provision equipment;
(d) suitable working environments;
(e) suitable inspection, measuring and test equipment, capable of the necessary accuracy and precision;
(f) the implementation of suitable monitoring, inspection or testing activities;
(g) provision for identifying the status of the product/service, with respect to required measurement and verification activities;
(h) suitable methods for release and delivery of products and/or services.

Where applicable, the organization needs to identify the product/service by suitable means throughout all processes. Where traceability is a requirement for the organization, there is a need to control the identification of product/service. There is also a need to ensure that, during internal processing and final delivery of the product/service, the identification, packaging, storage, preservation, and handling do not adversely affect conformity with the requirements. This applies equally to parts or components of a product and elements of a service.

Where the resulting output cannot be easily or economically verified by monitoring, inspection or testing, including where processing deficiencies may become apparent only after the product is in use or the service has been delivered, the organization needs to validate the production and service delivery processes to demonstrate their effectiveness and acceptability.

The arrangements for validation might include:

(a) processes being qualified prior to use;
(b) qualification of equipment or personnel;
(c) use of specific procedures or records.

Evidence of validated processes, equipment and personnel needs to be recorded and maintained, of course.

Post-delivery services

Where there is a requirement for the organization to provide support services, after delivery of the product or service, this needs to be planned and in line with the customer requirement.

Monitoring and measuring devices

There is a need to control, calibrate, maintain, handle and store the applicable measuring, inspection and test equipment to specified requirements. Measuring, inspection, and test equipment should be used in a way which ensures that any measurement uncertainty, including accuracy and precision, is known and is consistent with the required measurement capability. Any test equipment software should meet the applicable requirements for the design and development of the product.

The organization needs to:

(a) calibrate and adjust measuring, inspection and test equipment at specified intervals or prior to use, against equipment traceable to international or national standards. Where no standards exist, the basis used for calibration needs to be recorded;
(b) identify measuring, inspection and test equipment with a suitable indicator or approved identification record to show its calibration status;
(c) record the process for calibration of measuring, inspection and test equipment;
(d) ensure the environmental conditions are suitable for any calibrations, measurements, inspections and tests;
(e) safeguard measuring, inspection and test equipment from adjustments which would invalidate the calibration;
(f) verify validity of previous inspection and test results when equipment is found to be out of calibration;
(g) establish the action to be initiated when calibration verification results are unsatisfactory.

4 Measurement, analysis and improvement

Any organization needs to define and implement measurement, analysis and improvement processes to demonstrate that the products or services conform to the specified requirements. The type, location and timing of these measurements needs to be determined and the results recorded based on their importance. The results of data analysis and improvement activities should be an input to the management review process, of course.

Measurement and monitoring

There is a need to determine and establish processes for measurement of the quality management system performance. Customer satisfaction must be a primary measure of system output and the internal audits should be used as a primary tool for evaluating ongoing system compliance.

The organization needs to establish a process for obtaining and monitoring information and data on customer satisfaction. The methods and measures for obtaining customer satisfaction information and data and the nature and frequency of reviews need to be defined to demonstrate the level of customer confidence in the delivery of conforming product and/or service supplied by the organization. Suitable measures for establishing internal improvement need to be implemented and the effectiveness of the measures periodically evaluated.

The organization must establish a process for performing internal audits of the quality management system and related processes. The purpose of the internal audit is to determine whether:

(a) the quality management system established by the organization conforms to the requirements of the International Standard; or
(b) the quality management system has been effectively implemented and maintained.

The internal audit process should be based on the status and importance of the activities, areas or items to be audited, and the results of previous audits.

The internal audit process should include:

(a) planning and scheduling the specific activities, areas or items to be audited;
(b) assigning trained personnel independent of those performing the work being audited;
(c) assuring that a consistent basis for conducting audits is defined.

The results of internal audits should be recorded including:

(a) activities, areas, and processes audited;
(b) non-conformities or deficiencies found;
(c) status of commitments made as the result of previous audit, such as corrective actions or product audits;
(d) recommendations for improvement.

The results of the internal audits should be communicated to the area audited and the management personnel responsible need to take timely corrective action on the non-conformities recorded.

Suitable methods for the measurement of processes necessary to meet customer requirements need to be applied, including monitoring the output of the processes that control conformity of the product or service provided to customers. The measurement results then need to be used to determine opportunities for improvements.

The organization needs also to apply suitable methods for the measurement of the product or service to verify that the requirements have been met. Evidence from any inspection and testing activities and the acceptance criteria used need to be recorded. If there is an authority responsible for release of the product and/or service, this should also be recorded.

Products or services should not be dispatched until all the specified activities have been satisfactorily completed and the related documentation is available and authorized. The only exception to this is when the product or service is released under positive recall procedures.

Control of non-conforming products

Products and services which do not conform to requirements need to be controlled to prevent unplanned use, application or installation, and the organization needs to identify, record and review the nature and extent of the problem encountered, and determine the action to be taken. This needs to include how the non-conforming service will be:

(a) corrected or adjusted to conform to requirements; or
(b) accepted under concession, with or without correction; or
(c) reassigned for an alternative, valid application; or
(d) rejected as unsuitable.

The responsibility and authority for the review and resolving of non-conformities need to be defined, of course.

When required by the contract, the proposed use or repair of non-conforming product or a modified service needs to be reported for concession to the customer. The description of any corrections or adjustments, accepted non-conformities, product repairs or service modifications also need to be recorded. Where it is necessary to repair or rework a product or modify a service, verification requirements need to be determined and implemented.

Analysis of data

Analysis of data needs to be established as a means of determining where system improvements can be made. Data needs to be collected from relevant sources, including internal audits, corrective and preventive action, non-conforming product service, customer complaints and customer satisfaction results.

The organization should then analyze the data to provide information on:

(a) the effectiveness of the quality management system;
(b) process operation trends;

(c) customer satisfaction;
(d) conformance to customer requirements of the product/service;
(e) suppliers.

There is also a need to determine the statistical techniques to be used for analyzing data, including verifying process operations and product service characteristics. Of course the statistical techniques selected should be suitable and their use controlled and monitored.

Improvement

The organization needs to establish a process for eliminating the causes of non-conformity and preventing recurrence. Non-conformity reports, customer complaints and other suitable quality management system records are useful as inputs to the corrective action process. Responsibilities for corrective action need to be established together with the procedures for the corrective action process, which should include:

(a) identification of non-conformities of the products, services, processes, the quality management system, and customer complaints;
(b) investigation of causes of non-conformities, and recording results of investigations;
(c) determination of corrective actions needed to eliminate causes of non-conformities;
(d) implementation of corrective action;
(e) follow-up to ensure corrective action taken is effective and recorded.

Corrective actions also need to be implemented for products or services already delivered, but subsequently discovered to be non-conforming and customers need to be notified where possible.

The organization needs to establish a process for eliminating the causes of potential non-conformities to prevent their occurrence. Quality management system records and results from the analysis of data should be used as inputs for this and responsibilities for preventive action established. The process should include:

(a) identification of potential product, service and process non-conformities;
(b) investigation of the causes of potential non-conformities of products/services, process and the quality management system, and recording the results;
(c) determination of preventive action needed to eliminate causes of potential non-conformities;
(d) implementation of preventive action needed;
(e) follow-up to ensure preventive action taken is effective, recorded and submitted for management review.

Processes for the continual improvement of the quality management system need to be established including methods and measures suitable for the products/services.

Case study ▪▪▪ ▬▬▬▬▬▬▬▬▬▬▬▬▬▬▬▬▬

▪ Developing a business management
▪ system (BMS) in QinetiQ

QinetiQ is the result of a lengthy transformation from a set of independent government research and development establishments (previously known as DERA – Defence Evaluation & Research Agency) to a public limited company (plc), which is at the forefront of creating technology and its use in providing solutions for a variety of customers, to afford them a competitive edge in their market places. The transformation had many facets, mostly to do with the changing of attitudes and behaviors, and many initiatives and change programs were spawned to develop new skills and attitudes. This short case study examines one thread – the latter stages of the development of a business management system (BMS) based on the right processes to achieve the goals of the company.

The development of a single corporate management system began with an examination of the processes that needed to operate successfully. Using process-mapping techniques, a new process model was derived.

A benchmarking study and its recommendations were key in the development of a revision to the DERA BMS (see Chapter 9).

With the advent of Public Private Partnerships (PPP) and the splitting of DERA into QinetiQ and [dstl], with the majority of the assets going to QinetiQ as a plc, there was a clear need to remove unnecessary bureaucracy and enable entrepreneurship, while retaining the controls essential for good governance. Hence, sets of new principles and structure and concept were derived, as follows:

BMS Principles

- ▪ the board sets out the strategic intent for QinetiQ;
- ▪ we clearly and unambiguously set out the policies that we must follow in QinetiQ to achieve that strategic intent;
- ▪ we do this through a framework of processes;
- ▪ the processes describe the minimum set of mandatory steps and controls to deliver QinetiQ's objectives;
- ▪ at the delivery level, the principle of subsidiarity applies;
- ▪ we adopt a risk based approach to control;
- ▪ we employ people competent to carry out their tasks;

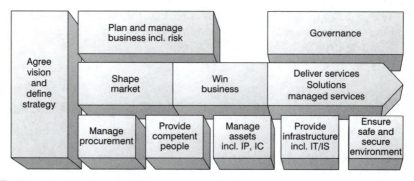

Figure 12.4   New QinetiQ BMS top level process framework

■ we demonstrate that the processes continue to:
  − be effective in delivering QinetiQ's needs; be efficient in their operation;
  − compare favorably with other world-class organizations.

Structure and concept of the BMS

■ Processes
  − policy
  − mandatory steps
■ Obligatory
  − essential 'rules' but not processes
■ Practice
  − good ways of working
■ Why
  − objectives
  − policies
■ What
  − process steps
  − corporate tools
■ How
  − plans
    ◆ calling on the practices as appropriate.

Here the processes describe the minimum mandatory steps to ensure consistency, if appropriate, and governance. The main assumption is that people are competent to do their job and, therefore, do not need to be told *how* to do it – merely *what* they must do. In this way QinetiQ ensure the right balance in 'the way we work'. The processes are contained in a new framework shown in some detail in Figure 12.4.

Continuous improvement is now embedded in the new QinetiQ through its process-based electronic business management system. This is the backbone of the organization's highly specialized operations and provides a route to sustained performance in the increasingly commercial and competitive world in which its scientists live.

Other management systems and models

Organizations of all kinds are increasingly concerned to achieve and demonstrate sound environmental performance. Many have undertaken environmental audits and reviews to assess this. To be effective these need to be conducted within a structured management system, which in turn is integrated with the overall management activities dealing with all aspects of desired environmental performance.

Such a system should establish processes for setting environmental policy and objectives, and achieving compliance to them. It should be designed to place emphasis on the prevention of adverse environmental effects, rather than on detection after occurrence. It should also identify and assess the environmental effects arising from the organization's existing or proposed activities, products, or services and from incidents, accidents, and potential emergency situations. The system must identify the relevant regulatory requirements, the priorities, and pertinent environmental objectives and targets. It needs also to facilitate planning, control, monitoring, auditing and review activities to ensure that the policy is complied with, that it remains relevant, and is capable of evolution to suit changing circumstances.

The International Standard ISO 14001 contains a specification for environmental management systems for ensuring and demonstrating compliance with stated policies and objectives. The standard is designed to enable any organization to establish an effective management system, as a foundation for both sound environmental performance and participation in environmental auditing schemes.

ISO 14001 shares common management system principles with the ISO 9001:2000 standard and organizations may elect to use an existing management system, developed in conformity with that standard, as a basis for environmental management. The ISO 14001 standard defines environmental policy, objectives, targets, effect, management, systems, manuals, evaluation, audits and reviews. It mirrors the ISO 9001:2000 standard requirements in many of its own requirements, and it includes a guide to these in an informative annex. EN ISO 9001:2000 also gives a useful annex of 'correspondence between ISO 9001:2000 and ISO 14001:1996' under each of the main headings of both standards. This shows that the quality management system standard has been aligned with requirements of the environmental management system standard in order to enhance the compatibility of the two for the benefit of the users. As such ISO 9001:2000 enables an organization to align and integrate its own quality management systems with other related management system requirements. In this way, it should be possible for any organization to adapt its existing management system(s) in order to establish one that complies with ISO 9001:2000.

ISO 9000:2000 also makes comments on the relationship between quality management systems and excellence models. It claims that the approaches set down in the ISO 9000 family of standards and in the various excellence models (see Chapter 2) are based on common principles and that the only differences are in the scope of application. Excellence models and ISO 9000:

(a) help organizations identify strengths and weaknesses;
(b) aid the evaluation of organizations;
(c) establish a basis for continuous improvement;
(d) allow and support external recognition.

It is also recognized that excellence models add value to the quality management system approach by providing criteria that allow comparative evaluation of organizational performance with other organizations.

Management systems are needed in all areas of activity, whether large or small businesses, manufacturing, service or public sector. The advantages of systems in manufacturing are obvious, but they are just as applicable in areas such as marketing, sales, personnel, finance, research and development, as well as in the service industries and public sectors. No matter where it is implemented a good management system will improve process control, reduce wastage, lower costs, increase market share (or funding), facilitate training, involve staff, and raise morale.

■ Improvements made to quality management ■ systems

One measure of the changes made between 1994 and 2000 in the ISO 9000 family of quality management system standards is in the language used. Table 12.2 shows a simple count of the number of times certain words are to be found in each of the versions of ISO 9001. These reflect the main thrusts of the new standard:

■ increased customer focus;
■ process approach (at the expense of procedural documentation);

Table 12.2 Change in vocabulary in ISO 9001:2000

Change in ISO 9001 vocabulary (text and notes)

Word	Occurrence	
	1994	2000
• Improvement	0	16
• Processes	21	39
• Procedures	36	9
• Customer	15	32
• Competency	0	4
• Capability	5	1
• Analyze(is)	0	5

- continuous improvement;
- skills-based approach to human resource management.

The adoption of a process approach reflects the changes in management thinking in recent years and supports the author's 'campaign' for improved process management to deliver better performance in most organizations. The standard now explicitly recognizes the process-oriented nature of many organizations and addresses all sorts of important areas – risk assessment, relationship and interactions between functions/departments – the internal customer/supplier chains of Chapter 1 – and the vital one of capability.

Hence, the revised standard is a major step forward from the earlier versions. It pays a lot of attention to the management of people, an area which many organizations have neglected in their past 'quality initiatives', especially those concerned with documenting quality systems.

Organizations making the transition from the 1994 version to ISO 9002:2001 are finding how vital are the management of skills, including those of the auditors – external as well as internal! A mindset change is required away from the procedures-based earlier approaches which led in some cases to mechanistic and even bureaucratic systems. With understanding, commitment and the right attitude, particularly in the senior management, a quality management system can support a dynamic organization and help it achieve its aims and objectives – 'whats' – through a thorough understanding of the 'hows'. The ISO 9000:2000 family of standards is now much more aligned with the EFQM Excellence Model in terms of the intuitively appealing: planning–process–people–performance structure.

Chapter highlights

■ ■ ■

Why a quality management system?

■ An appropriate quality management system will enable the objectives set out in the quality policy to be accomplished.

■ The International Organization for Standardization (ISO) 9000:2000 series set out methods by which a system can be implemented to ensure that the specified customer requirements are met.

■ A quality system may be defined as an assembly of components, such as the management responsibilities, process, and resources.

Quality management system design

■ Quality management systems should apply to and interact with processes in the organization. The activities are generally processing, communicating, and controlling. These should be documented in the form of a quality manual.

■ The system should follow the Plan, Do, Check, Act cycle, through documentation, implementation, audit and review.

■ The ISO 9000:2000 family together forms a coherent set of quality management system standards to facilitate mutual understanding across national and international trade.

Quality management system requirements

■ The general categories of the ISO 9001:2000 standard on quality management systems include: management responsibility, resource management, product realization, measurement analysis and improvement, which are detailed in the standard.

Other management systems and models

■ The International Standard ISO 14001 contains specifications for environmental management systems for ensuring and demonstrating compliance with the stated policies and objectives, and acting as a base for auditing and review schemes.

■ ISO 14000 shares common principles with the ISO 9001:2000 standard on quality management systems. The latter shows, by 'correspondence' between the two, under the main headings of both standards, that the quality standard has been aligned with the requirements of the environmental standard.

■ ISO 9000:2000 also makes comments on the relationship between quality management systems and excellence models. The two are based on the common principles of identifying strengths and weaknesses, evaluation, continuous improvement, external recognition.

Improvements made to quality management systems

- The language used in the 1994 and 2000 versions of the ISO 9000 family shows changes in emphasis to increased customer focus, a process rather than procedural approach, continuous improvement, and a skills-based approach to people management.
- The ISO 9000:2000 standard is a major step forward from the earlier versions, paying much more attention to human resource management and being in tune with the EFQM Excellence Model in terms of the direction–process–people–performance alignment.

Chapter 13

Continuous improvement

A systematic approach

In the never-ending quest for improvement in the ways processes are operated, numbers and information should always form the basis for understanding, decisions and actions; and a thorough data-gathering, recording and presentation system is essential:

Record data	– all processes can and should be measured
	– all measurements should be recorded.
Use data	– if data is recorded and not used it will be abused.
Analyze data	– data analysis should be carried out by means of some basic systematic tools.
Act on the results	– recording and analysis of data without action leads to frustration.

In addition to the basic elements of a quality management system that provide a framework for recording, there exists a set of methods the Japanese quality guru Ishikawa has called the 'seven basic tools'. These should be used to interpret and derive the maximum use from data. The simple methods listed below, of which there are clearly more than seven, will offer any organization a means of collecting, presenting, and analyzing most of its data:

■ Process flowcharting – what is done?
■ Check sheets/tally charts – how often is it done?
■ Histograms – what do overall variations look like?
■ Scatter diagrams – what are the relationships between factors?

- Stratification – how is the data made up?
- Pareto analysis – which are the big problems?
- Cause and effect analysis and brainstorming, including CEDAC (cause and effect diagrams with addition of cards), NGT (normal group technique), and the five whys – what causes the problems?
- Force field analysis – what will obstruct or help the change or solution?
- Emphasis curve – which are the most important factors?
- Control charts – which variations to control and how?

Sometimes more sophisticated techniques, such as analysis of variance, regression analysis, and design of experiments, need to be employed.

The effective use of the tools requires their application by the people who actually work on the processes. Their commitment to this will be possible only if they are assured that management cares about improving quality. Managers must show they are serious by establishing a systematic approach and providing the training and implementation support required.

Improvements cannot be achieved without specific opportunities, commonly called problems, being identified or recognized. A focus on improvement opportunities leads to the creation of teams whose membership is determined by their work on and detailed knowledge of the process, and their ability to take improvement action. The teams must then be provided with good leadership and the right tools to tackle the job.

The systematic approach shown in Figure 13.1 should lead to the use of factual information, collected and presented by means of proven techniques, to open a channel of communications not available to the many organizations that do not follow this or a similar structured approach to problem solving and improvement. Continuous improvements in the quality of products, services, and processes can often be obtained without major capital investment, if an organization marshals its resources, through an understanding and breakdown of its processes in this way.

By using reliable methods, creating a favorable environment for team-based problem solving, and continuing to improve using systematic techniques, the never-ending improvement helix (see Chapter 3) will be engaged. This approach demands the real time management of data, and actions focused on processes and inputs rather than outputs. It will require a change in the language of many organizations from percentage defects, percentage 'prime' product, and number of errors, to *process capability*. The climate must change from the traditional approach of 'If it meets the specification, there are no problems and no further improvements are necessary'. The driving force for this will be the need for better internal and external customer satisfaction levels, which will lead to the continuous improvement question, 'Could we do the job better?'

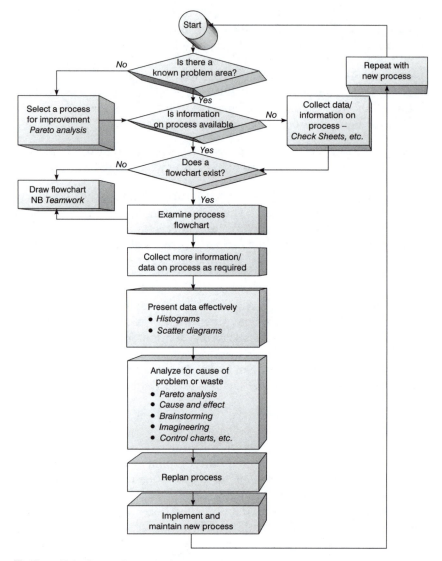

■ Some basic tools and techniques

Understanding processes so that they can be improved by means of the systematic approach requires knowledge of a simple kit of tools or techniques. What follows is a brief description of each technique, but a full description and further examples of some of them may be found in reference 1.

Process modeling and flowcharting _____

The use of these techniques, described in Chapter 10, ensures a full understanding of the inputs, outputs and flow of the process. Without that understanding, it is not possible to map or draw the correct flowchart of the process. It is important to remember that in all but the smallest tasks no single person is able to complete a process map or chart without help from others. This makes mapping and flowcharting powerful team forming exercises.

Check sheets or tally charts _____

A check sheet is a tool for data gathering, and a logical point to start in most process control or problem solving efforts. It is particularly useful for recording direct observations and helping to gather in facts rather than opinions about the process.

In the recording process it is essential to understand the difference between data and numbers. Data are pieces of information, including numerical, that are useful in solving problems, or providing knowledge about the state of a process. Numbers alone often represent meaningless measurements or counts, which tend to confuse rather than to enlighten. Numerical data on quality will arise either from counting or measurement.

Data from counting can occur only at definite points or in 'discrete' jumps. There can be only 0, 1, 2, etc., errors in an invoice page; there cannot be 2.46 errors. The number of pens that fail to write properly give rise to discrete data called ATTRIBUTES. As there is only a two-way classification to consider, right or wrong, present or not present, attributes give rise to counted data, which necessarily vary in jumps.

Data from measurement can occur anywhere at all on a continuous scale, and are called VARIABLE data. The weight of a capsule, the diameter of a piston, the tensile strength of a piece of rod, the time taken to process an insurance claim, are all variables, the measurement of which produces continuous data.

Check sheets are prepared by following four steps:

1 Select and agree on the exact event to be observed.
2 Decide on the data collection time period. This includes both how often the data are to be obtained (frequency) and for how long they will be collected (duration).

Observer: F. Oldsman		Computer No. 148	Date 26 June	
Number of observations: 95			Total	Percentage
Computer in use		┼┼┼┼ ┼┼┼┼ ┼┼┼┼ ┼┼┼┼ ┼┼┼┼ ┼┼┼┼ ┼┼┼┼ ┼┼┼┼ ┼┼┼┼ ┼┼┼┼ ┼┼┼┼	55	57.9
Computer idle	Repairs	┼┼┼┼	5	5.3
	No work	┼┼┼┼ ┼┼┼┼ ‖	12	12.6
	Operator absent	┼┼┼┼ ┼┼┼┼	10	10.5
	System failure	┼┼┼┼ ┼┼┼┼ ‖‖	13	13.7

■ **Figure 13.2** Activity sampling record in an office

3 Design a form that is simple, easy to use and large enough to record the information. Each column must be clearly labeled.
4 Collect the data and fill in the check sheet. Be honest in recording the information and allow enough time for it to be collected and recorded.

Follow up the recording by some analysis or presentation of the data.

The use of simple check sheets or tally charts aids the collection of data of the right type, in the right form, at the right time. The objectives of the data collection will determine the design of the record sheet used. Two examples of tally charts for different purposes are shown in Figures 13.2 and 13.3. These give rise to *frequency distributions.*

Histograms

Histograms show, in a very clear pictorial way, the frequency with which a certain value or group of values occurs. They can be used to display both attribute and variable data, and are an effective means of letting the people who operate the process know the results of their efforts. Data gathered on truck turn-round times is drawn as a histogram in Figure 13.4.

Scatter diagrams

Depending on the technology, it is frequently useful to establish the association, if any, between two parameters or factors. A technique to begin such an analysis is a simple X-Y plot of the two sets of data. The

Truck turn-round time (minutes – rounded to nearest 5)	Tally	Number of trucks (frequency)
10	I	1
15	III	3
20	ΗΗΤ I	6
25	ΗΗΤ IIII	9
30	ΗΗΤ ΗΗΤ ΗΗΤ ΗΗΤ ΗΗΤ ΗΗΤ ΗΗΤ ΗΗΤ II	42
35	ΗΗΤ ΗΗΤ ΗΗΤ ΗΗΤ ΗΗΤ ΗΗΤ ΗΗΤ ΗΗΤ ΗΗΤ ΗΗΤ ΗΗΤ ΗΗΤ ΗΗΤ ΗΗΤ ΗΗΤ ΗΗΤ ΗΗΤ ΗΗΤ ΗΗΤ ΗΗΤ ΗΗΤ II	107
40	ΗΗΤ ΗΗΤ ΗΗΤ ΗΗΤ ΗΗΤ ΗΗΤ ΗΗΤ ΗΗΤ ΗΗΤ ΗΗΤ ΗΗΤ ΗΗΤ ΗΗΤ ΗΗΤ ΗΗΤ ΗΗΤ ΗΗΤ ΗΗΤ ΗΗΤ ΗΗΤ ΗΗΤ ΗΗΤ ΗΗΤ ΗΗΤ ΗΗΤ ΗΗΤ ΗΗΤ ΗΗΤ ΗΗΤ ΗΗΤ ΗΗΤ ΗΗΤ ΗΗΤ ΗΗΤ	170
45	ΗΗΤ ΗΗΤ ΗΗΤ ΗΗΤ ΗΗΤ ΗΗΤ ΗΗΤ ΗΗΤ ΗΗΤ ΗΗΤ ΗΗΤ ΗΗΤ ΗΗΤ ΗΗΤ ΗΗΤ ΗΗΤ ΗΗΤ ΗΗΤ ΗΗΤ ΗΗΤ	100
50	ΗΗΤ ΗΗΤ ΗΗΤ ΗΗΤ ΗΗΤ ΗΗΤ ΗΗΤ III	38
55	ΗΗΤ ΗΗΤ ΗΗΤ I	16
60	ΗΗΤ	5
65	II	2
70	I	1
	Total	500

Figure 13.3 Tally chart and frequency distribution for truck turn-round times

resulting grouping of points on scatter diagrams (e.g. Figure 13.5) will reveal whether or not a strong or weak, positive or negative, correlation exists between the parameters. The diagrams are simple to construct and easy to interpret, and the absence of correlation can be as revealing as finding that a relationship exists.

Stratification

Stratification is simply dividing a set of data into meaningful groups. It can be used to great effect in combination with other techniques, including histograms and scatter diagrams. If, for example, three shift teams are responsible for the output described by the histogram (a) in Figure 13.6, 'stratifying' the data into the shift groups might produce

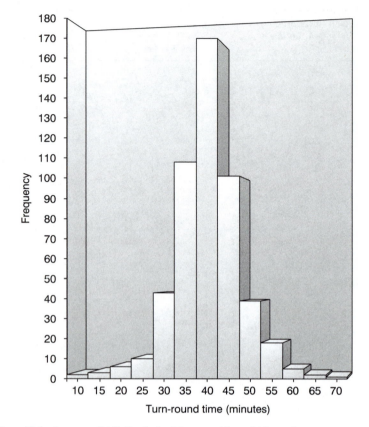

Figure 13.4   Frequency distribution for truck turn-round times (histogram)

Figure 13.5   Scatter diagram showing a negative correlation between two variables

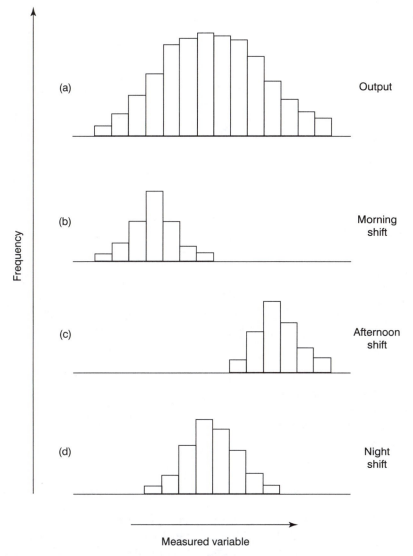

Figure 13.6 Stratification of data into shift teams

histograms (b), (c) and (d), and indicate process adjustments that were taking place at shift changeovers.

Figure 13.7 shows the scatter diagram relationship between advertising investment and revenue generated for all products. In diagram (a) all the data are plotted, and there seems to be no correlation. But if the data are stratified according to product, a correlation is seen to exist. Of course the reverse may be true, so the data should be kept together and plotted in different colors or symbols to ensure all possible interpretations are retained.

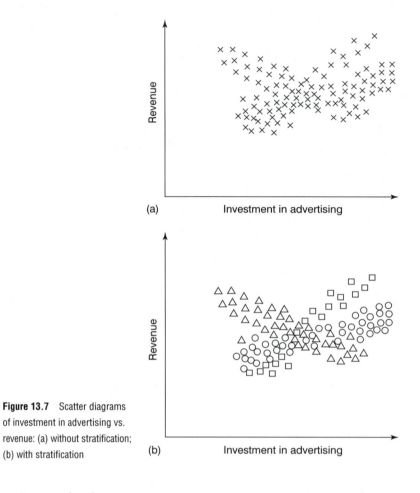

Figure 13.7 Scatter diagrams of investment in advertising vs. revenue: (a) without stratification; (b) with stratification

Pareto analysis

If the symptoms or causes of defective output or some other 'effect' are identified and recorded, it will be possible to determine what percentage can be attributed to any cause, and the probable results will be that the bulk (typically 80 percent) of the errors, waste, or 'effects', derive from a few of the causes (typically 20 percent). For example, Figure 13.8 shows a *ranked frequency distribution* of incidents in the distribution of a certain product. To improve the performance of the distribution process, therefore, the major incidents (broken bags/drums, truck scheduling, and temperature problems) should be tackled first. An analysis of data to identify the major problems is known as *Pareto analysis*, after the Italian economist who realized that approximately 90 percent of the wealth in his country was owned by approximately 10 percent of the people. Without an analysis of this sort, it is far too easy to devote resources to addressing one symptom only because its cause seems immediately apparent.

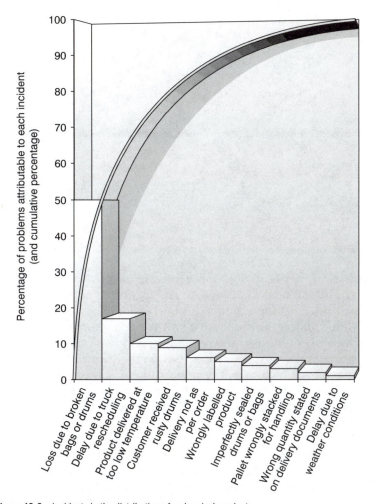

Figure 13.8   Incidents in the distribution of a chemical product

Cause and effect analysis and brainstorming _____

A useful way of mapping the inputs that affect quality is the *cause and effect diagram*, also known as the Ishikawa diagram (after its originator) or the fishbone diagram (after its appearance, Figure 13.9). The effect or incident being investigated is shown at the end of a horizontal arrow. Potential causes are then shown as labeled arrows entering the main cause arrow. Each arrow may have other arrows entering it as the principal factors or causes are reduced to their subcauses and subsubcauses by *brainstorming*.

Brainstorming is a technique used to generate a large number of ideas quickly, and may be used in a variety of situations. Each member of a

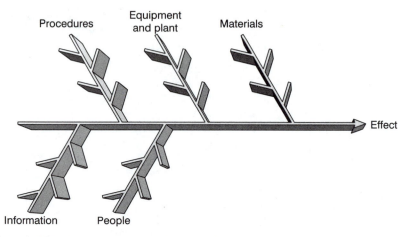

■ **Figure 13.9**   The cause and effect, Ishikawa or fishbone diagram

group, in turn, may be invited to put forward ideas concerning a problem under consideration. Wild ideas are safe to offer, as criticism or ridicule is not permitted during a brainstorming session. The people taking part do so with equal status to ensure this. The main objective is to create an atmosphere of enthusiasm and originality. All ideas offered are recorded for subsequent analysis. The process is continued until all the conceivable causes have been included. The proportion of non-conforming output attributable to each cause, for example, is then measured or estimated, and a simple Pareto analysis identifies the causes that are most worth investigating.

A useful variant on the technique is negative brainstorming. Here the group brainstorms all the things that would need to be done to ensure a negative outcome. For example, in the implementation of TQM, it might be useful for the senior management team to brainstorm what would be needed to make sure TQM *was not* implemented. Having identified in this way the potential roadblocks, it is easier to dismantle them (see also Force Field Analyses).

CEDAC

A variation on the cause and effect approach, which was developed at Sumitomo Electric and is now used by many major corporations across the world, is the cause and effect diagram with addition of cards (CEDAC).

The effect side of a CEDAC chart is a quantified description of the problem, with an agreed and visual quantified target and continually updated results on the progress of achieving it. The cause side of the CEDAC chart

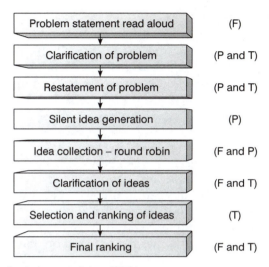

Problem statement read aloud	(F)
Clarification of problem	(P and T)
Restatement of problem	(P and T)
Silent idea generation	(P)
Idea collection – round robin	(F and P)
Clarification of ideas	(F and T)
Selection and ranking of ideas	(T)
Final ranking	(F and T)

Figure 13.10   Nominal group technique (NGT)

uses two different colored cards for writing facts and ideas. This ensures that the facts are collected and organized before solutions are devised. The basic diagram for CEDAC has the classic fishbone appearance.

Nominal group technique (NGT)

The nominal group technique (NGT) is a particular form of team brainstorming used to prevent domination by particular individuals. It has specific application for multi-level, multi-disciplined teams, where communication boundaries are potentially problematic.

In NGT a carefully prepared written statement of the problem to be tackled is read out by the facilitator (F). Clarification is obtained by questions and answers and then the individual participants (P) are asked to restate the problem in their own words. The group then discusses the problem until its formulation can be satisfactorily expressed by the team (T). The method is set out in Figure 13.10. NGT results in a set of ranked ideas that are close to a team consensus view obtained without domination by one or two individuals.

Even greater discipline may be brought to brainstorming by the use of 'soft systems methodology (SSM)', developed by Peter Checkland.[2] The component stages of SSM are gaining a 'rich understanding' through 'finding out', input/output diagrams, root definition (which includes the so-called CATWOE analysis: customers, 'actors', transformations, 'world-view', owners, environment), conceptualization, comparison, and recommendation.

Force field analyses

Force field analysis is a technique used to identify the forces that either obstruct or help a change that needs to be made. It is similar to negative brainstorming and cause/effect analysis and helps to plan how to overcome the barriers to change or improvement. It may also provide a measure of the difficulty in achieving the change.

The process begins with a team describing the desired change or improvement, and defining the objectives or solution. Having prepared the basic force field diagram, it identifies the favorable/positive/driving forces and the unfavorable/negative/restraining forces, by brainstorming. These forces are placed in opposition on the diagram and, if possible, rated for their potential influence on the ease of implementation. The results are evaluated. Then comes the preparation of an action plan to overcome some of the restraining forces, and increase the driving forces. Figure 13.11 shows a force field analysis produced by a senior management team considering the implementation of TQM in its organization, during a brainstorming session.

The emphasis curve

This is a technique for ranking a number of factors, each of which cannot be readily quantified in terms of cost, frequency of occurrence, etc., in priority order. It is almost impossible for the human brain to make a judgement of the relative importance of more than three or four non-quantifiable factors. It is, however, relatively easy to judge which is the most important of two factors, using some predetermined criteria.

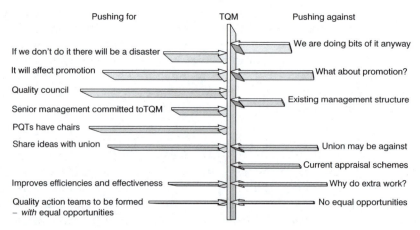

Figure 13.11 Force field analysis

The emphasis curve technique uses this fact by comparing only two factors at any one time.

The procedural steps for using the 'emphasis curve chart' are as follows:

1 List the factors for ranking under a heading 'Scope'.
2 Compare factor 1 with factor 2 and rank the most important. To assist in judging the relative importance of two factors, it may help to use weightings, e.g. degree of seriousness, capital investment, speed of completion, etc., on a scale of 1 to 10.
3 Compare factor 1 with 3, 1 with 4, 1 with 5 and so on – ringing the most important number in the matrix.
4 Having compared factor 1 against the total scope, proceed to compare factor 2 with 3, 2 with 4 and so on.
5 Count the number of 'ringed' number 1s in the matrix and put the total in a right-hand column against Number 1. Next count the total number of 2s in the matrix and put the total in a column against Number 2 and so on.
6 Add up the numbers in the column and check the total, using the formula $\{n(n - 1)\}/2$, where n is the number of entries in the column. This check ensures that all numbers have been 'ringed' in the matrix.
7 Proceed to rank the factors using the numbers in the column.

Generally the length of time to make a judgement between two factors does not significantly affect the outcome; therefore the rule is 'accept the first decision, record it and move quickly onto the next pair'.

Control charts

A control chart is a form of traffic signal whose operation is based on evidence from the small samples taken at random during a process. A green light is given when the process should be allowed to run. All too often processes are 'adjusted' on the basis of a single measurement, check or inspection, a practice that can make a process much more variable than it is already. The equivalent of an amber light appears when trouble is possibly imminent. The red light shows that there is practically no doubt that the process has changed in some way and that it must be investigated and corrected to prevent production of defective material or information. Clearly, such a scheme can be introduced only when the process is 'in control'. Since samples taken are usually small, there are risks of errors, but these are small, calculated risks and not blind ones. The risk calculations are based on various frequency distributions.

These charts should be made easy to understand and interpret and they can become, with experience, sensitive diagnostic tools to be used by operating staff and first-line supervision to prevent errors or defective output being produced. Time and effort spent to explain the working of the charts to all concerned are never wasted.

The most frequently used control charts are simple run charts, where the data is plotted on a graph against time or sample number. There are different types of control charts for variables and attribute data: for variables mean ($\bar{X}$) and range (**R**) charts are used together; number defective or **np** charts and proportion defective or **p** charts are the most common ones for attributes. Other charts found in use are moving average and range charts, numbers of defects (**c** and **u**) charts, and cumulative sum (cusum) charts. The latter offer very powerful management tools for the detection of trends or changes in attributes and variable data.

The cusum chart is a graph that takes a little longer to draw than the conventional control chart, but gives a lot more information. It is particularly useful for plotting the evolution of processes, because it presents data in a way that enables the eye to separate true changes from a background of random variation. Cusum charts can detect small changes in data very quickly, and may be used for the control of variables and attributes. In essence, a reference or 'target value' is subtracted from each successive sample observation, and the result accumulated. Values of this cumulative sum are plotted, and 'trend lines' may be drawn on the resulting graphs. If they are approximately horizontal, the value of the variable is

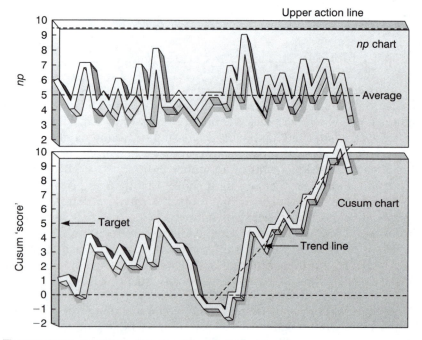

■ **Figure 13.12**   Comparison of cusum and *np* charts for the same data

about the same as the target value. A downward slope shows a value less than the target and an upward slope a value greater. The technique is very useful, for example, in comparing sales forecast with actual sales figures.

Figure 13.12 shows a comparison of an ordinary run chart and a cusum chart that have been plotted from the same data – errors in samples of 100 invoices. The change, which is immediately obvious on the cusum chart, is difficult to detect on the conventional control chart.

The range of type and use of control charts is now very wide, and within the present text it is not possible to indicate more than the basic principles underlying such charts.[1] All of them can be generated electronically using the various software tools available.

Statistical process control (SPC)

The responsibility for quality in any transformation process must lie with the operators of that process. To fulfill this responsibility, however, people must be provided with the tools necessary to:

- Know whether the process is capable of meeting the requirements.
- Know whether the process is meeting the requirements at any point in time.
- Make correct adjustment to the process or its inputs when it is not meeting the requirements.

The techniques of statistical process control (SPC) will greatly assist in these stages. To begin to monitor and analyze any process, it is necessary first of all to identify what the process is, and what the inputs and outputs are. Many processes are easily understood and relate to known procedures, e.g. drilling a hole, compressing tablets, filling cans with paint, polymerizing a chemical using catalysts. Others are less easily identifiable, e.g. servicing a customer, delivering a lecture, storing a product in a warehouse, inputting to a computer. In many situations it can be extremely difficult to define the process. For example, if the process is inputting data into a computer terminal, it is vital to know if the scope of the process includes obtaining and refining the data, as well as inputting. Process definition is so important because the inputs and outputs change with the scope of the process.

All processes can be monitored and brought 'under control' by gathering and using data – to measure the performance of the process and provide the feedback required for corrective action, where necessary. SPC methods, backed by management commitment and good organization, provide an objective means of *controlling* quality in any transformation process, whether used in the manufacture of artefacts, the provision of services, or the transfer of information.

SPC is not only a tool kit, it is a strategy for reducing variability, the cause of most quality problems: variation in products, in times of deliveries, in ways of doing things, in materials, in people's attitudes, in equipment and its use, in maintenance practices, in everything. Control by itself is not sufficient. Total quality management requires that the processes should be improved continually by reducing variability. This is brought about by studying all aspects of the process, using the basic question: 'Could we do this job more consistently and on target?' The answer drives the search for improvements. This significant feature of SPC means that it is not constrained to measuring conformance, and that it is intended to lead to action on processes that are operating within the 'specification' to minimize variability.

Process control is essential, and SPC forms a vital part of the TQM strategy. Incapable and inconsistent processes render the best design impotent and make supplier quality assurance irrelevant. Whatever process is being operated, it must be reliable and consistent. SPC can be used to achieve this objective.

In the application of SPC there is often an emphasis on techniques rather than on the implied wider managerial strategies. It is worth repeating that SPC is not only about plotting charts on the walls of a plant or office, it must become part of the company-wide adoption of TQM and act as the focal point of never-ending improvement. Changing an organization's environment into one in which SPC can operate properly may take several years rather than months. For many companies SPC will bring a new approach, a new 'philosophy', but the importance of the statistical techniques should not be disguised. Simple presentation of data using diagrams, graphs, and charts should become the means of communication concerning the state of control of processes. It is on this understanding that improvements will be based.

The SPC system

A systematic study of any process through answering the questions:

- Are we capable of doing the job correctly?
- Do we continue to do the job correctly?
- Have we done the job correctly?
- Could we do the job more consistently and on target?[3]

provides knowledge of the *process capability* and the sources of nonconforming outputs. This information can then be fed back quickly to marketing, design, and the 'technology' functions. Knowledge of the current state of a process also enables a more balanced judgement of equipment, both with regard to the tasks within its capability and its rational utilization.

Statistical process control procedures exist because there is variation in the characteristics of all material, articles, services, and people. The inherent variability in each transformation process causes the output from it to vary over a period of time. If this variability is considerable, it is impossible to predict the value of a characteristic of any single item or at any point in time. Using statistical methods, however, it is possible to take meagre knowledge of the output and turn it into meaningful statements that may then be used to describe the process itself. Hence, statistically based process control procedures are designed to divert the process itself. Hence, statistically based process control procedures are designed to divert attention from individual pieces of data and focus it on the process as a whole. SPC techniques may be sued to measure and control the degree of variation of any purchased materials, services, processes, and products, and to compare this, if required, to previously agreed specifications. In essence, SPC techniques select a representative, simple, random sample for the 'population', which can be an input to or an output from a process. From an analysis of the sample it is possible to make decisions regarding the current performance of the process.

Organizations that embrace the TQM concepts should recognize the value of SPC techniques in areas such as sales, purchasing, invoicing, finance, distribution, training, and in the service sector generally. These are outside the traditional areas for SPC use, but it needs to be seen as an organization-wide approach to reducing variation with the specific techniques integrated into a program of change throughout. A Pareto analysis, a histogram, a flowchart, or a control chart is a vehicle for communication. Data is data and whether the numbers represent defects or invoice errors, weights or delivery times, or the information relates to machine settings, process variables, prices, quantities, discounts, sales or supply points, is irrelevant – the techniques can always be used.

In the author's experience, some of the most exciting applications of SPC have emerged from organizations and departments which, when first introduced to the methods, could see little relevance in them to their own activities. Following appropriate training, however, they have learned how to, for example:

- *Pareto analyze* errors on invoices to customers and industry injury data.
- *Brainstorm and cause and effect analyze* reasons for late payment and poor purchase invoice matching.

- *Histogram* defects in invoice matching and arrival of trucks at certain times during the day.
- *Control chart* the weekly demand of a product.

Distribution staff have used control charts to monitor the proportion of late deliveries, and Pareto analysis and force field analysis to look at complaints about the distribution system. Bank operators have been seen using cause and effect analysis, NGT and histograms to represent errors in the output from their services. Moving average and cusum charts have immense potential for improving processes in the marketing area.

Those organizations that have made most progress in implementing continuous improvement have recognized at an early stage that SPC is for the whole organization. Restricting it to traditional manufacturing or operational activities means that a window of opportunity for improvement has been closed. Applying the methods and techniques outside manufacturing will make it easier, not harder, to gain maximum benefit from SPC.

■ Some additional techniques for process design ■ and improvement

Seven 'new' qualitative tools may be used as part of quality function deployment (see Chapter 6) or to improve processes. These do not replace the basic systematic tools described earlier in this chapter, neither are they extensions of these. The new tools are systems and documentation methods used to achieve success in design by identifying objectives and intermediate steps in the finest detail. The seven new tools are:

1 Affinity diagram.
2 Inter-relationship diagraph.
3 Tree diagram.
4 Matrix diagram or quality table.
5 Matrix data analysis.
6 Process decision program chart (PDPC).
7 Arrow diagram.

The tools are inter-related, as shown in Figure 13.13 and are summarized below.

1 Affinity diagram

This is used to gather large amounts of language data (ideas, issues, opinions) and organizes them into groupings based on the natural relationship between the items. In other words, it is a form of brainstorming.

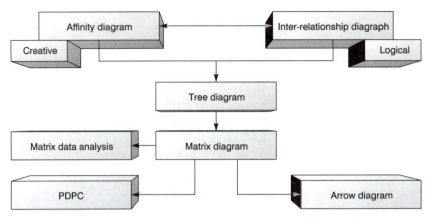

Figure 13.13   The seven new tools of quality design

One of the obstacles often encountered in the quest for improvement is past success or failure. It is assumed that what worked or failed in the past will continue to do so in the future. Although the lessons of the past should not be ignored, unvarying patterns of thought, which can limit progress, should not be enforced. This is especially true in QFD, where *new* logical patterns should always be explored.

The affinity diagram, like other brainstorming methods, is part of the creative process. It can be used to generate ideas and categories that can be used later with more strict, logic-based tools. This tool should be used to 'map the geography' of an issue when:

- Facts or thoughts are in chaos and the issues are too large or complex to define easily.
- Breakthroughs in traditional concepts are needed to replace old solutions and to expand a team's thinking.
- Support for a solution is essential for successful implementation.

The affinity diagram is not recommended when a problem is simple or requires a very quick solution.

The steps for generating an affinity diagram are as follows:

1 Assemble a group of people familiar with the problem of interest. Six to eight members in the group works best.
2 Phrase the issue to be considered. It should be vaguely stated so as not to prejudice the responses in a predetermined direction.
3 Give each member of the group a stack of cards and allow 5–10 minutes for everyone individually in the group to record ideas on the cards, writing down as many ideas as possible.

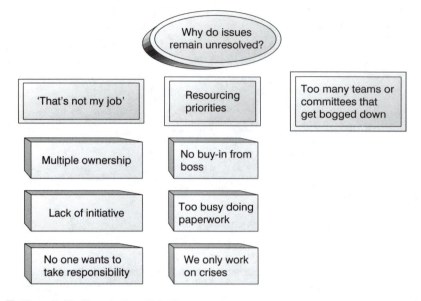

Figure 13.14   Example of an affinity diagram

4 At the end of the 5–10 minutes each member of the group, in turn, reads out one of his/her ideas and places it on the table for everyone to see, without criticism or justification.

5 When all ideas are presented, members of the group place together all cards with related ideas repeating the process until the ideas are in a few groups.

6 Look for one card in each group that captures the meaning of that group.

The output of this exercise is a compilation of a maximum number of ideas under a limited number of major headings (see, for example Figure 13.14). This data can then be used with other tools to define areas for attack. One of these tools is the inter-relationship diagraph.

2 Inter-relationship diagraph _____

This tool is designed to take a central idea, issue or problem, and map out the logical or sequential links among related factors. While this still requires a very creative process, the inter-relationship diagraph begins to draw the logical connections that surface in the affinity diagram. In designing, planning, and problem solving it is obviously not enough just to create an explosion of ideas. The affinity diagram allows some organized creative patterns to emerge but the inter-relationship dia-graph lets *logical* patterns become apparent.

The diagraph is based on a principle that the Japanese frequently apply regarding the natural emergence of ideas. This tool starts therefore from a central concept, leads to the generation of large quantities of ideas, and finally to the delineation of observed patterns. To some this may appear to be like reading tealeaves, but it works incredibly well. Like the affinity diagram, the inter-relationship diagram allows unanticipated ideas and connections to rise to the surface.

The inter-relationship diagraph is adaptable to both specific operational issues and general organizational questions. For example, a classic use of this tool at Toyota focused on all the factors behind the establishment of a 'billboard system' as part of their JIT program. On the other hand, it has also been used to deal with issues underlying the problem of getting top management support for TQM.

In summary, the inter-relationship diagraph should be used when:

(a) An issue is sufficiently complex that the inter-relationship between ideas is difficult to determine.
(b) The correct sequencing of management actions is critical.
(c) There is a feeling or suspicion that the problem under discussion is only a symptom.
(d) There is ample time to complete the required reiterative process and define cause and effect.

The inter-relationship diagraph can be used by itself, or it can be used after the affinity diagram, using data from the previous effort as input. The steps for using this tool are:

1 Clearly define one statement that describes the key issue to be discussed. Record this statement on a card and place it on the wall or a table, in the center of a large sheet of paper. Mark this card in some way so that it can be easily identified as the central idea, e.g. use a double circle around the text.
2 Generate related issues or problems. This may be done in wide open brainstorming, or may be taken directly from an affinity diagram. Place each of the ideas on a card and place the cards around the central idea card.
3 Use arrows to indicate which items are related and what leads to what. Look for possible relationships between all items.
4 Look for patterns of arrows to determine key factors or causes. For example, if one card has seven arrows coming from it to other issues, the idea on that card is a key factor or cause. Mark these key areas in some way, e.g. by using a double box.
5 Use the key factors in a tree diagram for further analysis.

Figure 13.15 gives an example of a simple inter-relationship diagraph.

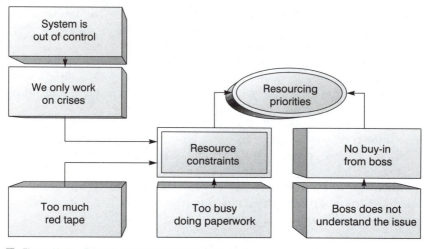

3 Systems flow/tree diagram

The systems flow/tree diagram (usually referred to as a tree diagram) is used to systematically map out the full range of activities that must be accomplished in order to reach a desired goal. It may also be used to identify all the factors contributing to a problem under consideration. As mentioned above, major factors identified by an inter-relationship diagraph can be used as inputs for a tree diagram. One of the strengths of this method is that it forces the user to examine the logical and chronological link between tasks. This assists in preventing a natural tendency to jump directly from goal or problem statement to solution (Ready … Fire … Aim!).

The tree diagram is indispensable when a thorough understanding of what needs to be accomplished is required, together with how it is to be achieved, and the relationships between these goals and methodologies. It has been found to be most helpful in situations when:

(a) Very ill-defined needs must be translated into operational characteristics, and to identify which characteristics can presently be controlled.
(b) All the possible causes of a problem need to be explored. This use is closest to the cause and effect diagram or fishbone chart.
(c) Identifying the first task that must be accomplished when aiming for a broad organizational goal.
(d) The issue under question has sufficient complexity and time available for solution.

Depending on the type of issue being addressed, the tree diagram will be similar to either a cause and effect diagram or a flowchart, although

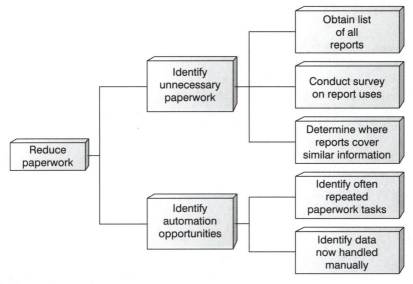

▓ **Figure 13.16**   An example of the tree diagram

it may be easier to interpret because of its clear linear layout. If a problem is being considered, each branch of the tree diagram will be similar to a cause and effect diagram. If a general objective is being considered, each branch may represent chronological activities, in which case the diagram will be similar to a flowchart. Although this tool is similar to other tools, suggestions on the stepwise procedure are included below. The procedure is based on trying to accomplish a goal, but it can be easily modified for use in problem solving:

1 Start with one statement that clearly and simply states the issue or goal. Write the idea on a card and place it on the left side of a flip chart or table.
2 Ask 'What method or task is needed to accomplish this goal or purpose?' Use the inter-relationship diagraph to find ideas that are most closely related to that statement, and place them directly to the right of the statement card.
3 Look at each of these 'second tier' ideas and ask the same question. Place these ideas to the right of the ones that they relate to. Continue this process until all the ideas are gone. Note: if none of the existing ideas on the inter-relationship diagraph can adequately answer the question, new ideas may be developed so as not to leave holes in the tree diagram.
4 Review the entire tree diagram by starting on the right and asking 'If this is done, will it lead to the accomplishment of the next idea or task?' The diagram produced will be similar to an organization chart.

An example is shown in Figure 13.16

4 Matrix diagrams _____

The matrix diagram is the heart of the seven new tools and the house of quality described in Chapter 6. The purpose of the matrix diagram is to outline the inter-relationships and correlations between tasks, functions or characteristics, and to show their relative importance. There are many versions of the matrix diagram, but the most widely used is a simple L-shaped matrix known as the *quality table*.

L-shaped matrix diagram

This is the most basic form of matrix diagram. In the L shape two inter-related groups of items are presented in line and row format. It is a simple two-dimensional representation that shows the intersection of related pairs of items as shown in Figure 13.17. It can be used to display relationships between items in *all* operational areas, including administration, manufacturing, personnel, R&D etc., to identify all the organizational tasks that need to be accomplished and how they should be allocated to individuals. In a QFD it is even more interesting if each person completes the matrix individually and then compares the coding with everyone in the work group.

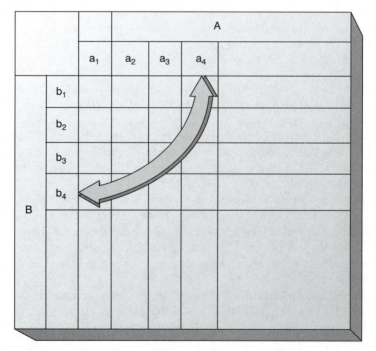

■ **Figure 13.17**   L-shaped matrix

Quality table

In a *quality table* customer demands (the whats) are analyzed with respect to substitute quality characteristics (the hows). See Figure 13.18. Correlations between the two are categorized as strong, moderate and possible. The customer demands shown on the left of the matrix are determined in co-operation with the customer. This effort requires a kind of a verbal 'ping-pong' with the customer to be truly effective: ask the customer what he wants, write it down, show it to him and ask him if that is what he meant, then revise and repeat the process as necessary. This should be done in a joint meeting with the customer, if at all possible. It is often of value to use a tree diagram to give structure to this effort.

The right side of the chart is often used to compare current performance to competitors' performance, company plan, and potential sales points with reference to the customer demands. Weights are given to these items to obtain a 'relative quality weight', which can be used to identify the key customer demands. The relative quality weight is then used with the correlations identified on the matrix to determine the key quality characteristics.

A modification that is added to create the house of quality table is a second matrix that explores the correlations between the quality characteristics. This is done so that errors caused by the manipulation of variables in a one-at-a-time fashion can be avoided. This also gives indications of where designed experiments would be of use in the design process. In the training required for use of this technique, several hours should be dedicated to a detailed explanation of the steps in the construction of a quality table, and the system to be used to compare numerically the various items.

T-shaped matrix diagram

A T-shaped matrix is nothing more than the combination of two L-shaped matrix diagrams. Figure 13.19 shows one application, the relationship between a set of courses in a curriculum and two important sets of

		Substitute quality characteristics							
	MFR	Ash	Importance	Current	Best competitor	Plan	IR	SP	RQW
No film breaks	◯ 17	▲ 6	4	4	4	4	1	◯	5.6
High rates	◉ 23		3	3	4	4	1.3		4.6
Low gauge variability	◉ 37	▲ 7	4	3	4	4	1.3	◯	7.3

(left label: Customer demands)

◉ Strong correlation IR Improvement ratio
◯ Some correlation SP Sales point
▲ Possible correlation RQW Relative quality weight

Figure 13.18 An example of the matrix diagram (quality table)

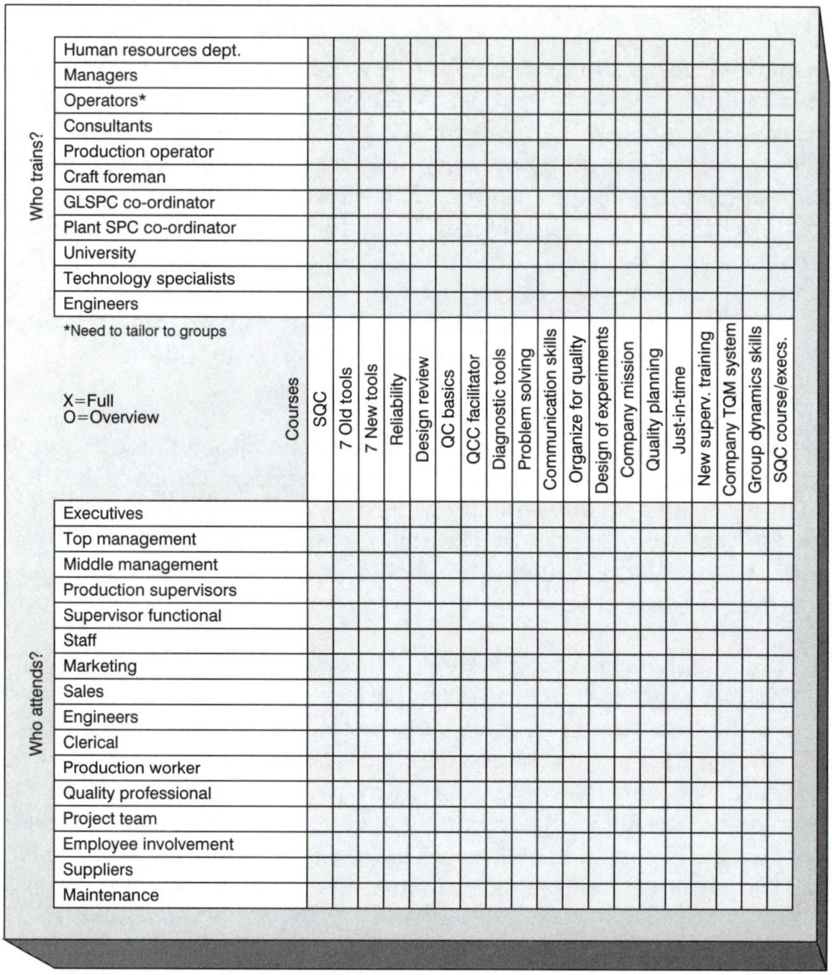

The matrix table shows "Who trains?" on the left (rows):
- Human resources dept.
- Managers
- Operators*
- Consultants
- Production operator
- Craft foreman
- GLSPC co-ordinator
- Plant SPC co-ordinator
- University
- Technology specialists
- Engineers

*Need to tailor to groups

X = Full
O = Overview

Courses (columns): SQC, 7 Old tools, 7 New tools, Reliability, Design review, QC basics, QCC facilitator, Diagnostic tools, Problem solving, Communication skills, Organize for quality, Design of experiments, Company mission, Quality planning, Just-in-time, New superv. training, Company TQM system, Group dynamics skills, SQC course/execs.

"Who attends?" (rows):
- Executives
- Top management
- Middle management
- Production supervisors
- Supervisor functional
- Staff
- Marketing
- Sales
- Engineers
- Clerical
- Production worker
- Quality professional
- Project team
- Employee involvement
- Suppliers
- Maintenance

Figure 13.19   T-matrix diagram on company-wide training

considerations: who should do the training for each course and which would be the most appropriate functions to attend each of the courses?

There are other matrices that deal with ideas such as product or service function, cost, failure modes, capabilities, etc., and there are at least 40 different types of matrix diagrams available.

5 Matrix data analysis

Matrix data analysis is used to take data displayed in a matrix diagram and arrange them so that they can be more easily viewed and show the strength of the relationship between variables. It is used most often in

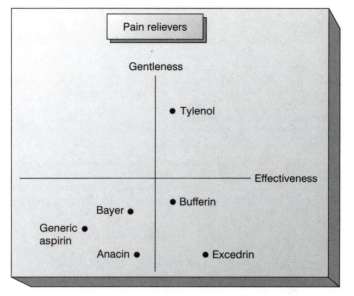

Figure 13.20   An example of matrix data analysis

marketing and product research. The concept behind matrix data analysis is fairly simple, but its execution (including data gathering) is complex.

A good idea of the uses and value of the construction of a chart for matrix data analysis may be shown in a simple example in which types of pain relievers are compared based on gentleness and effectiveness (Figure 13.20). This information could be used together with some type of demographic analysis to develop a marketing plan. Based on the information, advertising and product introduction could be effectively tailored for specific areas. New product development could also be carried out to attack specific niches in markets that would be profitable.

6 Process decision program chart

A process decision program chart (PDPC) is used to map out each event and contingency that can occur when progressing from a problem statement to its solution. The PDPC is used to anticipate the unexpected and plan for it. It includes plans for counter-measures on deviations. The PDPC is related to a failure mode and effect analysis and its structure is similar to that of a tree diagram. (An example of the PDPC is shown in Figure 13.21.) Suggested steps for constructing a PDPC are as follows:

1 Construct a tree diagram as described previously.
2 Take one major branch of the tree diagram (an item just to the right of the main goal or purpose). Ask 'What could go wrong at this step?' or 'What other path could this step take?'

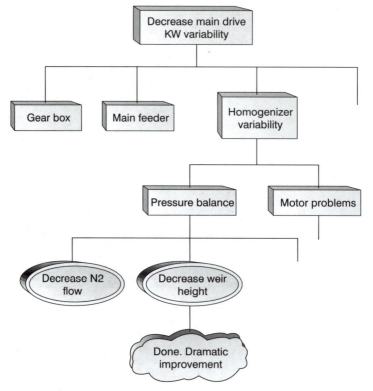

■ **Figure 13.21**   Process decision program chart

3 Answer the questions by branching off the original path in 'organization chart' manner.
4 Off to the side of each step, list actions or counter-measures that
could be taken.
5 Continue the process until the branch is exhausted.
6 Repeat with other main branches.

The PDPC is very simply an attempt to be proactive in the analysis of
failure and to construct, on paper, a 'dry run' of the process so that the
'check' part of the improvement cycle can be defined in advance. PDPC
is likely to enjoy widespread use because of increasing attention to
product liability.

7 Arrow diagram

The arrow diagram is used to plan or schedule a task. To use it, one
must know the sub-task sequence and duration. This tool is essentially
the same as the standard Gantt chart shown in Figure 13.22. Figure 13.23
is the same sequence shown as an arrow diagram. Although it is a simple

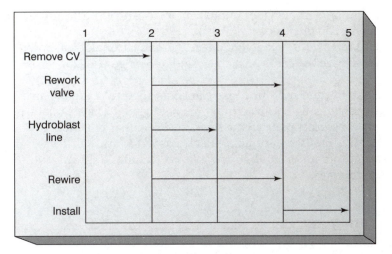

Figure 13.22 Gantt chart

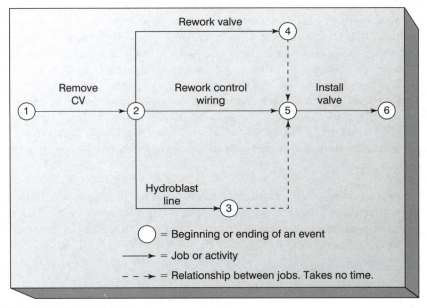

Figure 13.23 The arrow diagram

and well known tool for planning work, it is surprising how often it is ignored. The arrow diagram is useful in analyzing a repetitive job in order to make it more efficient. Some suggestions on constructing arrow diagrams are:

1 Use a team of people working on a job or project, e.g. a QFD team, to list all the tasks necessary to complete the job, and write them on

individual cards. On the bottom half of the card, write the time required to complete the task.

2 Place related tasks together. Place them in chronological order.

3 Summarize the cards on a chart similar to Figure 13.23.

Refinements and modifications can be applied to make the arrow diagram more detailed or to account for contingency. The technique is used widely on project planning, where it is known as critical path analysis (CPA). It is easily computerized and has led to further developments, such as program evaluation and review technique (PERT) and PRINZ methodologies.

Summary

What has been described in this section is a system for improving the design of products, processes, and services by means of seven 'new' qualitative tools. For the most part the seven tools are neither new nor revolutionary, but rather a compilation and modification of some methods that have been around for a long time. These tools do not replace statistical methods or other techniques, but they are meant to be used together as part of continuous process improvement.

The tools work best when representatives from all parts of an organization take part in their use and execution of the results. Besides the structure that the tools provide, the co-operation between functions or departments that is required will help break down barriers within the organizations.

While design, marketing and operations people will see the most direct applications for these tools, proper use of the 'philosophy' behind them requires participation from all parts of an organization. In addition, some of the seven new tools can be used in direct problem-solving activities.

■ Taguchi methods for process improvement

Genichi Taguchi was a noted Japanese engineering specialist who advanced 'quality engineering' as a technology to reduce costs and improve quality simultaneously. The popularity of Taguchi methods today testifies to the merit of his philosophies on quality. The basic elements of Taguchi's ideas, which have been extended here to all aspects of product, service and process quality, may be considered under four main headings.

1 Total loss function _____

An important aspect of the quality of a product or service is the total loss to society that it generates. Taguchi's definition of product quality as 'the loss imparted to society from the time a product is shipped is rather strange, since the word *loss* denotes the very opposite of what is normally conveyed by using the word *quality*. The essence of his definition is that the smaller the loss generated by a product or service from the time it is transferred to the customer, the more desirable it is.

The main advantage of this idea is that it encourages a new way of thinking about investment in quality improvement projects, which become attractive when the resulting savings to customers are greater than the cost of improvements.

Taguchi claims with some justification that any variation about a target value for a product or process parameter causes loss to the customer. The loss may be some simple inconvenience, but it can represent actual cash losses, owing to rework or badly fitting parts, and it may well appear as loss of customer goodwill and eventually market share. The loss (or cost) increases exponentially as the parameter value moves away from the target, and is at a minimum when the product or service is at the target value.

2 Design of products, services and processes _____

In any product or service development, three states may be identified: product or service design, process design, and production or operations. Each of these overlapping stages has many steps, the output of one often being the input to others. The output/input transfer points between steps clearly affect the quality and cost of the final product or service. The complexity of many modern products and services demands that the crucial role of design be recognized. Indeed the performance of the quality products from the Japanese automotive, banking, camera, and machine tool industries can be traced to the robustness of their product and process designs.

The prevention of problems in using products or services under varying operations and environmental conditions must be built in at the design stage. Equally, the costs during production or operation are determined very much by the actual manufacturing or operating process. Controls, including SPC methods, added to processes to reduce imperfections at the operational stage are expensive, and the need for controls *and* the production of non-conformance can be reduced by correct initial designs of the process itself.

Taguchi distinguished between *off-line* and *on-line* quality control methods, 'quality control' being used here in the very broad sense to include quality planning, analysis and improvement. Off-line QC uses technical aids in the *design* of products and processes, whereas on-line methods are technical aids for controlling quality and costs in the *production* of products or services. Too often the off-line QC methods focus on evaluation rather than improvement. The belief by some people (often based on experience!) that it is unwise to buy a new model or a motor car 'until the problems have been sorted out' testifies to the fact that insufficient attention is given to improvement at the product and process design stages. In other words, the bugs should be removed *before* not after product launch. This may be achieved in some organizations by replacing detailed quality and reliability evaluation methods with approximate estimates, and using the liberated resources to make improvements.

3 Reduction of variation

The objective of a continuous quality improvement program is to reduce the variation of key products' performance characteristics about their target values. The widespread practice of setting specifications in terms of simple upper and lower limits conveys the wrong idea that the customer is satisfied with all values inside the specification band, but is suddenly not satisfied when a value slips outside one of the limits. The practice of stating specifications as tolerance intervals only can lead manufacturers to produce and despatch goods whose parameters are just inside the specification band. Owing to the interdependence of many parameters of component parts and assemblies, this is likely to lead to quality problems.

The target value should be stated and specified as the ideal, with known variability about the mean. For those performance characteristics that cannot be measured on the continuous scale, the next best thing is an ordered categorical scale such as excellent, very good, good, fair, unsatisfactory, very poor, rather than the binary classification of 'good' or 'bad' that provides meagre information with which the variation reduction process can operate.

Taguchi has introduced a three-step approach to assigning nominal values and tolerances for product and process parameters:

(a) System design – the application of scientific engineering and technical knowledge to produce a basic functional prototype design. This requires a fundamental understanding of the needs of the customer *and* the production environment.

(b) Parameter design – the identification of the settings of product or process parameters that reduce the sensitivity of the designs to sources of variation. This requires a study of the whole process system design to achieve the most robust operational settings, in terms of tolerance to ranges of the input variables.
(c) Tolerance design – the determination of tolerance around the nominal settings identified by parameter design. This requires a tradeoff between the customer's loss due to performance variation and the increase in production or operational costs.

4 Statistically planned experiments

Taguchi has pointed out that statistically planned experiments should be used to identify the settings of product and process parameters that will reduce variation in performance. He classifies the variables that affect the performance into two categories: design parameters and sources of 'noise'. As we have seen earlier, the nominal settings of the *design parameters* define the specification for the product or process. The *sources of noise* are all the variables that cause the performance characteristics to deviate from the target values. The *key* noise factors are those that represent the major sources of variability, and these should be identified and included in the experiments to design the parameters at which the effect of the noise factors on the performance is minimum. This is done by systematically varying the design parameter settings and comparing the effect of the noise factors for each experimental run.

Statistically planned experiments may be used to identify:

(a) The design parameters that have a large influence on the product or performance characteristic.
(b) The design parameters that have no influence on the performance characteristics (the tolerances of these parameters may be relaxed).
(c) The settings of design parameters at which the effect of the sources of noise on the performance characteristic is minimal.
(d) The settings of design parameters that will reduce cost without adversely affecting quality.[1]

Taguchi methods have stimulated a great deal of interest in the application of statistically planned experiments to product and process designs. The use of 'design of experiments' to improve industrial products and processes is not new – Tippett used these techniques in the textile industry more than 50 years ago. What Taguchi has done, however, is to acquaint us with the scope of these techniques in off-line quality control.

Taguchi's methods, like all others, should not be used in isolation, but be an integral part of continuous improvement.

■ Six sigma

Since the early 1980s, most of the world has been in what the author has called a 'quality revolution'. Based on the simple premise that organizations of all kinds exist mainly to serve the needs of the customers of their products or services, good quality management has assumed great importance. Competitive pressures on companies and Government demands on the public sector have driven the need to find more effective and efficient approaches to managing businesses and non-profit-making organizations.

In the early days of the realization that improved quality was vital to the survival of many companies, especially in manufacturing, senior managers were made aware, through national campaigns and award programs, that the basic elements had to be right. They learned through adoption of quality management systems, the involvement of improvement teams and the use of quality tools, that improved business performance could be achieved only through better planning, capable processes and the involvement of people. These are the basic elements of a TQM approach and this has not changed no matter how many sophisticated approaches and techniques come along.

The development of TQM has seen the introduction and adoption of many dialects and components, including quality circles, international systems and standards, statistical process control, business process re-engineering, lean manufacturing, continuous improvement, benchmarking and business excellence.

An approach finding favor in some companies is six sigma, most famously used in Motorola, General Electric, and Allied Signal. This operationalized TQM into a project-based system, based on delivering tangible business benefits, often directly to the bottom line. Strange combinations of the various approaches have led to Lean Sigma and other company specific acronyms such as 'Statistically Based Continuous Improvement (SBCI)'.

The six-sigma improvement model _____

There are five fundamental phases or stages in applying the six-sigma approach to improving performance in a process: Define, Measure, Analyze, Improve, and Control (DMAIC). These form an improvement cycle grounded in Deming's original Plan, Do, Check, Act (PDCA), (Figure 13.24). In the six-sigma approach, DMAIC provides a breakthrough strategy and disciplined methods of using rigorous data

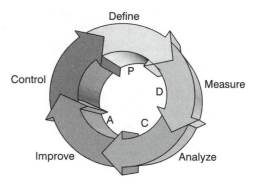

Figure 13.24
The six-sigma improvement
model – DMAIC

Table 13.1 The DMAIC steps

D	Define the scope and goals of the improvement project in terms of customer requirements and the process that delivers these requirements – inputs, outputs, controls and resources.
M	Measure the current process performance – input, output and process – and calculate the short- and longer-term process capability – the sigma value.
A	Analyze the gap between the current and desired performance, prioritize problems and identify root causes of problems. Benchmarking the process outputs, products or services against recognized benchmark standards of performance may also be carried out.
I	Generate the improvement solutions to fix the problems and prevent them from recurring so that the required financial and other performance goals are met.
C	This phase involves implementing the improved process in a way that 'holds the gains'. Standards of operation will be documented in systems such as ISO 9000 and standards of performance will be established using techniques such as statistical process control (SPC).

gathering and statistically based analysis to identify sources of errors and ways of eliminating them. It has become increasingly common in so-called 'six-sigma organizations' for people to refer to 'DMAIC projects'. These revolve around the three major strategies for processes to bring about rapid bottom-line achievements – design/redesign, management and improvement.

Table 13.1 shows the outline of the DMAIC steps.

Building a six-sigma organization and culture

Six-sigma approaches question many aspects of business, including its organization and the cultures created. The goal of most commercial organizations is to make money through the production of saleable goods or services and, in many, the traditional measures used are capacity or throughput based. As people tend to respond to the way they are being measured, the management of an organization tends to get what it measures. Hence, throughput measures may create work-in-progress and finished goods inventory thus draining the business of cash and working capital. Clearly, supreme care is needed when defining what and how to measure.

Six-sigma organizations focus on:

■ understanding their customers' requirements;
■ identifying and focusing on core-critical processes that add value to customers;
■ driving continuous improvement by involving all employees;
■ being very responsive to change;
■ basing managing on factual data and appropriate metrics;
■ obtaining outstanding results, both internally and externally.

The key is to identify and eliminate variation in processes. Every process can be viewed as a chain of independent events and, with each event subject to variation, variation accumulates in the finished product or service. Because of this, research suggests that most businesses operate somewhere between the three- and four-sigma level. At this level of performance, the real cost of quality is about 25–40 percent of sales revenue. Companies that adopt a six-sigma strategy can readily reach the five sigma level and reduce the cost of quality to 10 percent of sales. They often reach a plateau here and to improve to six-sigma performance and 1 percent cost of quality takes a major rethink.

Properly implemented six-sigma strategies involve:

■ leadership involvement and sponsorship;
■ whole organization training;
■ project selection tools and analysis;
■ improvement methods and tools for implementation;
■ measurement of financial benefits;
■ communication;
■ control and sustained improvement.

One highly publicized aspect of the six-sigma movement, especially in its application in companies such as General Electric (GE), Motorola, Allied Signal and GE Capital in Europe, is the establishment of process improvement experts, known variously as 'Master Black Belts', 'Black Belts' and 'Green Belts'. In addition to these martial arts-related characters, who perform the training, lead teams and do the improvements, are other roles which the organization may consider, depending on the seriousness with which they adopt the six-sigma discipline. These include the:

- leadership group or council/steering committee;
- sponsors and/or champions/process owners;
- implementation leaders or directors – often master black belts;
- six-sigma coaches – master black belts or black belts;
- team leaders or project leaders – black belts or green belts;
- team members – usually green belts.

Many of these terms will be familiar from TQM and continuous improvement activities. The 'Black Belts' reflect the finely honed skill and discipline associated with the six-sigma approaches and techniques. The different levels of Green, Black and Master Black Belts recognize the depth of training and expertise.

Mature six-sigma programs, such as at GE, Johnson & Johnson and Allied Signal, have about 1 percent of the workforce as full-time Black Belts. There is typically one Master Black Belt to every ten Black Belts or about one to every 1000 employees. A Black Belt typically oversees/ completes five to seven projects per year, which are led by Green Belts who are not employed full time on six-sigma projects (Figure 13.25).

The means of achieving six-sigma capability are, of course, the key. At Motorola this included millions of dollars spent on a company-wide education program, documented quality systems linked to quality goals, formal processes for planning and achieving continuous improvements, individual QA organizations acting as the customer's advocate in all areas of the business, a corporate quality council for co-ordination, promotion, rigorous measurement and review of the various quality systems/ programs to facilitate achievement of the policy.

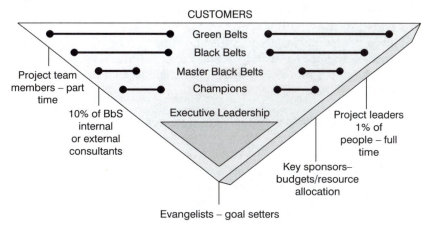

Figure 13.25 A six-sigma company

Ensuring the financial success of six-sigma projects _____

Six-sigma approaches are not looking for incremental or 'virtual' improvements, but breakthroughs. This is where six sigma has the potential to outperform other improvement initiatives. An intrinsic part of implementation is to connect improvement to bottom-line benefits and projects should not be started unless they plan to deliver significantly to the bottom line.

Estimated cost savings vary from project to project, but reported average results range from $100 000 to $150 000 per project, which typically last four months. The average Black Belt will generate $500 000–1 000 000 benefits per annum, and large savings are claimed by the leading exponents of six sigma. For example, GE has claimed returns of $1.2 bn from its investment of $450 m.

Six-sigma project selection takes on different faces in different organizations. While the overall goal of any six-sigma project should be to improve customer results and business results, some projects will focus on production/service delivery processes, and others will focus on business/commercial processes. Whichever they are, all six-sigma projects must be linked to the highest levels of strategy in the organization and be in direct support of specific business objectives. The projects selected to improve business performance must be agreed upon by both the business and operational leadership, and someone must be assigned to 'own' or be accountable for the projects, as well as someone to execute them.

At the business level, projects should be selected based on the organization's strategic goals and direction. Specific projects should be aimed at improving such things as customer results, non-value add, growth, cost and cash flow. At the operations level, six-sigma projects should still tie to the overall strategic goals and direction but directly involve the process/operational management. Projects at this level then should focus on key operational and technical problems that link to strategic goals and objectives.

When it comes to selecting six-sigma projects, key questions which must be addressed include:

- what is the nature of the projects being considered?
- what is the scope of the projects being considered?
- how many projects should be identified?
- what are the criteria for selecting projects?
- what types of results may be expected from six-sigma projects?

Project selection can rely on a 'top-down' or 'bottom-up' approach. The top-down approach considers a company's major business issues and objectives and then assigns a champion – a senior manager most affected by these business issues – to broadly define the improvement objectives, establish performance measures, and propose strategic improvement projects with specific and measurable goals that can be met in a given time period. Following this, teams identify processes and critical-to-quality characteristics, conduct process baselining, and identify opportunities for improvement. This is the favored approach and the best way to align 'localized' business needs with corporate goals.

At the process level, six-sigma projects should focus on those processes and critical-to-quality characteristics that offer the greatest financial and customer results potential. Each project should address at least one element of the organization's key business objectives, and be properly planned.

Concluding observations and links with TQM, SPC excellence, etc.

Six sigma is not a new technique, its roots can be found in TQM and SPC but it is more than TQM or SPC rebadged. It is a framework within which powerful TQM and SPC tools can be allowed to flourish and reach their full improvement potential. With the TQM philosophy, many practitioners promised long-term benefits over 5–10 years, as the programs began to change hearts and minds. Six sigma is about delivering breakthrough benefits in the short term and is distinguished from TQM by the intensity of the intervention and pace of change.

Excellence approaches such as the EFQM Excellence Model and six sigma are complementary vehicles for achieving better organizational performance. The Excellence Model can play a key role in the baselining phase of strategic improvement, while the six-sigma breakthrough strategy is a delivery vehicle for achieving excellence through:

1 Committed leadership.
2 Integration with top level strategy.
3 A cadre of change agents – Black Belts.
4 Customer and Market focus.
5 Bottom-line impact.
6 Business process focus.
7 Obsession with measurement.
8 Continuous innovation.
9 Organizational learning.
10 Continuous reinforcement.

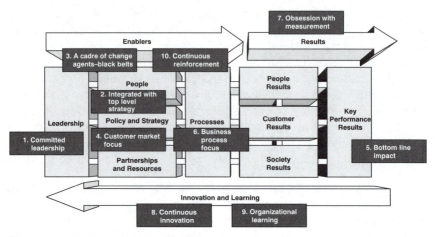

■ **Figure 13.26**   The Excellence Model and six sigma

These are 'mapped' onto the Excellence Model in Figure 13.26 (see also 'Six Sigma Excellence' by Les Porter, *Quality World*, April 2002, pp. 12–15[4]).

There is a whole literature and many conferences have been held on the subject of six sigma and it is not possible here to do justice to the great deal of thought that has gone into the structure of these approaches. As with Taguchi methods the major contribution of six sigma has not been in the creation of new technology or methodologies, but in bringing to the attention of senior management the need for a disciplined, structured approach and their commitment, if real performance and bottom-line improvements are to be achieved.

Technical note

Sigma is a statistical unit of measurement that describes the distribution about the mean of any process or procedure. A process or procedure that can achieve plus or minus *six-sigma* capability can be expected to have a defect rate of no more than a few parts per million, even allowing for some shift in the mean. In statistical terms, this approaches *zero defects*.

In a process in which the characteristic of interest is a variable, defects are usually defined as the values which fall outside the specification limits. Assuming and using a normal distribution of the variable, the percentage and/or parts per million defects can be found. For example, in a centered process with a specification set average ±3 sigma there will be 0.27 percent or 2700 ppm defects. This may be referred to as 'an unshifted ±3 sigma process' and the quality called '±3 sigma quality'. In an 'unshifted ±6 sigma process', the specification range is average 6 sigma and it produces only 0.002 ppm defects.

It is difficult in the real world, however, to control a process so that the mean is always set at the nominal target value – in the center of the specification. Some shift in the process mean is expected. The ppm defects produced by such a 'shifted process' are the sum of the ppm outside each specification limit, which can be obtained from the normal distribution.

■ The 'DRIVE' framework for continuous improvement

The author and his colleagues have developed a framework for a structured approach to problem solving in teams, the *DRIVE* model. The mnemonic provides landmarks to keep the team on track and in the right direction.

<u>D</u>efine   the problem. *Output*: written definition of the task and its success criteria.

<u>R</u>eview   the information. *Output*: presentation of known data and action plan for further data.

<u>I</u>nvestigate   the problem. *Output*: documented proposals for improvement and action plans.

<u>V</u>erify   the solution. *Output*: proposed improvements that meet success criteria.

<u>E</u>xecute   the change. *Output*: task achieved and improved process documented.

The various stages are discussed in detail below. Some of the steps may be omitted if they have already been answered or are clearly not relevant to a particular situation.

Define

At this stage the improvement team is concerned with gaining a common understanding and agreement within the groups of the task that it faces, in terms of the problem to be solved and the boundaries of the process or processes that contain it. It is necessary to generate at the outset a means of knowing when the team has succeeded. There is no concern at this stage with solutions. The key steps are:

1 *Look at the task*
   Typical questions:
   (a) What is the brief?
   (b) Is it understood?
   (c) Is there agreement with it?
   (d) Is it sufficiently explicit?
   (e) Is it achievable?

There may be a need for clarification with the 'sponsor' at this stage, and possibly some redefinition of the task.

2 *Understand the process*
   (a) What processes 'contain' the problem?
   (b) What is wrong at present?
   (c) Brainstorm – ideas for improvement.
   (d) Perhaps draw a rough flowchart to focus thinking.

3 *Prioritize*
   (a) Set boundaries to the investigation.
   (b) Make use of ranking, Pareto, matrix analysis, etc., as appropriate.
   (c) Review and gain agreement in the team of what is 'do-able'.

4 *Define the task*
   (a) Produce a written description of the process or problem area that can be confirmed with the team's sponsor.
   (b) Confirm agreement in the team.
   (c) This step may generate further questions for clarification by the sponsor of the process.

5 *Agree success criteria*
   (a) List possible success criteria. How will the team know when it has succeeded?
   (b) Choose and agree success criteria in the team.
   (c) Discuss and agree time scales for the project.
   (d) Agree with 'sponsor'.
   (e) Document the task definition, success criteria and time scale for the complete project.

Review

This stage is concerned with finding out what information is already available, gathering it together, structuring it, identifying what further information might be needed, and agreeing in the team WHAT is needed, HOW it is going to be obtained, and WHO is going to get it.

1 *Gather existing information*
   (a) Locate sources – verbal inputs, existing files, charts, quality records, etc.
   (b) Go and collect, ask, investigate.

2 *Structure information*
   Information may be available but not in the right format.

3 *Define gaps*
   (a) Is enough information available?
   (b) What further information is needed?
   (c) What equipment is affected?

(d) Is the product/service from one plant or area?

(e) How is the product/service at fault?

4 *Plan further data collection*

(a) Use any data already being collected.

(b) Draw up check sheet(s).

(c) Agree data-collection tasks in the team – WHO, WHAT, HOW, WHEN.

(d) Seek to consult others, where appropriate. Who actually has the information? Who really understands the process?

(e) This is a good opportunity to start to 'extend the team' in preparation for the *Execute* stage later on.

Investigate

This stage is concerned with analyzing all the data, considering all possible improvements, and prioritizing these to come up with one or more solutions to the problem, or improvements to the process, which can be verified as being the answer which meets the success criteria.

1 *Implement data-collection action plan*

Check at an early stage that the plan is satisfying the requirements.

2 *Analyze data*

(a) What picture is the data painting?

(b) What conclusions can be drawn?

(c) Use all appropriate tools to give a clearer picture of the process.

3 *Generate potential improvements*

(a) Brainstorm improvements.

(b) Discuss all possible solutions.

(c) Write down all suggestions (have there been any from outside the team?).

4 *Agree proposed improvements*

(a) Prioritize possible proposals.

(b) Decide what is achievable in what time scales.

(c) Work out how to test proposed solution(s) or improvement(s).

(d) Design check sheets to collect all necessary data.

(e) Build a checking/verifying plan of action.

Verify

This stage is concerned with testing the plans and proposals to make sure that they work before any commitment to major process changes. This may require a relatively short discussion round a table in a meeting or lengthy pilot trials in a laboratory, office or even a main operations area or production plant.

1 *Implement action plan*
   Carry out the agreed tests on the proposals.
2 *Collect data*
   (a) Consider the use of questionnaires, if appropriate.
   (b) Make sure the check sheets are accumulating the data properly.
3 *Analyze data*
4 *Verify that success criteria are met*
   (a) Compare performance of new or changed process with success
       criteria from *Define* stage.
   (b) If success criteria are not met, return to appropriate stage in
       drive model (usually the *Investigate*) stage.
   (c) Continue until the success criteria have been met. For difficult
       problems, it may be necessary to go a number of times round
       this loop.

Execute

This stage is concerned with selling the solution or process improvement to others, e.g. the process owner, who may not have taken part in
the investigation but whose commitment is vital to ensure success. Part
of this stage may well be the need to address the existing documented
quality management system.

1 *Develop implementation plan to gain commitment*
   (a) Is there commitment from others? Consider all possible impacts.
   (b) Actions?
   (c) Timing?
   (d) Selling required?
   (e) Training required for new or modified process?
2 *Review appropriate system paperwork/documentation*
   (a) Who should do this? The team? The activity/process owner?
   (b) What are the implications for other systems?
   (c) What controlled documents are affected?
3 *Gain agreement to all facets of the execution plan from the process owner.*
4 *Implement the plan.*
5 *Monitor success*
   (a) Extent of original team involvement? Initially perhaps and then
       at intervals?
   (b) 'Delegate' to process owner/department concerned? At what
       stage?
6 *Responsibility*
   Balance between team taking responsibility for meeting its agreed
   project success criteria and ownership within the organization of

Table 13.2 Likely tools in the DRIVE model

	Define	Review	Investigate	Verify	Execute
Brainstorming	●	●	●		
Cause and effect diagram	●				
Pareto	●	●		●	●
Matrix analysis	●	●	●		
Check sheets		●	●	●	●
Flowcharts		●	●		●
Force field		●			●
Scatter diagram		●		●	
Histograms		●			
Charts		●	●	●	●
Project bar chart				●	●

processes and continuous improvement. In the case of registered firms, responsibility for continued monitoring can be delegated to the quality management.

The problem-solving tools most likely to be used at each of the DRIVE stages are shown in Table 13.2.

An example of the DRIVE model used in practice ____

The example below shows how the DRIVE model for a particular project worked out in practice. The sort of responses made by the team are given in quotes thus ' '.

Stated task: 'We lose orders because our response time in making quotations is too long. We must significantly reduce our response time.'

1 *Define stage*
   (a) Look at the task:
       (i)   Can we accept the problem as stated? 'Yes, this is generally known to be a problem area.'
       (ii) Does this apply to all customers, or to specific product lines? 'All customers.'
       (iii) Does anyone measure response time at the moment? 'There was a one-off assessment a long time ago but it is not routinely measured.'
       (iv) Have we the right expertise in our team to tackle it? 'Not really sure until we understand the problem.'

(v) Can we succeed with this problem? 'Yes, if we don't let it get too big.'

(b) Understand the process:

(i) What process 'contain' this problem? 'Customer visits by sales reps, telephone enquiries system, telex/fax enquiries, development department (technical vetting and provision of samples) pricing department', etc.

(ii) What is wrong at present? 'No real liaison between departments, labs have other priorities, visit reports are not explicit.'

(c) Prioritize:

(i) What should be the boundaries of our investigation? 'Enquiries arising from direct sales visits to customers compose about 70 percent of all enquiries. We will restrict our project to this area initially.'

(d) Define the task

(e) Agree success criteria

(f) In response to (d) and (e) the team finally documented:

'Customer quotations project:

Our task is to investigate and reduce delays in the handling of those customer enquiries which arise from direct visits by our sales force. The project will be conducted in three phases:

1 (a) To establish the average time (in days) between the salesperson's visit and receipt by the customer of our quotation.

(b) To agree a target reduction in the average response time to enquiries.

2 To make recommendations that will enable the target reduction in response time to be achieved.

3 To implement the recommendations and monitor response times on a sample basis to demonstrate that the desired reduction has been achieved.

Milestones for the completion of each phase, measured from the formal go-ahead date for these proposals, are
Phase 1–1 month
Phase 2–4 months
Phase 3–8 months'

2 *Review stage*

The team was not able to locate the original study report, but did discover a memo that gave the following summary:

■ Average time to process a quotation request: 17 days
■ Average time of quotations judged 'too late': 20 days
■ Percentage of quotations 'too late': 30 percent

From this, the team concluded that the average time would have to be reduced to about nine days to give a frequency of exceeding 20 days of only 1 in 100, i.e. only 1 percent of quotations 'too late'.

3 *Investigate stage*

The team constructed flowcharts of the various stages of the process. Major 'gray areas' occurred in dealing with quotations for new(er) products or new customers where more technical letting by the laboratory was required, because (a) customer requirements were not clear, and (b) salesmen were not authorized to offer new specifications without checking with the technical department. It was in these areas that the sales visit reports were not sufficiently specific.

The flowchart of the process was changed to include a path giving early warning to the technical department of requirements from 'new business areas', by identifying specifically named customers and product types. Quotations in these areas were treated with priority.

A check sheet to measure quotation turn-round times was designed.

4 *Verify stage*
- The modified process was implemented.
- The check sheet was used to gather data.
- A 'c chart' was used to monitor the average turn-round time in days for each week's orders, with the new procedure introduced at week 10. Action and warning lines for the chart were based on a target average of nine days. The ensuing chart is shown in Figure 13.27.

5 *Execute stage*
- The above data was presented to a meeting of the sales and technical departments.
- The changed procedures were agreed, documented and circulated, including the list of current customers and products for special vetting.
- A procedure, which was documented, called for *all* quotations for new customers and products to be added to the special list and

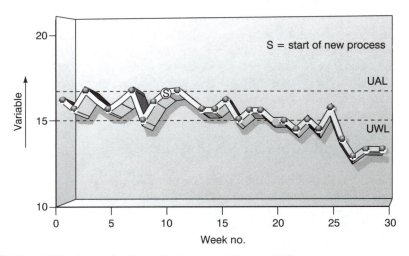

Figure 13.27   Charting the effects of the improvement through DRIVE

retained until familiarity enabled them to become 'standard'. They were then removed from the list for special attention.

■ Continued monitoring, using the chart in Figure 13.16, showed the average turn-round time reduce eventually to ten days. Only 2 percent of quotations were then taking longer than 20 days.

The DRIVE framework has been found invaluable in structuring the thinking and approach by numerous situations including large project management. Like the six-sigma DMAIC model, it helps groups of people cover all aspects of the continuous improvement cycle.

Case study ■■■ ▬▬▬▬▬▬▬▬▬▬▬▬▬▬▬▬▬

■ Quality approach to major change in BT Retail

A highlight of the quality program in BT Retail has been the development of the 'Performance Accelerator' 'a clear systematic ten step framework for complex change management (Figure 13.28). Developed specifically for BT Retail, Performance Accelerator is a unique methodology drawing on BT's deep understanding of business excellence and supports this by integrating elements of other proven effective change methodologies such as six sigma. A suite of quality tools underpins each of the ten steps.

The ten steps require a high degree of rigor with a clear, fact-based approach that ensures the program thoroughly takes account of customer, business and employee views (together with benchmarks) in defining its goals. A comprehensive analysis of root cause is also required. The approach is used on all of BT Retail's key programs.

■ SQF tools and techniques

In order to achieve a sustained and consistent approach in using the Shell Quality Framework (SQF) it was recognized that a set of supporting tools and techniques was required, which together would allow the full benefits to be realized in achieving improved business performance. This had to be in place in order to begin to address complex, outdated and broken processes across the organization. This was the genesis of the Business Improvement System shown in Figure 13.29.

Improvement tools and techniques _____

In order to operationalize this system a structured approach to process improvement was developed around the DRIVER methodology. This provided the tools and techniques to analyze business problems in detail. It also provided a systematic way of identifying and implementing solutions and supported the use of tools such as affinity diagrams, cause and effect analysis, force field analysis, metaplanning, and Pareto analysis. The six steps within DRIVER are outlined in Table 13.3.

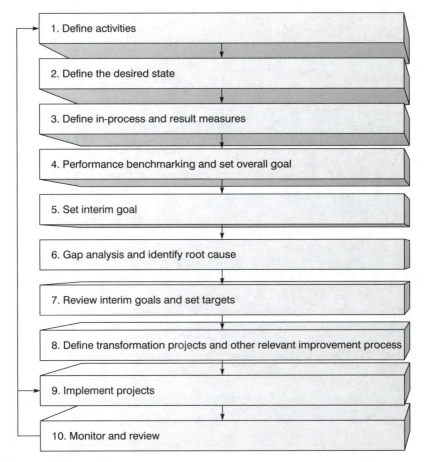

1. Define activities

2. Define the desired state

3. Define in-process and result measures

4. Performance benchmarking and set overall goal

5. Set interim goal

6. Gap analysis and identify root cause

7. Review interim goals and set targets

8. Define transformation projects and other relevant improvement process

9. Implement projects

10. Monitor and review

Figure 13.28   BT's 'Performance Accelerator'

Process mapping guidelines

This provided the ability to understand a process by graphical analysis. This was seen as necessary to build a common understanding of issues affecting the process in question, as well as a way of dealing with purpose, inputs, ouputs, resources, controls and interfaces. Such maps were invaluable to understand the 'as is' position when process redesign is appropriate.

Process management guide

This is where the SQF really came into its own. At level 4 it provided a clear definition of process operation reflecting best practice, and essentially provided the bedrock on which new processes were defined. This forced issues to be considered such as the operation of processes under controlled conditions, the concept of process ownership, and starting to think about performance measurement.

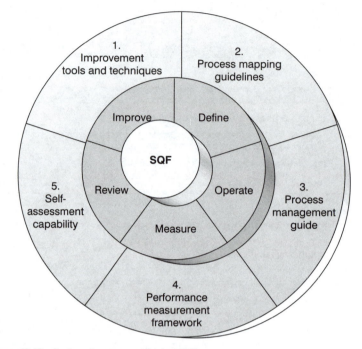

Figure 13.29   Business Improvement System (BIS)

Table 13.3

Stage		Output
D	Define scope of process improvement	Purpose, scope and success criteria
R	Review 'as is' process	Defined process, problems and performance
I	Identify 'should be' process	Defined improvement (process, problem prevention and performance)
V	Verify improvements	Prioritized and verified improvements to deliver benefit
E	Execute improvement delivery	Deployed improvements
R	Review implementation	Measure of improvement benefit

Performance measurement framework

This was probably the most difficult aspect of the whole Business Improvement System to implement. A set of project templates were developed that allowed the selection of outcome metrics for any standard business process. Properly assembled this provided, for the first time, a regime of measurement and tracking for key processes, aimed at achieving the 'perfect transaction' in spirit if not in practice. Considering the results of the original baseline study referred to above, this was the area most in need of improvement.

Self-assessment capability

The intention of this component was twofold: first, to provide an ongoing self-assessment capability where templates, questionnaires, scoring sheets and action plans were provided, based on the SQF; second, to allow those parts of the organization for whom accreditation and recognition was important to use the SQF and the Business Improvement System to test their readiness for achieving their goals.

Once these tools and techniques had been developed and integrated into the Business Improvement System, two pilots were undertaken to validate and fine tune the approach. The first was improving the accounting processes around joint ventures in Exploration & Production. The result here was a simplified and shortened end-to end process, resources released for more value-added tasks and higher customer satisfaction levels. The second pilot addressed problem management in the IS service delivery organization. Again, by looking afresh at the process, particularly at performance tracking, the end result was a global, stable process with defined service levels and performance tracking – and happier customers.

References

1. Oakland, J.S., *Statistical Process Control*, 5th edition, Butterworth-Heinemann, Oxford, 2002.
2. Checkland, P., *Soft Systems Methodology in Action*, Wiley, 1990.
3. This system for process capability and control is based on Frank Price's very practical framework for thinking about quality in manufacturing:
   Can we make it OK?
   Are we making it OK?
   Have we made it OK?
   Could we make it better?
   Which he presented in this excellent book *Right First Time*. Gower, London, 1985.
4. Porter, L., 'Six Sigma Excellence', *Quality World* (IQA–London), April 2002, pp. 12–15.

Chapter highlights

A systematic approach

- Numbers and information will form the basis for understanding, decisions, and actions in never-ending improvement – record data, use/analyze data, act on results.

■ A set of simple tools is needed to interpret fully and derive maximum use from data. More sophisticated techniques may need to be employed occasionally.

■ The effective use of the tools requires the commitment of the people who work on the processes. This in turn needs management support and the provision of training.

Some basic tools and techniques

■ The basic tools and the questions answered are:

Process flowcharting	– what is done?
Check/tally charts	– how often is it done?
Histograms	– what do variations look like?
Scatter diagrams	– what are the relationships between factors?
Stratification	– how is the data made up?
Pareto analysis	– which are the big problems?
Cause and effect analysis and brainstorming (also CEDAC and NGT)	– what causes the problem?
Force field analysis	– what will obstruct or help the change or solution?
Emphasis curve	– which are the most important factors?
Control charts (including cusum)	– which variations to control and how?

Statistical process control

■ People operating a process must know whether it is capable of meeting the requirements, know whether it is actually doing so at any time, and make correct adjustments when it is not. SPC techniques will help here.

■ Before using SPC, it is necessary to identify what the process is, what the inputs/outputs are, and how the suppliers and customers and their requirements are defined. The most difficult areas for this can be in non-manufacturing.

■ All processes can be monitored and brought 'under control' by gathering and using data. SPC methods, with management commitment, provide objective means of controlling quality in any transformation process.

■ SPC is not only a tool kit, it is a strategy for reducing variability, part of never-ending improvement. This is achieved by answering the following questions:

Are we capable of doing the job correctly?

Do we continue to do the job correctly?

Have we done the job correctly?

Could we do the job more consistently and on target?

- SPC provides knowledge and control of process capability.
- SPC techniques have value in the service sector and in the non-manufacturing areas, such as marketing and sales, purchasing, invoicing, finance, distribution, training and personnel.

Some additional techniques for process design and improvement

- Seven 'new tools' may be used as part of quality function deployment (QFD, see Chapter 6) or to improve processes. These are systems and documentation methods for identifying objectives and intermediate steps in the finest detail.
- The seven new tools are: affinity diagram, inter-relationship digraph, tree diagram, matrix diagrams or quality table, matrix data analysis, process decision program chart (PDPC), and arrow diagram.
- The tools are inter-related and their promotion and use should lead to better designs in less time. They work best when people from all parts of an organization are using them. Some of the tools can be used in activities related to problem solving and design.

Taguchi methods for process improvement

- Genichi Taguchi has advanced 'quality engineering' as a technology to reduce costs and make improvements.
- Taguchi's approach may be classified under four headings: total loss function; design of products, services and processes; reduction in variation; and statistically planned experiments.
- Taguchi methods, like all others, should not be used in isolation, but as an integral part of continuous improvement.

Six sigma

- Six sigma is not a new technique – its origins may be found in TQM and SPC. It is a framework through which powerful TQM and SPC tools flourish and reach their full potential. It delivers breakthrough benefits in the short term through the intensity and speed of change. The Excellence Model is a useful framework for mapping the key six-sigma breakthrough strategies.
- A process that can achieve six-sigma capability (where sigma is the statistical measure of variation) can be expected to have a defect rate of a few parts per million, even allowing for some drift in the process setting.
- Six sigma is a disciplined approach for improving performance by focusing on enhancing value for the customer and eliminating costs which add no value.

- There are five fundamental phases/stages in applying the six-sigma approach: Define, Measure, Analyze, Improve, and Control (DMAIC). These form an improvement cycle similar to Deming's Plan, Do, Check, Act (PDCA), to deliver the strategies of process design/redesign, management and improvement, leading to bottom-line achievements.
- Six-sigma approaches question organizational cultures and the measures used. Six-sigma organizations, in addition to focusing on understanding customer requirements, identify core processes, involve all employees in continuous improvement, are responsive to change, base management on fact and metrics, and obtain outstanding results.
- Properly implemented six-sigma strategies involve: leadership involvement and sponsorship, organization-wide training, project selection tools and analysis, improvement methods and tools for implementation, measurement of financial benefits, communication, control and sustained improvement.
- Six-sigma process improvement experts, named after martial arts – Master Black Belts, Black Belts and Green Belts – perform the training, lead teams and carry out the improvements. Mature six-sigma programs have about 1 percent of the workforce as Black Belts.

The 'DRIVE' framework for continuous improvement

- A structured approach to problem solving is provided by the DRIVE model: Define the problem, Review the information, Investigate the problem, Verify the solution, and Execute the change.
- After initial problems are solved, others should be tackled – successful solutions motivating new teams. In all cases teams should follow a disciplined approach to problem solving, using proven techniques.
- The DRIVE model helps structure – thinking and approaches and like the six-sigma DMAIC framework, can be invaluable in all aspects of continuous improvement.

Part 5

People

The people are the masters.

Edmund Burke, 1729–1797

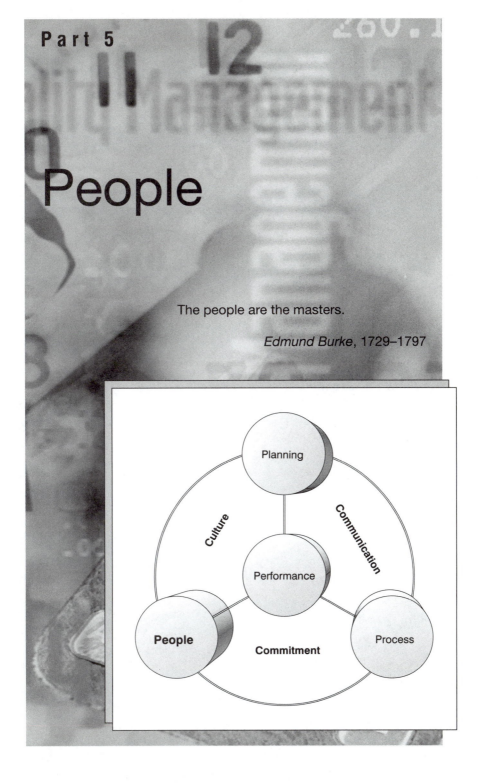

Chapter 14

■ Human resource management

■ Introduction

In recent years, the way in which people are managed and developed at work has come to be recognized as one of the primary keys to improved organizational performance. This is reflected by popular idioms such as 'people are our most important asset' or 'people make the difference'. Indeed, such axioms now appear in the media and on corporate public relations documents with such regularity, that the accuracy and integrity of such assertions has begun to be questioned (see references 1 and 2). This chapter draws on some of the research undertaken by the European Centre for Business Excellence (EC*for*BE), the research and education division of Oakland Consulting, which focused on world class, successful and, in many cases, quality award-winning organizations. It describes the main people management activities that are currently being used in leading edge companies.

There is an overwhelming amount of evidence that successful companies pay much more than lip service to the claim that people are our most important resource. On a general level, successful organizations share a fundamental philosophy to value and invest in their employees. More specifically, world class companies value and invest in their people through the following activities:

■ Strategic alignment of human resource management (HRM) policies.
■ Effective communication.

- Employee empowerment and involvement.
- Training and development.
- Teams and teamwork.
- Review and continuous improvement.

■ Strategic alignment of HRM policies

It is clear that leading edge organizations adopt a common approach or plan, illustrated in Figure 14.1, to align their HR policies to the overall business strategy.

Key elements of the HR strategy (e.g. skills, recruitment and selection, health and safety, appraisal, employee benefits, remuneration, training, etc.) are first identified, usually by an HR director who then reports regularly to the board, and the HR plan, typically spanning three years, is aligned with the overall business objectives and is an integral part of company strategy. For example, if a business objective is to expand at a particular site, then the HR plan provides the necessary additional manpower with the appropriate skills profile and training support. The HR plan is revised as part of the overall strategic planning process. Divisional boards then liaise with the HR director to ensure that the HR plan supports and is aligned with overall policy.

In addition, the HR director holds regular meetings with key personnel from employee relations, health and safety, training and recruitment, etc. to review and monitor the HR plan, drawing upon published data and benchmarking activities in all relevant areas of policy and practice.

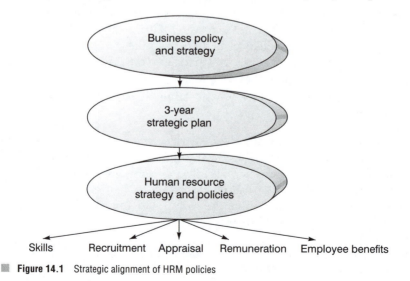

■ **Figure 14.1**   Strategic alignment of HRM policies

Divisional managing directors and the HR director report progress on how the HR plan is supporting the business to the quality committee or board. An overview of this human resource process is illustrated in Figure 14.2.

Although it is beyond the scope of this chapter to make a detailed examination of HR policy, it is prudent to outline briefly some of the common practices that emerged from the identified best practice relating to selection and recruitment, skills and competencies, appraisal, and employee reward, recognition and benefits.

Selection and recruitment

The following practices are common among the organizations studied regarding selection and recruitment:

1 Ensure fairness by using standard tools and practices for job descriptions and job evaluations.
2 Enhance 'transparency' and communication through jargon-free booklets that provide detailed information to new recruits about performance, appraisal, job conditions and so on.

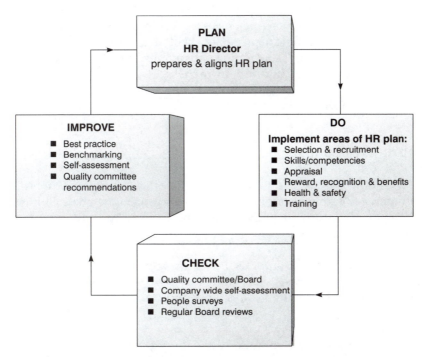

PLAN
HR Director
prepares & aligns HR plan

IMPROVE
■ Best practice
■ Benchmarking
■ Self-assessment
■ Quality committee recommendations

DO
Implement areas of HR plan:
■ Selection & recruitment
■ Skills/competencies
■ Appraisal
■ Reward, recognition & benefits
■ Health & safety
■ Training

CHECK
■ Quality committee/Board
■ Company wide self-assessment
■ People surveys
■ Regular Board reviews

Figure 14.2   Human resource process

3 Ensure that job descriptions are responsibility rather than task oriented.
4 Train all managers and supervisors in interviewing and other selection techniques.
5 Align job descriptions and competencies so that people with the appropriate skills and attributes for the job are identified.
6 Compare the organization's employment terms and conditions (on a regular basis) with published data on best practice and documents to ensure the highest standards are being met.
7 Review HR policies regularly to ensure that they fully reflect legislative and regulatory changes together with known best practice.

Skills/competencies

Since the publication of *The Competent Manager*,[3] the terms competence and competency have been widely used and underpin the work of the specific bodies in any country associated with vocational qualifications and occupational standards. In line with this, good organizations have skills/competence-based human resource management policies underpinning selection and recruitment, training and development, promotion and appraisal.

Although numerous lists of generic management competencies have been published, in essence they are all very similar and are closely allied to the core management competencies underpinning HR policies: leadership, motivation, people management skills, team working skills, comprehensive job knowledge, planning and organizational skills, customer focus, commercial and business awareness, effective communication skills – oral and written – and change management skills, coupled with a drive for continuous improvement.

Appraisal process

As with other HR policies, the main thrust of the appraisal process is alignment – alignment of personal, team and corporate goals coupled with appraisals to help individuals achieve their full potential (see Figure 14.3).

Without exception, the appraisal systems described in world class organizations were based on agreed objectives, time-based so that completion dates provided the opportunity for an automatic review process.

Typically, employees are appraised annually and the managers conducting appraisals attend training in appraisal skills. Before each appraisal,

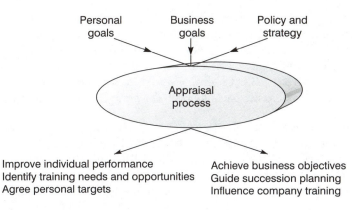

Figure 14.3 The appraisal process

the appraisee and appraiser each complete preparation forms thus making the interview a two-way discussion on performance against objectives during, say, the previous 12 months. Training and development work to achieve the objectives is agreed and, if necessary, additional help is available in the form of advice and counseling.

Employee reward, recognition and benefits _____

Reward and recognition

Although an in-depth study of the policies and practices relating to financial reward and recognition was beyond the scope of the research, it is possible to highlight the following activities that were common among the organizations:

- Rewards are based on consistent, quality-based performance.
- Awards are given to employees but also to customers, suppliers, universities, colleges, students, etc.
- Financial incentives are offered for company-wide suggestions and new idea schemes.
- Internal promotion, for example, from non-supervisory roles to divisional managing directors encourage a highly motivated workforce and enhance job security.
- Commendations include ad hoc recognition for length of service, outstanding contributions, etc.
- Recognition is given through performance feedback mechanisms, development opportunities, pay progressions and bonuses.
- Recognition systems operate at all levels of the organization but with particular emphasis on informal recognition ranging from a personal 'thank you' to recognition at team meetings and events.

With regard to employee benefits, it is well documented that benefits are seen as a tangible expression of the psychological bond between employers and employees. However, to maximize effectiveness, benefits packages should be able to be selected on the basis of what is good for the employee as well as the employer. Moreover, when employees can design their own benefits package both they and the company benefit.

Leading edge organizations favor a 'cafeteria' approach to employee benefits and in recent years there has been increasing interest in this idea of cafeteria benefits to maximize flexibility and choice, particularly in the area of fringe benefits, which can make up a high proportion of the total remuneration package. Under this scheme, the company provides a core package of benefits to all employees (including salary) and a 'menu' of other costed benefits (e.g. personal medical care, dental care, company car, health insurance, etc.) from which the employee can select their personal package.

Some of the ideas underpinning cafeteria benefits sit well with the literature on motivation – to emphasize that different individuals have different needs and expectations from work. Moreover, through communicating the benefits package and providing employees with benefit flexibility, the positive impact is further increased; not only are employees more likely to get what benefits they want, but also communication makes them more aware of the benefits they are gaining, thus informing and increasing morale.

■ Effective communication

Effective communication emerges from the research as an essential facet of people management – be it communication of the organization's goals, vision, strategy and policies or the communication of facts, information and data. For business success, regular, two-way communication, particularly face to face with employees, is an important factor in establishing trust and a feeling of being valued. Two-way communication is regarded as both a core management competency and as a key management responsibility. For example, a typical list of management responsibilities for effective communication is to:

- Regularly meet all their people.
- Ensure people are briefed on key issues in a language free of technical jargon.
- Communicate honestly and as fully as possible on all issues which affect their people.
- Encourage team members to discuss company issues and give upward feedback.

■ Ensure issues from team members are fed back to senior managers and timely replies given.

Regular two-way communication also involves customers, shareholders, financial communities and the general public.

Communications process

Successful organizations follow a systematic process for ensuring effective communications as shown in Figure 14.4.

Plan

Typically, the HR function, e.g. the HR director, is responsible for the communication process. He/she assesses the communication needs of the organization and liaises with divisional directors, managers or local management teams to ensure that the communication plans are in alignment with overall policy and strategy. A communication program accompanies any major changes on organization policy or objectives.

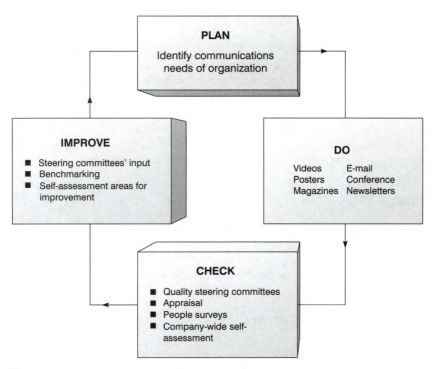

PLAN
Identify communications needs of organization

IMPROVE
■ Steering committees' input
■ Benchmarking
■ Self-assessment areas for improvement

DO
Videos   E-mail
Posters   Conference
Magazines Newsletters

CHECK
■ Quality steering committees
■ Appraisal
■ People surveys
■ Company-wide self-assessment

■ **Figure 14.4**   Best practice communications process

Do

A comprehensive mix of diverse media is used to support effective communication throughout their organization. These include:

Videos	Posters	Open-door policies
Surveys	Campaigns	E-mail
Magazines	Briefings	Notice boards
Newsletters	Conferences	Internet/Intranet
Appraisals	Meetings	Focus groups

It is evident that the introduction of electronic systems has brought about radical changes in communications. Typically employees are able to access databases, spreadsheets, word processing, e-mail and diary facilities. Information on business performance, market intelligence and quality issues can also be easily and quickly cascaded. Further, video conferencing is used to facilitate internal face-to-face communications with major customers across the world, resulting in substantial savings on travel and associated costs. Furthermore, provision is made for depots, units, regions, divisions, departments, etc. to hold 'virtual' meetings and conferences. Feedback questionnaires then check that events are valuable and help the planning of future events.

Check

Quality steering or review committees, people surveys, appraisal and company-wide self-assessment are used to review the effectiveness of the communications process. Appraisal and staff survey data are analyzed to ensure that the communications process is continuing to deliver effective upward, downward and lateral communications. Reports are then made quarterly, six-monthly and/or annually to the chief executive and/or the most senior team on the effectiveness and relevance of the communications process. The people survey data is also used to ascertain employee perceptions and to keep in touch with current opinion.

Improve

The results of the various review processes highlight areas for improvement and results are verified by benchmarking against, for instance, a national survey. Quality steering committees then put forward recommendations for future planning and continuous improvements.

Communications structure

Successful organizations place great emphasis on communication channels that enable people at all levels in the organization to feel able to

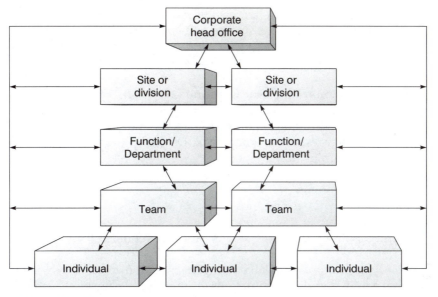

Figure 14.5   Multi-directional communications structure

talk to each other. Consequently, managers are not only trained but 'are committed to being open-minded, honest, more visible and approachable' (field research). Many formal and informal communication mechanisms exist, all designed to foster an environment of open dialog, shared knowledge, information and trust in an effective upward, downward, lateral and cross-functional structure such as the one illustrated in Figure 14.5.

TNT, for example, have a regional structure that provides a link between local and central management which ensures that the chain of communication is complete so that information can cascade down and rise up. Briefings of senior managers by executives are followed by briefings of middle managers and so on all the way down the line. The same chain also works in reverse to facilitate bottom-up communications.

(See also Chapter 15 for more detail on the communication process.)

Employee empowerment and involvement

To encourage employee commitment and involvement, successful organizations place great importance on empowering their employees. The positive effects of employee empowerment are well documented but the notion has been challenged with some writers claiming that it is not possible to empower people – rather, it is possible only to create a climate and a structure in which people will take responsibility. Nonetheless, it is clear that the organizations studied in the research considered empowerment to be a key issue *and* made efforts to create a working environment that was conducive to the employees taking responsibility.

The Dana Commercial Credit Corporation (DCC), for example, subscribe to the importance of employee empowerment for their impressive customer satisfaction scores that are two points higher than the industry average (on a five-point scale). DCC empower their people by encouraging employees to:

- Set their own goals.
- Judge their own performance.
- Take ownership of their actions.
- Identify with DCC (e.g. to become stock shareholders).

Similarly, TNT report that 'all employees are empowered to respond to normal and extraordinary situations without further recourse' and that they have 'worked hard to create a no blame culture where our people are empowered to take decisions to achieve their objectives'.

People empowerment is also a key issue in Texas Instruments. Here empowerment is built into the TI operational approach, organized around processes, by stimulating creativity and encouraging quality teams. Along the same lines, Hewlett-Packard advocate teamwork and high levels of empowerment combined with a strong setting of objectives and freedom for employees to achieve them. Similarly, the Eastman Chemical Company describe how they focus on employee empowerment and have learnt that a company cannot empower employees who do not care, do not have authority and do not have the appropriate skills.

To address the above issues, management map out processes to provide employees with the necessary authority and skills. In addressing the issue of not caring, employee surveys reveal that appraisal systems can be a major roadblock and the appraisal process may need to be revised.

Common initiatives

There are three common initiatives in successful organizations which place great store by:

1 *Corporate employee suggestion schemes* – these provide a formalized mechanism for promoting participative management, empowerment and employee involvement.

2 *Company-wide culture change programs* – in the form of workshops, ceremonies and events used to raise awareness and to empower individuals and teams to practice continuous improvement.
3 *Measurement of key performance indicators (KPIs)* – whereby the effectiveness of staff involvement and empowerment is measured by improvements in human resource key performance indicators (KPIs), such as labour turnover, accident rate, absenteeism, and lost time through accidents. Typically, KPI measurements, coupled with appraisal feedback and survey results, are regularly reviewed by the HR director who uses the information as the basis for reports and suggestions for improvements to the board.

On a more general level, successful organizations increase commitment by empowering and involving more and more of their employees in formulating plans that shape the business vision. As more people understand the business and where it is planned to go, the more they become involved in and committed to developing the organization's goals and objectives.

Case study ▪▪▪ ▬▬▬▬▬▬▬▬▬▬▬▬

▪ Involving the workforce in BT Retail

At local operational level employees are encouraged to be innovative in their approach to day-to-day problem solving and put forward reasoned ideas based on their own insight of the business, to improve elements of BT Retail's operations. One simple example of how this can work comes from the conferencing team. A group of front-line employees are brought together monthly in a forum called the 'customer listening post'. Ideas and issues that are stimulated by day-to-day contact with customers are discussed and then actions agreed. In the last year, this team initiated 72 improvements ranging from product enhancements to training in recently launched products.

In addition a number of approaches have been introduced to ensure that employees are fully involved in the design and development of major change. For example, a program which is transforming call-center operations successfully involved about 1500 people in the design stage. This input was combined with data from customers and internal performance measures to produce a design that reflected the views of all key stakeholders.

BT Retail places a high value on involvement and experience suggests that it has to be deployed in a way that meets local requirements. An organization of 50 000 people can benefit massively from involvement of its people, but that number of people also represents a major challenge in continually maintaining their involvement.

BT Retail places a high value on product and process innovation. To support this, in a business regarded by many as traditional and conservative, required specific action, and a detailed innovation strategy was developed in 2001. At a strategic level a number of approaches have been introduced to focus on delivering an innovative strategy. These include a much broader involvement of people in strategy development workshops specifically focused on coming up with innovative business and process proposals together with the senior management team investing more time on strategy with increased external stimulus.

A Corporate Venturing unit has also been formed to generate, select and exploit radical business concepts throughout BT Retail. The Venturing unit delivers significant value by rapidly developing innovations that provide new revenue opportunities but which would be difficult and diverting for mainstream units to evolve. In its first year the venturing approach established over six new businesses each of which has the potential to generate significant revenue (at least £10 m – c. $15 m).

■ Training and development

The training and development of people at work has increasingly come to be recognized as an important part of human resource management. Through the 1980s, major changes in many organizations resulted in increasing workloads, the introduction of new technology and wider ranges of tasks, all of which required training provision. During the 1990s, initiatives such as ISO 9000, TQM, Investors in People (IiP), benchmarking and self-assessment against frameworks such as the EFQM Excellence Model have further highlighted the need for properly trained employees.

It is widely acknowledged that many writers and practicing managers sing the praises of training, saying it is a 'symbol of the employers' commitment to staff', or that it shows an organization's strategy is based on 'adding value rather than lowering costs'. However, others claim that a lack of effective training predominates in many organizations today and that serious doubts remain as to whether or not management actually *does* invest in the training of their human resources.

It is perhaps not surprising then that research on successful and quality award-winning organizations revealed an ongoing commitment to investing in the provision of planned, relevant and appropriate training. Training was found to be carefully planned through training needs analysis processes that linked the training needs with those of the organization, groups, departments, divisions, and individuals. To maintain training relevancy and currency, databases of training courses are widely available and, to encourage diversification, employees are able to realize their full potential by training in quality, job skills, general education, health and safety and so on, through exams, qualifications, assessor training, etc. Typically, training strategies in these organizations require managers to:

- Play an active role in training delivery (cascade training) and support (including quality tools and techniques).
- Receive training and development based on personal development plans.
- Fund training and improvement activities to allow autonomy at 'local' levels for short payback investments.
- Co-ordinate discussions and peer assessments to develop tailored training plans for individuals.

TI Europe, for instance, has a development and performance management (DPM) process that uses discussion and peer assessment to help create individually tailored training plans with business objectives through policy deployment. Similarly, Trident Precision Manufacturing manages and motivates its employees through training. Training is provided in quality, skills related to the job, general education and safety and since 1989, they have invested an average of c. 5 percent of the payroll on training – a figure that represents almost three times the average for all US industries. In addition, Trident encourages employees to diversify their abilities and, typically, 80 percent of their employees are trained in at least two job functions. As a result of their investments these companies boast business benefits such as:

- increases in sales volumes;
- not losing customers to competitors;
- low employee turnover.

What is particularly noteworthy about the training activities identified is that they are almost identical to those processes and activities commonly found in the management literature on the theory of training. Many writers have developed models of the training process which can be summarized into the four phases shown in Figure 14.6.

The **Assessment Phase** identifies what is needed (the content of the training) at the organizational, group and individual levels. This may involve some overlap, e.g. an individual's poor sales performance may be a symptom of production problems at the group level, as well as a need for more product innovation at the organizational level. The Assessment Phase thus involves identifying training needs by assessing the gaps between future requirements of a job and the current skills, knowledge or attitudes of the person in the job. So the organization looks at what is presently happening and what should or could be

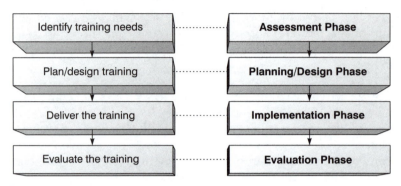

happening. Any differences between the two will give some indications of training needs.

The **Planning/Design Phase** identifies where and when the training will take place and involves such questions as:

■ Who needs to be trained?
■ What competencies are required?
■ How long will training take?
■ What are the expected benefits of training?
■ Who/How many will undertake the training?
■ What resources are needed, e.g. money, equipment, accommodation, etc.?

Typically, the organizations involved in the studies planned their training programs (e.g. annually) according to the needs of the business and circulated lists of available courses well in advance of the training dates. The lists ensure all managers are aware of what is provided so that they are able to schedule attendance by staff. The strategic training plan is supported by an annual budgeting and planning system with quarterly meetings to monitor and review performance. The budgeting and planning process with its integral HR element is cascaded throughout the organization to teams at all locations.

The **Implementation Phase** involves the actual delivery of the training. This might be on site or away from the premises and will include training techniques such as: simulators; business games; case studies; coaching and mentoring; planned experience; computer-assisted instruction. Demonstration or 'sitting next to Nellie' is another commonly used training technique, the effectiveness of which is dependent upon Nellie!

Induction and devolved training form an integral part of the training implementation phase. New employees attend induction courses and are issued with personal development documents giving details of what training and assessment will take place in the first few months of employment, as well as copies of the vision and mission statements. At regular intervals throughout the induction period (e.g. every three or four weeks) new employees are then reviewed to identify training and development needs for the remainder of the year.

In addition, much of the training is devolved to line managers through facilitation and facilitator packs. This requires the training and development of all levels of managers and supervisors in facilitation skills. Line managers then identify team members to be trained as facilitators. Adopting this approach is said to create an environment in which everyone is aware of training and development issues for themselves and their colleagues.

The **Evaluation Phase** is widely acknowledged as one of the most critical steps in the training process and can take many forms such as observation, questionnaires, interviews, etc. For example, in this phase the overall effectiveness of training is evaluated and this provides feedback for the trainers, for future improvements to the program, for senior managers and the trainees themselves. Providing trainees with a set of training objectives will help them know what they need to learn and give them feedback on their progress. This will then influence their attitudes towards future training and even the company itself.

In sum, it seems that successful organizations approach training and development in a planned and systematic way involving training needs analysis, assessment of training content, carefully planned implementation and continuous evaluation and review – a convincing argument for the value of theory when it is put into practice.

(See also Chapter 15 for more detail on the training and development process.)

Teams and teamwork

It is clear that leading edge organizations place great emphasis on the value of people working together in teams. This is hardly surprising as a great deal of theory and research indicates that people are motivated and work better when they are part of a team. Teams can also achieve more through integrated efforts and problem solving.

Teams are a management tool and are most effective when team activity is clearly linked to organizational strategy. For this, the strategy must be communicated to influence team direction, which then links to the production of team mission statements and the use of team agendas and scorecards. Importantly, though, many people emphasize the value of cross-functional teams which proved to be a common feature in many of the organizations studied. Here, teams which have originally evolved out of the old functional departments or units within an organization gain experience and benefit from team building and become cross-functional. For example, in the Eastman company: 'Over the past decade, the Eastman Corporation has developed a quality focus which has expanded from individuals to the concept of interlocking teams.' Each team is required to identify its customers, the customer requirements and what measures need to be used to ensure that those requirements are being satisfied.

Cross-functional teams are also an important feature in TI Europe where almost every employee belongs to at least one team, ranging from managers on quality steering teams, to operators on quality improvement teams, to fully empowered, self-directed work teams. Cross-functional teams are used to address the entire process.

(See also Chapter 15 for more detail on teams and teamwork.)

■ Review, continuous improvement and conclusions

In all the organizations involved in the research, processes for reviewing performance and continuous improvement exist at the individual, team, departmental/divisional and organizational levels. These include such processes as:

- Annual staff surveys and subsequent actions, which are viewed as the cornerstones of continuous improvement. The people surveys are also critically reviewed against data from other world class organizations and benchmarks to determine best practice and feed into the continuous improvement processes.
- Quality committees, the HR department and cross-functional teams drawn from depots, regions and divisions review feedback from surveys as well as the format of the surveys.
- Ongoing performance feedback and development through on-the-job coaching plus regular one-to-one individual and team reviews.

This chapter so far has highlighted the main people management activities that are currently being used in some world class organizations. A general conclusion from the research supporting this is that successful organizations pay much more than lip service to the popular idiom 'people are our most important asset'. Indeed successful organizations value and invest in their people in a never-ending quest for effective management and development of their employees. This involves rigorous planning of processes, skillful implementation, regular review of processes, and continuous improvement practices.

From a theoretical viewpoint, these findings about people management activities in successful organizations are hardly surprising, since the management literature is strewn with examples of the benefits of systematic planning, followed by strategic implementation, regular review and continuous improvement. Nonetheless, from a practical viewpoint, the real value of the findings is that they flesh out in some detail those people management activities that are being used to good effect in some world class organizations which are reaping the benefits of putting theory into practice.

Organizing people for quality management

In many organizations management systems are still viewed in terms of the internal dynamics between marketing, design, sales, production/ operations, distribution, accounting, etc. A change is required from this to a larger process-based system that encompasses and integrates the business interests of customers and suppliers. Management needs to develop an in-depth understanding of these relationships and how they may be used to cement the partnership concept. A quality function can be the organization's focal point in this respect, and should be equipped to gauge internal and external customers' expectations and degree of satisfaction. It should also identify deficiencies in all business functions, and promote improvements.

The role of the quality function is to make quality an inseparable aspect of every employees' performance and responsibility. The transition in many companies from quality departments with line functions will require careful planning, direction, and monitoring. Quality professionals have developed numerous techniques and skills, focused on product or service quality. In many cases there is a need to adapt these to broader, process applications. The first objectives for many 'quality managers' will be to gradually disengage themselves from line activities, which will then need to be dispersed throughout the appropriate operating departments. This should allow quality management to be understood as a 'process' at a senior level, and to be concerned with the following throughout the organization:

- Encouraging and facilitating improvement.
- Monitoring and evaluating the progress of improvement.

- Promoting the 'partnership' in relationships with customers and suppliers.
- Planning, managing, auditing, and reviewing quality management systems.
- Planning and providing training and counselling or consultancy.
- Giving advice to management on:
  - establishment of process management and control;
  - relevant statutory/legislation requirements with respect to quality;
  - quality and process improvement programs;
  - inclusion of quality elements in all processes, job instructions and procedures.

Quality directors and managers may have an initial task, however, to help those who control the means to implement this concept – the leaders of industry and commerce – to really believe that quality must become an integral part of all the organization's operations.

The author has a vision of quality as a strategic business management function that will help organizations to develop or change their cultures. To make this vision a reality, quality professionals must expand the application of quality concepts and techniques to all business processes and functions, and develop new forms of providing assurance of quality at every supplier/customer interface. They will need to know the entire cycle of products or services, from concept to the *ultimate* end user. An example of this was observed in the case of a company manufacturing pharmaceutical seals, whose customer expressed concern about excess aluminum projecting below and round a particular type of seal. This was considered a cosmetic defect by the immediate customer, the Health Service, but a safety hazard by a blind patient – the *customer's customer*. The prevention of this 'curling' of excess metal meant changing practices at the mill that rolled the aluminum – at the *supplier's supplier*. Clearly, the quality professional dealing with this problem needed to understand the supplier's processes and the ultimate customer's needs, in order to judge whether the product was indeed capable of meeting the requirements.

The shift in 'philosophy' will require considerable staff education in many organizations. Not only must people in other functions acquire quality management skills, but quality personnel must change old attitudes and acquire new skills – replacing the inspection, calibration, specification-writing mentality with knowledge of defect prevention, wide-ranging quality management systems design and audit. Clearly, the challenge for many quality professionals is not so much making changes in their organization, as recognizing the changes required in themselves. It is more than an overnight job to change the attitudes of an inspection police force into those of a consultative, team-oriented improvement resource. This emphasis on prevention and improvement-based systems elevates the role of quality professionals from a technical one to that of general management. A narrow departmental view of quality is totally out of place in an organization aspiring to TQM, and many quality directors and managers will need to widen their perspective and increase their knowledge to encompass all facets of the organization.

To introduce the concepts of process management will require not only a determination to implement change, but sensitivity and skills in inter-personal relations. This will depend very much, of course, on the climate within the organization. Those whose management is truly concerned with co-operation and concerned for the people will engage strong employee support for the quality manager or director in his catalytic role in process improvement. Those with aggressive, confrontational management will create for the quality professional impossible difficulties in obtaining support from the 'rank and file'.

Quality appointments

Many organizations have realized the importance of the contribution a senior, qualified director of quality can make to the prevention strategy. Smaller organizations may well feel that the cost of employing a full-time quality manager is not justified, other than in certain very high-risk areas. In these cases a member of the management team may be appointed to operate on a part-time basis, performing the quality management function in addition to his/her other duties. To obtain the best results from a quality director/manager, he/she should be given sufficient authority to take necessary action to secure the implementation of the organization's quality policy, and must have the personality to be able to communicate the message to all employees, including staff, management and directors. Occasionally the quality director/manager may require some guidance and help on specific technical quality matters, and one of the major attributes required is the knowledge and wherewithal to acquire the necessary information and assistance.

In large organizations, then, it may be necessary to make several specific appointments or to assign details to certain managers. The following actions may be deemed to be necessary.

Assign a quality director, manager or co-ordinator

This person will be responsible for the planning and implementation of quality management. He or she will be chosen first for process, project and people management abilities rather than detailed knowledge of quality assurance matters. Depending on the size and complexity of the organization, and its previous activities in quality management, the position may be either full or part time, but it must report directly to the chief executive.

Appoint a quality manager adviser

A professional expert on quality may be required to advise on the 'technical' aspects of planning and implementing quality management. This is a consultancy role, and may be provided from within or without the organization, full or part time. This person needs to be a persuader, philosopher, teacher, adviser, facilitator, reporter and motivator. He or she must clearly understand the organization, its processes and interfaces, be conversant with the key functional languages used in the business, and be comfortable operating at many organizational levels. On a more general level this person must fully understand and be an effective advocate and teacher of quality management, be flexible and become an efficient agent of change.

Steering committees and teams _____

Devising and implementing quality management in an organization takes considerable time and ability. It must be given the status of a senior executive project. The creation of cost-effective performance improvement is difficult, because of the need for full integration with the organization's strategy, operating philosophy and management systems. It may require an extensive review and substantial revision of existing systems of management and ways of operating. Fundamental questions may have to be asked, such as 'do the managers have the necessary authority, capability, and time to carry this through?'

Any review of existing management and operating systems will inevitably 'open many cans of worms' and uncover problems that have been successfully buried and smoothed over – perhaps for years. Authority must be given to those charged with following quality management through with actions that they consider necessary to achieve the goals. The commitment will be continually questioned and will be weakened, perhaps destroyed, by failure to delegate authoritatively.

The following steps are suggested in general terms. Clearly, different types of organization will have need to make adjustments to the detail, but the component parts are the basic requirements.

A disciplined and systematic approach to continuous improvement may be established in a quality or business excellence 'steering committee or council' (Figure 14.7). The committee/council should meet at least monthly to review strategy, implementation progress, and improvement.

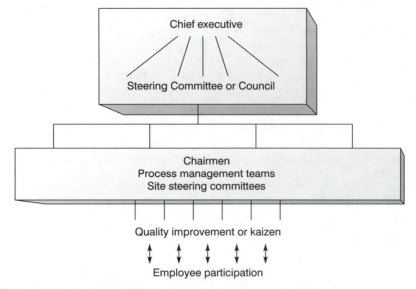

Figure 14.7 Figure 14.7   Employee participation through the team structures for quality management

It should be chaired by the chief executive, who must attend every meeting – only death or serious illness should prevent him/her being there. Clearly, postponement may be necessary occasionally, but the council should not carry on meeting without the chief executive present. The council members should include the top management team and the chairmen of any 'site' steering committees or process management teams, depending on the size of the organization. The objectives of the council are to:

- Provide strategic direction on quality for the organization.
- Establish plans for quality on each 'site'.
- Set up and review the process teams that will own the key or critical business processes.
- Review and revise quality plans for implementation.

The process management teams and any site steering committees should also meet monthly, shortly before the senior steering committee/council meetings. Every senior manager should be a member of at least one such team. This system provides the 'top-down' support for employee participation in process management and development. It also ensures that the commitment to quality at the top is communicated effectively through the organization.

The three-tier approach of steering committee, process management teams and quality improvement teams allows the first to concentrate on quality strategy, rather than become a senior problem-solving group.

Progress is assured if the team chairmen are required to present a status report at each meeting.

The process management teams or steering committees control all the quality improvement teams and have responsibility for:

- The selection of projects for the teams.
- Providing an outline and scope for each project to give to the teams.
- The appointment of team members and leaders.
- Monitoring and reviewing the progress and results from each team project.

As the focus of this work will be the selection of projects, some attention will need to be given to the sources of nominations. Projects may be suggested by:

- Steering committee/council members representing their own departments, process management teams, their suppliers or their customers, internal and external.
- Quality improvement teams.
- Kaizen teams or quality circles (if in existence).
- Suppliers.
- Customers.

The process team members must be given the responsibility and authority to represent their part of the organization in the process. The members must also feel that they represent the team to the rest of the organization. In this way the team will gain knowledge and respect and be seen to have the authority to act in the best interests of the organization, with respect to their process.

■ Quality circles or kaizen teams

No book on quality management would be complete without a mention of kaizen and quality circles. Kaizen is a philosophy of continuous improvement of all the employees in an organization so that they perform their tasks a little better each day. It is a never-ending journey centered on the concept of starting anew each day with the principle that methods can always be improved.

Kaizen teian is a Japanese system for generating and implementing employee ideas. Japanese suggestion schemes have helped companies to improve quality and productivity, and reduced prices to increase market share. They concentrate on participation and the rates of implementation, rather than on the 'quality' or value of the suggestion. The emphasis is on encouraging everyone to make improvements.

Kaizen teian suggestions are usually small-scale ones, in the worker's own area, and are easy and cheap to implement. Key points are that the rewards given are small, and implementation is rapid, which results in many small improvements that accumulate to massive total savings and improvements.

One of the most publicized aspects of the Japanese approach to quality has been these quality circles or kaizen teams. The quality circle may be defined then as a group of workers doing similar work who meet:

- Voluntarily.
- Regularly.
- In normal working time.
- Under the leadership of their 'supervisor'.
- To identify, analyze, and solve work-related problems.
- To recommend solutions to management.

Where possible quality circle members should implement the solutions themselves.

The quality circle concept first originated in Japan in the early 1960s, following a post-war reconstruction period during which the Japanese placed a great deal of emphasis on improving and perfecting their quality control techniques. As a direct result of work carried out to train foremen during that period, the first quality circles were conceived, and the first three circles registered with the Union of Japanese Scientists and Engineers (JUSE) in 1962. Since that time the growth rate has been phenomenal. The concept has spread to Taiwan, the USA and Europe, and circles in many countries have been successful. Many others have failed.

It is very easy to regard quality circles as the magic ointment to be rubbed on the affected spot, and unfortunately many managers in the West first saw them as a panacea for all ills. There are no panaceas, and to place this concept into perspective, Juran, who has been an important influence in Japan's improvement in quality, has stated that quality circles represent only 5–10 percent of the canvas of the Japanese success. The rest is concerned with understanding quality, its related costs and the organization, systems and techniques necessary for achieving customer satisfaction.

Given the right sort of commitment by top management, introduction, and environment in which to operate, quality circles can produce the 'shop floor' motivation to achieve quality performance at that level.

Circles should develop out of an understanding and knowledge of quality on the part of senior management. They must not be introduced as a desperate attempt to do something about poor quality. The term 'quality circle' may be replaced with a number of acronyms but the basic concepts and operational aspects may be found in many organizations.

The structure of a quality circle organization

The unique feature about quality circles or kaizen teams is that people are asked to join and not told to do so. Consequently, it is difficult to be specific about the structure of such a concept. It is, however, possible to identify four elements in a circle organization:

- Members.
- Leaders.
- Facilitators or co-ordinators.
- Management.

Members form the prime element of the concept. They will have been taught the basic problem-solving and process control techniques and, hence, possess the ability to identify and solve work-related problems.

Leaders are usually the immediate supervisors or foremen of the members. They will have been trained to lead a circle and bear the responsibility of its success. A good leader, one who develops the abilities of the circle members, will benefit directly by receiving valuable assistance in tackling nagging problems.

Facilitators are the managers of the quality circle programs. They, more than anyone else, will be responsible for the success of the concept, particularly within an organization. The facilitators must co-ordinate the meetings, the training and energies of the leaders and members, and form the link between the circles and the rest of the organization. Ideally the facilitator will be an innovative industrial teacher, capable of communicating with all levels and with all departments within the organization.

Management support and commitment are necessary to quality circles or, like any other concept, they will not succeed. Management must retain its prerogatives, particularly regarding acceptance or non-acceptance of recommendations from circles, but the quickest way to kill a program is to ignore a proposal arising from it. One of the most difficult facts for management to accept, and yet one forming the cornerstone of the quality circle philosophy, is that the real 'experts' on performing a task are those who do it day after day.

Training quality circles

The training of quality circle/kaizen leaders and members is the foundation of all successful programs. The whole basis of the training operation is that the ideas must be easy to take in and be put across in a way that facilitates understanding. Simplicity must be the key word, with emphasis being given to the basic techniques. Essentially there are eight segments of training:

1 Introduction to quality circles.
2 Brainstorming.
3 Data gathering and histograms.
4 Cause and effect analysis.
5 Pareto analysis.
6 Sampling and stratification.
7 Control charts.
8 Presentation techniques.

Managers should also be exposed to some training in the part they are required to play in the quality circle philosophy. A quality circle program can only be effective if management believes in it and is supportive and, since changes in management style may be necessary, managers' training is essential.

Operation of quality circles/kaizen teams

There are no formal rules governing the size of a quality circle/kaizen team. Membership usually varies from three to 15 people, with an average of seven to eight. It is worth remembering that, as the circle becomes larger, it becomes increasingly difficult for all members of the circle to participate.

Meetings should be held away from the work area, so that members are free from interruptions, and are mentally and physically at ease. The room should be arranged in a manner conducive to open discussion, and any situation that physically emphasizes the leader's position should be avoided.

Meeting length and frequency are variable, but new circles meet for approximately one hour once per week. Thereafter, when training is complete, many circles continue to meet weekly; others extend the interval to two or three weeks. To a large extent the nature of the problems selected will determine the interval between meetings, but this should never extend to more than one month, otherwise members will lose interest and the circle will cease to function.

Great care is needed to ensure that every meeting is productive, no matter how long it lasts or how frequently it is held. Any of the following activities may take place during a circle meeting:

- Training – initial or refresher.
- Problem identification.
- Problem analysis.
- Preparation and recommendation for problem solution.
- Management presentations.
- Quality circle administration.

A quality circle usually selects a project to work on through discussion within the circle. The leader then advises management of this choice and, assuming that no objections are raised, the circle proceeds with the work. Other suggestions for projects come from management, quality assurance staff, the maintenance department, various staff personnel, and other circles.

It is sometimes necessary for quality circles to contact experts in a particular field, e.g. engineers, quality experts, safety officers, maintenance personnel. This communication should be strongly encouraged, and the normal company channels should be used to invite specialists to attend meetings and offer advice. The experts may be considered to be 'consultants', the quality circle retaining responsibility for improving a process or solving the particular problem. The overriding purpose of quality circles or kaizen teams is to provide the powerful motivation of allowing people to take some part in deciding their own actions and futures.

Case study ■ ■ ■

■ My SQF

Once the Shell Quality Framework (SQF) had been introduced for organizational improvements and it was accepted as a tool for improved decision making, there was another, perhaps more important aspect yet to be developed – using the SQF as a tool to help people in their own personal work planning, and maybe even to start to address the issues of work/life balance.

Putting the SQF into a small leaflet format (Figure 14.8) that could be distributed to each member of staff proved immensely valuable. It helped in discussions between staff and supervisors, where real issues around their workload, priorities and challenges could be discussed in a way that took the heat and emotion out of the debate. For an individual reviewing work priorities, it was important to be absolutely clear about the purpose of a task or project, as well as having a good understanding of what was required to achieve success in implementation. It also meant having the confidence and a fair basis for saying No! to some activities.

Figure 14.8
'MySQF' leaflet

During the two years of development, testing and implementation of the SQF, it was never found wanting in terms of an area of the business where it could not be applied. One of the most difficult aspects was getting people to accept the simplicity of the construct, and not to look for complexity – as if it were perhaps a test of intellectual rigor. The SQF was deliberately predicated around the fact that it would outlast business fads and theories – keeping it down to earth meant that it was bound to be successful over time and not thrown away when the next idea came along. Managers will keep using the SQF to provide an anchor to business improvement – it costs little, is difficult to argue against and helps the most important asset in the organization – the people!

Acknowledgement

The author is grateful to Dr Susan Oakland for the contribution she made to this chapter.

References

1. Maguire, S., 'Learning to change', *European Quality*, Vol. 2, No. 6, p. 8, 1995.

2. Marchington, M. and Wilkinson, A., *Core Personnel and Development*, London, Institute of Personnel and Development, 1997.
3. Boyatsis, R., *The Competent Manager: A Model for Effective Performance*, New York, Wiley, 1982.

Chapter highlights

■ ■ ■

Introduction

■ In recent years the way people are managed has been recognized as a key to improving performance. Research (EC*for*BE) on world class, award-winning organizations has identified the main people management activities used in leading edge organizations.

■ World class organizations value and invest in people through: strategic alignment of HRM policies, effective communications, employee empowerment and involvement, training and development, teams and teamwork and review and continuous improvement.

Strategic alignment of HRM policies

■ Leading edge organizations adopt a common approach to aligning HR policies with business strategy. Key elements of policy, such as skills, recruitment and selection, training, health and safety, appraisal, employee benefits and remuneration, are first identified. The HR plan is then devised as part of the strategic planning process, following a Plan, Do, Check, Improve (PDCI) cycle.

Effective communication

■ Regular two-way communication, particularly face to face, is essential for success.

■ Again the PDCI cycle provides a systematic process for ensuring effective communications, which uses benchmarking and self-assessment as part of the improvement effort.

Employee empowerment and involvement

■ To encourage employee commitment and involvement, successful organizations place great importance on empowering employees. This can include people setting own goals, judging own performance, taking ownership of actions, and identifying with the organization itself (perhaps as shareholders).

■ Common initiatives include: employee suggestion schemes, culture change programs and measurement of KPIs. Generally commitment is increased by involving more employees in planning and shaping the vision.

Training and development

- Training and development has been highlighted by many initiatives as a critical success factor although lack of effective training still predominates in many organizations.
- In successful organizations, training is planned through needs analysis, use of databases, training delivery at local levels, and peer assessments for evaluation.

Teams and teamwork

- Leading edge organizations place great value in people working in teams, because this motivates and causes them to work better.
- Teams are most effective when their activities are clearly linked to the strategy, which in turn is communicated to influence direction. Cross-functional teams are particularly important to address end-to-end processes.

Review, continuous improvement and conclusions

- Effective organizations use processes for reviewing performance and continuous improvement at the individual, team, divisional/departmental and organizational levels. These include surveys of staff, committees/teams, and ongoing performance feedback.

Organizing people for quality management

- The quality function should be the organization's focal point of the integration of the business interests of customers and suppliers into the internal dynamics of the organization.
- Its role is to encourage and facilitate quality and process improvement; monitor and evaluate progress; promote the quality chains; plan, manage, audit and review systems; plan and provide quality training, counseling and consultancy; and give advice to management.
- In larger organizations a quality director will contribute to the prevention strategy. Smaller organizations may appoint a member of the management team to this task on a part-time basis. An external TQM adviser is usually required.
- In devising and implementing TQM for an organization, it may be useful to ask first if the managers have the authority, capability and time to carry it through.
- A disciplined and systematic approach to continuous improvement may be established in a steering committee/council, whose members are the senior management team.

▨ Reporting to the steering committee are the process management teams or any site steering committees, which in turn control the quality improvement or kaizen teams and quality circles.

Quality circles or kaizen teams

▨ Kaizen is a philosophy of small step continuous improvement, by all employees. In kaizen teams the suggestions and rewards are small but the implementation is rapid.

▨ A quality circle or kaizen team is a group of people who do similar work meeting voluntarily, regularly, in normal working time, to identify, analyze and solve work-related problems, under the leadership of their supervisor. They make recommendations to management. Alternative names may be given to the teams, other than 'quality circles'.

Chapter 15

Culture change through people and teamwork

The need for teamwork

The complexity of most of the processes that are operated in industry, commerce and the services places them beyond the control of any one individual. The only really efficient way to tackle process management and improvement is through the use of some form of teamwork which has many advantages over allowing individuals to work separately:

- A greater variety of complex processes and problems may be tackled – those beyond the capability of any one individual or even one department – by the pooling of expertise and resources.
- Processes and problems are exposed to a greater diversity of knowledge, skill, experience, and are solved more efficiently.
- The approach is more satisfying to team members, and boosts morale and ownership through participation in process management, problem solving and decision making.
- Processes and problems that cross departmental or functional boundaries can be dealt with more easily, and the potential/actual conflicts are more likely to be identified and solved.
- The recommendations are more likely to be implemented than individual suggestions, as the quality of decision making in *good teams* is high.

Most of these factors rely on the premise that people are willing to support any effort in which they have taken part or helped to develop.

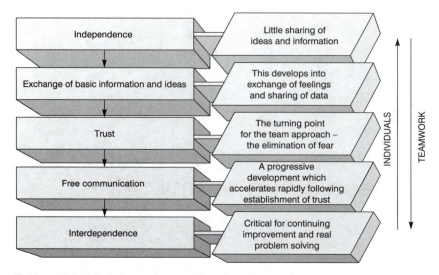

Figure 15.1   Independence to interdependence through teamwork

When properly managed and developed, teams improve the process of problem solving, producing results quickly and economically. Teamwork throughout any organization is an essential component of quality management, for it builds trust, improves communications and develops interdependence. Much of what has been taught previously in management has perhaps led to a culture in the West of independence, with little sharing of ideas and information. Knowledge is very much like organic manure – if it is spread around it will fertilize and encourage growth, if it is kept closed in, it will eventually fester and become useless.

Good teamwork changes the independence to interdependence through improved communications, trust and the free exchange of ideas, knowledge, data and information (Figure 15.1). The use of the face-to-face interaction method of communication, with a common goal, develops over time the sense of dependence on each other. This forms a key part of any quality improvement process, and provides a methodology for employee recognition and participation, through active encouragement in group activities.

Teamwork provides an environment in which people can grow and use all the resources effectively and efficiently to make continuous improvements. As individuals grow, the organization grows. It is worth pointing out, however, that employees will not be motivated towards continual improvement in the absence of:

- Commitment from top management.
- The right organizational 'climate'.
- A mechanism for enabling individual contributions to be effective.

All these are focused essentially at enabling people to feel, accept, and discharge responsibility. More than one organization has made this part of their strategy – to 'empower people to act'. If one hears from employees comments such as 'We know this is not the best way to do this job, but if that is the way management want us to do it, that is the way we will do it', then it is clear that the expertise existing at the point of operation has not been harnessed and the people do not feel responsible for the outcome of their actions. Responsibility and accountability foster pride, job satisfaction, and better work.

Empowerment is very easy to express conceptually, but it requires real effort and commitment on the part of all managers and supervisors to put into practice. Recognition that only partially successful but good ideas or attempts are to be applauded and not criticized is a good way to start. Encouragement of ideas and suggestions from the workforce, particularly through their part in team or group activities, requires investment. The ultimate rewards are total commitment, both inside the organization and outside through the supplier and customer chains.

Teamwork to support quality management and improvement has several components. It is driven by a strategy, needs a structure, and must be implemented thoughtfully and effectively. The strategy that drives the improvement comprises the:

- vision and mission of the organization;
- critical success factors;
- core process framework.

These components have been dealt with in other chapters. The structural and implementation aspects of teamwork are the subject of the remainder of this chapter.

Running process management and quality improvement teams

Process management and quality improvement teams are groups of people with the appropriate knowledge, skills, and experience who are brought together specifically by management to improve processes and/or tackle and solve particular problems, usually on a project basis. They are cross-functional and often multidisciplinary.

The 'task force' has long been a part of the culture of many organizations at the 'technology' and management levels. But quality and process teams go a step further; they expand the traditional definition of 'process' to cover the entire end-to-end operating system. This includes technology, paperwork, communication and other units, operating procedures, and the process equipment itself. By taking this broader view, the teams can address new problems. The actual running of teams calls several factors into play:

- team selection and leadership;
- team objectives;
- team meetings;
- team assignments;
- team dynamics;
- team results and reviews.

Team selection and leadership

The most important element of a quality process team is its members. People with knowledge and experience relevant to the process or solving the problem are clearly required. However, there should be a limit of five to ten members to keep the team small enough to be manageable but allow a good exchange of ideas. Membership should include appropriate people from groups outside the operational and technical areas directly 'responsible' for the process, if their presence is relevant or essential. In the selection of team members it is often useful to start with just one or two people who are clearly concerned directly with the process. If they try to draw maps or flowcharts (see Chapter 10) of the relevant processes, the requirement to include other people, in order to understand the process and complete the maps/charts, will aid the team selection. This method will also ensure that all those who can make a significant contribution to the process and its improvement are represented.

The process owner has a primary responsibility for team leadership, management and maintenance, and his/her selection and training is crucial to success. The leader need not be the highest ranking person in the team, but must be concerned about accomplishing the team objectives (this is sometimes described as 'task concern') and the needs of the members (often termed 'people concern'). Weakness in either of these areas will lessen the effectiveness of the team in solving problems or making breakthroughs. Team leadership training should be directed at correcting deficiencies in these crucial aspects.

Team objectives

At the beginning of any quality or process improvement project and at the start of every meeting the objectives should be stated as clearly as possible by the leader. This can take a simple form: 'This meeting is to continue the discussion from last Tuesday on the provision of current price data from salesmen to invoice preparation, and to generate suggestions for improvement in its quality'. Project and/or meeting objectives enable the team members to focus thoughts and efforts on the aims, which may need to be restated if the team becomes distracted by other issues.

Team meetings

An agenda should be prepared by the leader and distributed to each team member before every meeting. It should include the following information:

- Meeting place, time and how long it will be.
- A list of members (and co-opted members) expected to attend.
- Any preparatory assignments for individual members or groups.
- Any supporting material to be discussed at the meeting.

Early in a project the leader should orient the team members in terms of the approach, methods, and techniques they will use to solve the problem. This may require a review of the:

- Systematic approach (Chapter 13).
- Procedures and rules for using some of the basic tools, e.g. brainstorming – no judgement of initial ideas.
- Role of the team in the continuous improvement process.
- Authority of the team.

A team secretary should be appointed to take the minutes of meeting and distribute them to members as soon as possible after each meeting. The minutes should not be formal, but reflect decisions and carry a clear statement of the action plans, together with assignments of tasks. They may be hand-written initially, copied and given to team members at the end of the meeting, to be followed later by a more formal document that will be seen by any member of staff interested in knowing the outcome of the meeting. In this way the minutes form an important part of the communication system, supplying information to other teams or people needing to know what is going on.

Team assignments

It is never possible to solve problems by meetings alone. What must come out of those meetings is a series of action plans that assign specific tasks to team members. This is the responsibility of the team leader. Agreement must be reached regarding the responsibilities for individual assignments, together with the time scale, and this must be made clear in the minutes. Task assignments must be decided while the team is together and not by separate individuals in after-meeting discussions.

Team dynamics

In any team activity the interactions between the members are vital to success. If solutions to problems are to be found, the meetings and ensuing assignments should assist and harness the creative thinking process. This is easier said than done, because many people have either not learned or been encouraged to be innovative. The team leader clearly has a role here to:

- Create a 'climate' for creativity.
- Encourage all team members to speak out and contribute their own ideas or build on others.
- Allow differing points of view and ideas to emerge.
- Remove barriers to idea generation, e.g. incorrect preconceptions that are usually destroyed by asking 'Why?'
- Support all team members in their attempts to become creative.

In addition to the team leader's responsibilities, the members should:

- Prepare themselves well for meetings, by collecting appropriate data or information (*facts*) pertaining to a particular problem.
- Share ideas and opinions.
- Encourage other points of view.
- Listen 'openly' for alternative approaches to a problem or issue.
- Help the team determine the best solutions.
- Reserve judgement until all the arguments have been heard *and* fully understood.
- Accept individual responsibility for assignments and group responsibility for the efforts of the team.

Team results and reviews

A process approach to improvement and problem solving is most effective when the results of the work are communicated and acted upon. Regular feedback to the teams, via their leaders, will assist them to focus on objectives, and review progress.

Reviews also help to deal with certain problems that may arise in teamwork. For example, certain members may be concerned more with their own personal objectives than those of the team. This may result in some manipulation of the problem-solving process to achieve different goals, resulting in the team splitting apart through self-interest. If recognized, the review can correct this effect and demand greater openness and honesty.

A different type of problem is the failure of certain members to contribute and take their share of individual and group responsibility. Allowing other people to do their work results in an uneven distribution of effort, and leads to bitterness. The review should make sure that all members have assigned and specific tasks, and perhaps lead to the documentation of duties in the minutes. A team roster may even help.

A third area of difficulty, which may be improved by reviewing progress, is the ready-fire-aim syndrome of action before analysis. This often results from team leaders being too anxious to deal with a problem. A review should allow the problem to be redefined adequately and expose the real cause(s). This will release the trap the team may be in of having to do something before they really know what should be done. The review will provide the opportunity to rehearse the steps in the systematic approach.

Teamwork and action-centered leadership

Over the years there has been much academic work on the psychology of teams and on the leadership of teams. Three points on which all authors are in agreement are that teams develop a personality and culture of their own, respond to leadership, and are motivated according to criteria usually applied to individuals.

Key figures in the field of human relations, like Douglas McGregor (Theories X & Y), Abraham Maslow (Hierarchy of Needs) and Fred Hertzberg (Motivators and Hygiene Factors), all changed their opinions on group dynamics over time as they came to realize that groups are not the democratic entity that everyone would like them to be, but respond to individual, strong, well-directed leadership, both from without and within the group, just like individuals.

Action-centered leadership

During the 1960s John Adair, senior lecturer in Military History and the Leadership Training Adviser at the Military Academy, Sandhurst, and later assistant director of the Industrial Society, developed what he called the action-centered leadership model, based on his experiences at Sandhurst, where he had the responsibility to ensure that results in the cadet training did not fall below a certain standard. He had observed that some instructors frequently achieved well above average

results, owing to their own natural ability with groups and their enthusiasm. He developed this further into a team model, which is the basis of the approach of the author and his colleagues to this subject.

In developing his model for teamwork and leadership, Adair made it clear that for any group or team, big or small, to respond to leadership, they need a clearly defined *task*, and the response and achievement of that task are interrelated to the needs of the *team* and the separate needs of the *individual members* of the team (Figure 15.2).

The value of the overlapping circles is that it emphasizes the unity of leadership and the interdependence and multifunctional reaction to single decisions affecting any of the three areas.

Leadership tasks

Drawing upon the discipline of social psychology, Adair developed and applied to training the functional view of leadership. The essence of this he distilled into the three interrelated but distinctive requirements of a leader. These are to define and achieve the job or task, to build up and co-ordinate a team to do this, and to develop and satisfy the individuals within the team (Figure 15.3).

1 *Task needs*. The difference between a team and a random crowd is that a team has some common purpose, goal or objective, e.g. a football team. If a work team does not achieve the required results or meaningful results, it will become frustrated. Organizations have to make a profit, to provide a service, or even to survive. So anyone who manages others has to achieve results; in production, marketing, selling or whatever. Achieving objectives is a major criterion of success.

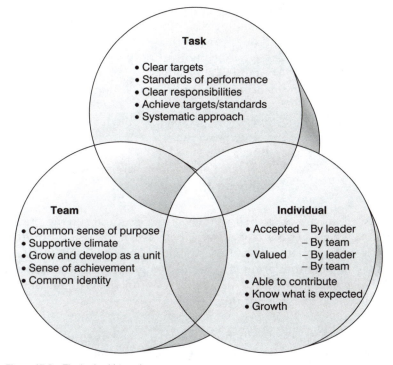

Task

- Clear targets
- Standards of performance
- Clear responsibilities
- Achieve targets/standards
- Systematic approach

Team

- Common sense of purpose
- Supportive climate
- Grow and develop as a unit
- Sense of achievement
- Common identity

Individual

- Accepted – By leader
           – By team
- Valued   – By leader
           – By team
- Able to contribute
- Know what is expected
- Growth

■ **Figure 15.3** The leadership needs

2 *Team needs*. To achieve these objectives, the group needs to be held together. People need to be working in a co-ordinated fashion in the same direction. Teamwork will ensure that the team's contribution is greater than the sum of its parts. Conflict within the team must be used effectively; arguments can lead to ideas or to tension and lack of co-operation.

3 *Individual needs*. Within working groups, individuals also have their own set of needs. They need to know what their responsibilities are, how they will be needed, how well they are performing. They need an opportunity to show their potential, take on responsibility and receive recognition for good work.

The task, team and individual functions for the leader are as follows:

(a) *Task functions*   Defining the task
                      Making a plan
                      Allocating work and resources
                      Controlling quality and tempo of work
                      Checking performance against the plan
                      Adjusting the plan

(b) *Team functions* Setting standards
Maintaining discipline
Building team spirit
Encouraging, motivating, giving a sense of purpose
Appointing subleaders
Ensuring communication within the group
Training the group

(c) *Individual* Attending to personal problems
    functions Praising individuals
Giving status
Recognizing and using individual abilities
Training the individual

The team leader's or facilitator's task is to concentrate on the small central area where all three circles overlap. In a business that is introducing quality management this is the 'action to change' area, where the leaders may be attempting to manage the change from *business as usual* to *TQM equals business as usual*, using the cross-functional quality improvement teams at the strategic interfaces.

In the action area the facilitator's or leader's task is similar to the task outlined by John Adair. It is to try to satisfy all three areas of need by achieving the task, building the team, and satisfying individual needs. If a leader concentrates on the task, e.g. in going all out for production schedules, while neglecting the training, encouragement and motivation of the team and individuals, (s)he may do very well in the short term. Eventually, however, the team members will give less effort than they are capable of. Similarly, a leader who concentrates only on creating team spirit, while neglecting the task and the individuals, will not receive maximum contribution from the people. They may enjoy working in the team but they will lack the real sense of achievement that comes from accomplishing a task to the utmost of the collective ability.

So the leader/facilitator must try to achieve a balance by acting in all three areas of overlapping need. It is always wise to work out a list of required functions within the context of any given situation, based on a general agreement on the essentials. Here is Adair's original Sandhurst list, on which one's own adaptation may be based:

■ *Planning*, e.g. seeking all available information.
Defining group task, purpose or goal.
Making a workable plan (in right decision-making framework).
■ *Initiating*, e.g. briefing group on the aims and the plan.
Explaining why aim or plan is necessary.

Allocating tasks to group members.
- *Controlling*, e.g. maintaining group standard.
Influencing tempo.
Ensuring all actions are taken towards objectives.
Keeping discussions relevant.
Prodding group to action/decision.
- *Supporting*, e.g. expressing acceptance of persons and their contribution.
Encouraging group/individuals.
Disciplining group/individuals.
Creating team spirit.
Relieving tension with humor.
Reconciling disagreements or getting others to explore them.
- *Informing*, e.g. clarifying task and plan.
Giving new information to the group, i.e. keeping them 'in the picture'.
Receiving information from the group.
Summarizing suggestions and ideas coherently.
- *Evaluating*, e.g. checking feasibility of an idea.
Testing the consequences of a proposed solution.
Evaluating group performance.
Helping the group to evaluate its own performance against standards.

A checklist (Table 15.1) should assist the team leader to measure the progress against the required functions of fulfilling the task, maintaining the team and growing the people.

Team processes

The team process is like any other process; it has inputs and outputs. High-performing teams have three main attributes: high task fulfillment, high team maintenance, and low self-orientation. These may be subdivided as follows:

TASK FULFILLMENT

INITIATING	Ideas, solutions, defining problems, suggesting procedures, proposing tasks or goals.
INFORMATION-SEEKING	Facts, opinions, suggestions, ideas.
INFORMATION-GIVING	Facts, opinions, suggestions, ideas.
CLARIFYING	Analyzing implications of information or ideas, interpreting, defining terms, indicating alternatives.
SUMMARIZING	Reviewing, drawing together ideas/information.
TESTING FOR CONSENSUS	Checking readiness for decision, checking agreements.

■ **Table 15.1** Task – team – individual checklist

Task	1	Are the targets clearly set out?
	2	Are there clear standards of performance?
	3	Are available resources defined?
	4	Are responsibilities clear?
	5	Are resources fully utilized?
	6	Are targets/standards being defined?
	7	Is a systematic approach being used?
Team	1	Is there a common sense of purpose?
	2	Is there a supportive climate?
	3	Is the unit growing and developing?
	4	Is there a sense of corporate achievement?
	5	Is there a common identity?
	6	Does the team know and respond to the leader's vision?
Individual	1	Is each individual accepted by the leader/team?
	2	Is each individual made part of the team by leader/team?
	3	Is each individual able to contribute?
	4	Does each individual know what is expected in relation to the task and by the team?
	5	Does each individual feel a part of the team?
	6	Does each individual feel valued by the team?
	7	Is there evidence of individual growth?

TEAM MAINTENANCE

ENCOURAGING	Supporting contributions, stimulating contributions, being friendly and responsive.
SETTING STANDARDS	Suggesting standards for group working, reviewing against these, evaluating group success.
EXPRESSING GROUP FEELINGS	Observing, understanding and expressing group emotions, reducing tension, mediating, recognizing conflicts and encouraging exploration of differences.
COMPROMISING	Giving weight to others' views, commitment to best solution.
GATEKEEPING	Keeping communications open, facilitating participation, suggesting procedures for sharing discussion.

SELF-ORIENTATION

BLOCKING	Interposing a difficulty without alternative or reasoning.
AGGRESSIVENESS	Attacking, overpainting the picture to stir up feelings, exaggerating.

Table 15.2 Team meeting review

Good points				Bad points
	10	5	0	
1 Goal clear and agreed				Goal unclear
2 Previous agreements complete				Partially or not at all
3 We listen to each other				No awareness of listening
4 Right people present				Team not correctly composed
5 Leadership needs creatively met				Drifting or dominating
6 Open and trusting atmosphere				Distrust and defensiveness
7 Time used efficiently				Time wasted
8 Systematic tools used				Lack of systematic approach
9 Agreements reached (what/who/when) and documented				Verbal agreements or none
10 Consensus decisions				Authoritarian or other
11 I was able to express my opinion				No opportunity
12 Opinions could be questioned				Opinions untouchable
13 Opinions distinguished from facts				Mixed and not aware of it
14 Everyone participating				Some not participating
15 Challenging, rewarding, committed atmosphere				Flat and lifeless

UNACCEPTABLE	0
MUST BE IMPROVED	1, 2, 3
FAIR	4, 5, 6
GOOD	7, 8, 9
EXCELLENT	10

DOMINATING	Asserting authority or superiority in manipulating group, refusing to budge.
FORMING CLIQUES	Forming sub-groups for protection or support.
SPECIAL PLEADING	Speaking for special interests as a cover for personal interest.
SEEKING SYMPATHY	Drawing attention, attempting to gain sympathy.
WITHDRAWING	Opting out or getting behind stronger members.
WASTING TIME	Various diversions for self-orientated reasons.
NOT LISTENING	Ignoring suggestions or closing off hearing when others are speaking.

These may be used in various ways to construct a 'team behavior checklist', which may be used by team facilitators or observers to rate the team performance. A second review document (see Table 15.2) may be used by individuals to rate the various aspects of a team meeting.

Situational leadership _____

In dealing with the task, the team, and with any individual in the team, a style of leadership appropriate to the situation must be adopted. The teams and the individuals within them will, to some extent, start 'cold', but they will develop and grow in both strength and experience. The interface with the leader must also change with the change in the team, according to the Tannenbaum and Schmidt model (Figure 15.4).[1]

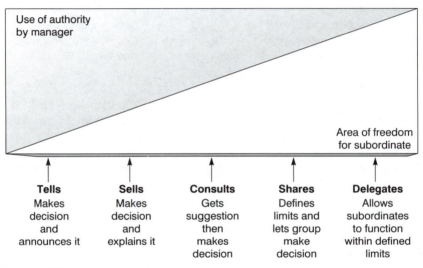

Tells	**Sells**	**Consults**	**Shares**	**Delegates**
Makes	Makes	Gets	Defines	Allows
decision	decision	suggestion	limits and	subordinates
and	and	then	lets group	to function
announces it	explains it	makes	make	within defined
		decision	decision	limits

■ **Figure 15.4**   Continuum of leadership behavior

Initially a very directive approach may be appropriate, giving clear instructions to meet agreed goals. Gradually, as the teams become more experienced and have some success, the facilitating team leader will move through coaching and support to a less directing and eventually less supporting and less directive approach – as the more interdependent style permeates the whole organization.

This equates to the modified Blanchard model[1] in Figure 15.5, where directive behavior moves from high to low as people develop and are more easily empowered. When this is coupled with the appropriate level of supportive behavior, a directing style of leadership can move through coaching and supporting to a delegating style. It must be stressed, however, that effective delegation is only possible with developed 'followers', who can be fully empowered.

One of the great mistakes in recent years has been the expectation by management that teams can be put together with virtually no training or development (S1 in Figure 15.5) and that they will perform as a mature team (S4). The Blanchard model emphasizes that there is no

The four leadership styles

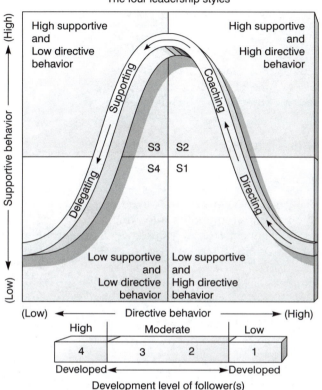

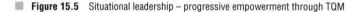

Figure 15.5   Situational leadership – progressive empowerment through TQM

quick and easy 'tunnel' from S1 to S4. The only route is the laborious climb through S2 and S3.

Stages of team development

Original work by Tuckman[1] suggested that when teams are put together, there are four main stages of team development, the so-called forming (awareness), storming (conflict), norming (co-operation), and performing (productivity). The characteristics of each stage and some key aspects to look out for in the early stages are given below.

Forming – awareness

Characteristics:

- Feelings, weaknesses and mistakes are covered up.
- People conform to established lines.

- Little care is shown for others' values and views.
- There is no shared understanding of what needs to be done.

Watch out for:

- Increasing bureaucracy and paperwork.
- People confining themselves to defined jobs.
- The 'boss' is ruling with a firm hand.

Storming – conflict

Characteristics:

- More risky, personal issues are opened up.
- The team becomes more inward looking.
- There is more concern for the values, views and problems of others in the team.

Watch out for:

- The team becomes more open, but lacks the capacity to act in a unified, economic, and effective way.

Norming – co-operation

Characteristics:

- Confidence and trust to look at how the team is operating.
- A more systematic and open approach, leading to a clearer and more methodical way of working.
- Greater valuing of people for their differences.
- Clarification of purpose and establishing of objectives.
- Systematic collection of information.
- Considering all options.
- Preparing detailed plans.
- Reviewing progress to make improvements.

Performing – productivity

Characteristics:

- Flexibility.
- Leadership decided by situations, not protocols.
- Everyone's energies utilized.
- Basic principles and social aspects of the organization's decisions considered.

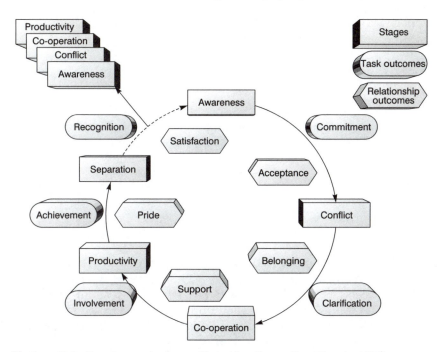

Figure 15.6   Team stages and outcomes. (Derived from Kormanski and Mozenter, 1987[1])

The team stages, the task outcomes and the relationship outcomes are shown together in Figure 15.6. This model, which has been modified from Kormanski,[2] may be used as a framework for the assessment of team performance. The issues to look for are:

1 How is leadership exercised in the team?
2 How is decision making accomplished?
3 How are team resources utilized?
4 How are new members integrated into the team?

Teams, which go through these stages successfully, should become effective teams and display the following attributes.

Attributes of successful teams

Clear objectives and agreed goals

No group of people can be effective unless they know what they want to achieve, but it is more than knowing what the objectives are. People are only likely to be committed to them if they can identify with and have ownership of them – in other words, objectives and goals are agreed by team members.

Often this agreement is difficult to achieve but experience shows that it is an essential prerequisite for the effective group.

Openness and confrontation

If a team is to be effective, then the members of it need to be able to state their views, their differences of opinion, interests and problems, without fear of ridicule or retaliation. No teams work effectively if there is a cut-throat atmosphere, where members become less willing or able to express themselves openly; then much energy, effort and creativity are lost.

Support and trust

Support naturally implies trust among team members. Where individual group members do not feel they have to protect their territory or job, and feel able to talk straight to other members, about both 'nice' and 'nasty' things, then there is an opportunity for trust to be shown. Based on this trust, people can talk freely about their fears and problems and receive from others the help they need to be more effective.

Co-operation and conflict

When there is an atmosphere of trust, members are more ready to participate and are committed. Information is shared rather than hidden. Individuals listen to the ideas of others and build on them. People find ways of being more helpful to each other and the group generally. Co-operation causes high morale – individuals accept each other's strengths and weaknesses and draw on their pool of knowledge or skill. All abilities, knowledge and experience are fully utilized by the group; individuals have no inhibitions about using other people's abilities to help solve their problems, which are shared.

Allied to this, conflicts are seen as a necessary and useful part of the organizational life. The effective team works through issues of conflict and uses the results to help objectives. Conflict prevents teams from becoming complacent and lazy, and often generates new ideas.

Good decision making

As mentioned earlier, objectives need to be clearly and completely understood by all members before good decision making can begin. In making decisions effective, teams develop the ability to collect information quickly then discuss the alternatives openly. They become committed to their decisions and ensure quick action.

Appropriate leadership

Effective teams have a leader whose responsibility it is to achieve results through the efforts of a number of people. Power and authority can be applied in many ways, and team members often differ on the style of leadership they prefer. Collectively, teams may come to different views of leadership but, whatever their view, the effective team usually sorts through the alternatives in an open and honest way.

Review of the team processes

Effective teams understand not only the group's character and its role in the organization, but how it makes decisions, deals with conflicts, etc. The team process allows the team to learn from experience and consciously to improve teamwork. There are numerous ways of looking at team processes – use of an observer, by a team member giving feedback, or by the whole group discussing members' performance.

Sound intergroup relationships

No human being or group is an island; they need the help of others. An organization will not achieve maximum benefit from a collection of quality improvement teams that are effective within themselves but fight among each other.

Individual development opportunities

Effective teams seek to pool the skills of individuals, and it necessarily follows that they pay attention to development of individual skills and try to provide opportunities for individuals to grow and learn, and of course have FUN.

Once again, these ideas are not new but are very applicable and useful in the management of teams for quality improvement, just as Newton's theories on gravity still apply!

Personality types and the MBTI

No one person has a monopoly of 'good characteristics'. Attempts to list the qualities of the ideal manager, for example, demonstrate why that paragon cannot exist. This is because many of the qualities are mutually exclusive, for example.

Highly intelligent	*v.*	Not *too* clever
Forceful and driving	*v.*	Sensitive to people's feelings
Dynamic	*v.*	Patient
Fluent communicator	*v.*	Good listener
Decisive	*v.*	Reflective

Although no individual can possess all these and more desirable qualities, a team often does.

A powerful aid to team development is the use of the Myers-Briggs Type Indicator (MBTI).[1] This is based on an individual's preferences on four scales for:

- giving and receiving 'energy';
- gathering information;
- making decisions;
- handling the outer world.

Its aim is to help individuals understand and value themselves and others, in terms of their differences as well as their similarities. It is well researched and non-threatening when used appropriately.

The four MBTI preference scales, which are based on Jung's theories of psychological types, represent two opposite preferences:

- *Extroversion – Introversion* – how we prefer to give/receive energy or focus our attention.
- *Sensing – Intuition* – how we prefer to gather information.
- *Thinking – Feeling* – how we prefer to make decisions.
- *Judgement – Perception* – how we prefer to handle the outer world.

To understand what is meant by preferences, the analogy of left- and right-handedness is useful. Most people have a preference to write with either their left or their right hand. When using the preferred hand, they tend not to think about it, it is done naturally. When writing with the other hand, however, it takes longer, needs careful concentration, seems more difficult, but with practice would no doubt become easier. Most people *can* write with and use both hands, but tend to prefer one over the other. This is similar to the MBTI psychological preferences: most people are able to use both preferences at different times, but will indicate a preference on each of the scales.

In all, there are eight possible preferences – E or I, S or N, T or F, J or P, i.e. two opposites for each of the four scales. An individual's *type* is the combination and interaction of the four preferences. It can be assessed initially by completion of a simple questionnaire. Hence, if each preference is represented by its letter, a person's type may be shown by a four letter code – there are 16 in all. For example, ESTJ represents a preference for *extroversion* (E) who prefers to gather information with *sensing* (S), prefers to make decisions by *thinking* (T) and has a *judging* (J) attitude towards the world, i.e. prefers to make decisions rather than continue to collect information. The person with opposite preferences on all four scales would be an INFP, a preference for *introversion* (I) who prefers intuition for perceiving, feelings or values for making decisions, and likes to maintain a perceiving attitude towards the outer world.

The questionnaire, its analysis and feedback must be administered by a qualified MBTI practitioner, who may also act as external facilitator to the team in its forming and storming stages.

Type and teamwork

With regard to teamwork, the preference types and their interpretation are extremely powerful. The *'extrovert'* prefers action and the outer world, while the *'introvert'* prefers ideas and the inner world.

Sensing–thinking types are interested in facts, analyze facts impersonally, and use a step-by-step process from cause to effect, premise to conclusion. The *sensing–feeling* combinations, however, are interested in facts, analyze facts personally, and are concerned about how things matter to themselves and others.

Intuition–thinking types are interested in possibilities, analyze possibilities impersonally, and have theoretical, technical, or executive abilities. On the other hand, the *intuition–feeling* combinations are interested in possibilities, analyze possibilities personally, and prefer new projects, new truths, things not yet apparent.

Judging types are decisive and planful, they live in an orderly fashion, and like to regulate and control. *Perceivers*, on the other hand, are flexible, live spontaneously, and understand and adapt readily.

As we have seen, an individual's type is the combination of four preferences on each of the scales. There are 16 combinations of the preference scales and these may be displayed on a *type table* (Figure 15.7). If the individuals within a team are prepared to share with each other their MBTI preferences, this can dramatically increase understanding

ISTJ	ISFJ	INFJ	INTJ
ISTP	ISFP	INFP	INTP
ESTP	ESFP	ENFP	ENTP
ESTJ	ESFJ	ENFJ	ENTJ

Figure 15.7 MBTI type table form. (Source: Isabel Myers-Briggs, Introduction to Type[2])

and frequently is of great assistance in team development and good team working. The similarities and differences in behavior and personality can be identified. The assistance of a qualified MBTI practitioner is absolutely essential in the initial stages of this work.

Problem solving using type preferences

The MBTI preferences may be used by an individual or a team in a step-by-step process for problem solving. The problem solving model, represented in Figure 15.8, is straightforward, but can be difficult to use, because people tend to skip over those steps that require them to use their non-preferences. For example, information tends to be gathered by the preferred function (S or N) and decisions made by the preferred function (T or F). A strongly ST type will spend much time gathering facts (S) and thinking logically through the decision process (T), with perhaps insufficient attention being given to other possibilities (N) and the impact on people (F). If the size of each letter represents a unit of time, the ST's problem solving method may be represented in Figure 15.9, in which the Z pattern of the model is not followed. Problems, solutions, and decisions are likely to be improved if *all* the preferences are used. Until individuals master the process of spending time in their non-preferred functions, i.e. type development, it may be wise to consult others of opposite preference when tackling important problems or making vital decisions.

Clearly, this has great implications for teamwork and requires that team members share their MBTI preferences or types.

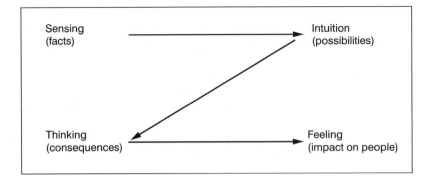

Figure 15.8   Team task analysis: the problem-solving model. Source: Sandra Krebs Hirsh and Jean M. Kummerow, *Introduction to Type in Organisational Settings*[2]

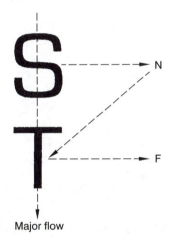

Figure 15.9
Relative time spent on each aspect of
problem solving model by 'ST type'

Major flow

Interpersonal relations – FIRO-B and the Elements

The FIRO-B (Fundamental Interpersonal Relations Orientation– Behavior) is a powerful psychological instrument which can be used to give valuable insights into the needs individuals bring to their relationships with other people. The instrument assesses needs for **inclusion**, **control** and **openness** and therefore offers a framework for understanding the dynamics of interpersonal relationships.

Use of the FIRO instrument helps individuals to be more aware of how they relate to others and to become more flexible in this behavior. Consequently it enables people to build more productive teams through better working relationships.

Since its creation by William Schutz in the 1950s, to predict how military personnel would work together in groups, the FIRO-B instrument has been used throughout the world by managers and professionals to look at management and decision-making styles. Through its ability to predict areas of probable tension and compatibility between individuals, the FIRO-B is a highly effective team building tool which can aid in the creation of the positive environment in which people thrive and achieve improvements in performance.

The theory underlying the FIRO-B incorporates ideas from the work of Adomo, Fromm and Bion and it was first fully described in Schutz's book, *FIRO: A Three Dimensional Theory of Personal Behaviour* (1958). In his more recent book *The Human Element*, Schutz developed the instrument into a series of 'elements', B, F, S, etc. and offers strategies for heightening our awareness of ourselves and others.

The FIRO-B takes the form of a simple-to-complete questionnaire the analysis of which provides scores that estimate the levels of behavior with which the individual is comfortable, with regard to his/her needs for inclusion, control and openness. Schutz described these three dimensions in the form of the decisions we make in our relationships regarding whether we want to be:

■ 'in' or 'out' – inclusion;
■ 'up' or 'down' – control;
■ 'close' or 'distant' – openness.

The FIRO-B estimates our unique level of needs for each of these dimensions of interpersonal interaction.

The instrument further divides each of these dimensions into:

■ the behavior we feel most comfortable **exhibiting towards** other people – **expressed** behaviors; and
■ the behavior we **want from** others – **wanted** behaviors.

Hence, the FIRO-B 'measures', on a scale of 0–9, each of the three interpersonal dimensions in two aspects (Table 15.3).

■ **Table 15.3** The FIRO-B interpersonal dimensions and aspects

	Inclusion	Control	Openness
Expressed behavior	Expressed inclusion	Expressed control	Expressed openness
Wanted behavior	Wanted inclusion	Wanted control	Wanted openness

Modified from: W. Schutz (1978) *FIRO Awareness Scales Manual*, Palo Alto, CA, Consulting Psychologists Press.

The **expressed** aspect of each dimension indicates the level of behavior the individual is most comfortable with towards others, so high scores for the expressed dimensions would be associated with:

High scored-expressed behaviors

Inclusion Makes efforts to include other people in his/her activities – tries to belong to or join groups and to be with people as much as possible.

Control tries to exert control and influence over people and tell them what to do.

Openness Makes efforts to become close to people – expresses friendly open feelings, tries to be personal and even intimate.

Low scores would be associated with the opposite expressed behavior.

The **wanted** aspect of each dimension indicates the behavior the individual prefers others to adopt towards him/her, so high scores for the wanted dimensions would be associated with:

High scored-wanted behaviors

Inclusion Wants other people to include him/her in their activities – to be invited to belong to or join groups (even if no effort is made by the individual to be included).

Control Wants others to control and influence him/her and be told what to do.

Openness Wants others to become close to him/her and express friendly, open, even affectionate feelings.

Low scores would be associated with the opposite wanted behaviors.

It is interesting to look at typical manager FIRO-B profiles, based on their scores for the six dimensions/aspects in Table 15.3. Figure 15.10 shows the average of a sample of 700 middle and senior managers in the UK with boundaries at one sigma, plotted on expressed/wanted scales for the three dimensions.

On average the managers show a higher level of expressed inclusion – including people in his/her activities – than wanted inclusion. Similarly, and not surprisingly perhaps, expressed control – trying to exert influence and control over others – is higher in managers than wanted control. When it comes to openness, the managers tend to want others to be open, rather than be open themselves.

It is even more interesting to contemplate these results when one considers the demands of some of the recent popular management programs,

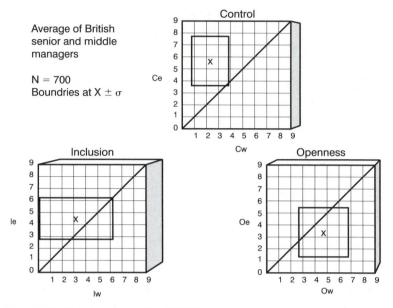

▨ **Figure 15.10**   Typical manager profiles (FIRO-B)

such as total quality management, employee involvement, and self-directed teams. These tend to require from managers certain behaviors, for example lower levels of expressed control and higher levels of wanted control, so that the people feel empowered. Similarly, managers are encouraged to be more open. These, however, are **opposite** to the apparent behaviors of the sample of managers shown graphically in Figure 15.10. It is not surprising then that TQM has failed in some organizations where managers were being asked to empower employees and be more open – and who can argue against that – yet their basic underlying needs caused them to behave in the opposite way.

Understanding what drives these behaviors is outside the scope of this book but other FIRO and Element instruments can help individuals to further develop understanding of themselves and others. FIRO and Schutz's Elements instruments for measuring **feelings** (F) and **self-concept** (S) can deepen the awareness of what lies behind our behaviors with respect to inclusion, control and openness. The reader is advised to undertake further reading and seek guidance from a trained administrator of these instruments, but the overall relationship between the B and F instruments is given below:

Behaviors related to:	Feelings related to:
Inclusion	Significance
Control	Competence
Openness	Likeability

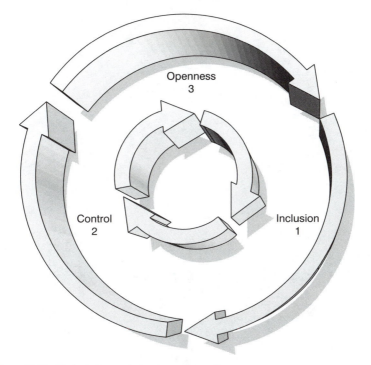

Figure 15.11   The inclusion, control and openness cycle

Issues around control behavior then may arise because of underlying feelings about competence. Similarly, underlying feelings concerning significance may lead to certain inclusion behaviors.

FIRO-B in the workplace

The inclusion, control and openness (I–C–O) dimensions form a cycle (Figure 15.11), which can help groups of people to understand how their individual and joint behavior develops as teams are formed. Given in Table 15.4 are the considerations, questions and outcomes under each dimension. If inclusion issues are resolved first it is possible to progress to dealing with the control issues, which in turn must be resolved if the openness issues are to be dealt with successfully. As a team develops, it travels around the inclusion, control and openness cycle time and time again. If the issues are not resolved in each dimension, further progress in the next dimension will be hindered – it is difficult to deal with issues of control if unresolved inclusion issues are still around and people do not know whether they are 'in' or 'out' of the group. Similarly it is difficult to be open if it is not clear where the power base is in the group.

Table 15.4 Considerations, questions and outcomes for the FIRO-B dimensions

Dimension	Considerations	Some typical questions	If resolved we get:	If not resolved we get:
Inclusion	Involvement – how much you want to include other people in your life and how much attention and recognition you want.	Do I care about this? Do I want to be involved? Does this fit with my values? Do I matter to this group? Can I be committed? … leading to … Am I 'in' or 'out'?	A feeling of belonging A sense of being recognized and valued Willingness to become committed	A feeling of alienation A sense of personal insignificance No desire for commitment or involvement
Control	Authority, responsibility, decision making, influence.	Who is in charge here? Do I have power to make decisions? What is the plan? When do we start? What support do I have? What resources do I have? … leading to … Am I 'up' or 'down'?	Confidence in self and others Comfort with level of responsibility Willingness to belong	Lack of confidence in leadership Discomfort with level of responsibility – fear of too much – frustration with too little 'Griping' between individuals
Openness	How much are we prepared to express our true thoughts and feelings with other individuals.	Does she like me? Will my work be recognized? Is he being honest with me? How should I show appreciation? Do I appear aloof? … leading to … Am I 'open' or 'closed'?	Lively and relaxed atmosphere Good-humored interactions Open and trusting relationships	Tense and suspicious atmosphere Flippant or malicious humor Individuals isolated

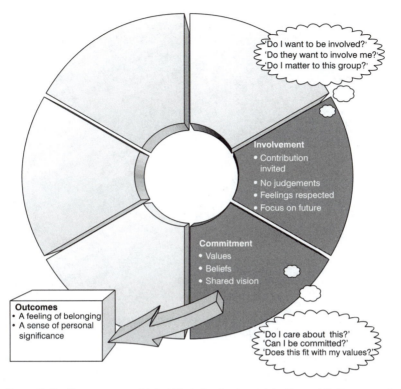

Figure 15.12   The openness model, Part I Inclusion: involvement, inviting contribution, responding

This I–C–O cycle has led to the development by the author and his colleagues of an 'openness model' which is in three parts. Part 1 is based on the premise that to participate productively in a team individuals must first be involved and then committed. Figure 15.12 shows some of the questions which need to be answered and the outcomes from this stage. Part 2 deals with the control aspects of empowerment and management and Figure 15.13 summarizes the questions and outcomes. Finally Part 3, summarized in Figure 15.14, ensures openness through acknowledgement and trust. The full openness cycle (Figure 15.15) operates in a clockwise direction so that trust leads to more involvement, further commitment, increased empowerment, etc. Of course, if progress is not made round the cycle and trust is replaced by fear, it is possible to send the whole process into reverse – a negative cycle of suspicion fault-finding, abdication and confusion (Figure 15.16). Unfortunately this will be recognized as the culture in some organizations where the focus of enquiry is 'what has gone wrong?' leading to 'whose fault was it?'

Fortunately, organizations and individuals seem keen to learn ways to change these negative communications that sour relationships, dampen personal satisfaction and reduce productivity. The inclusion,

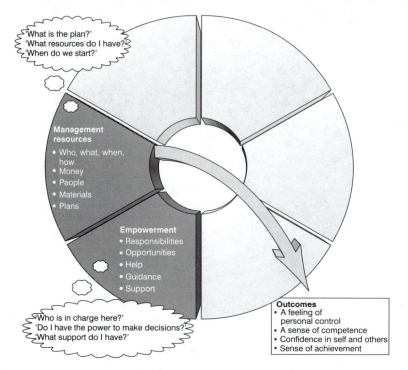

Figure 15.13   The openness model, Part 2 Control: choice, influence, power

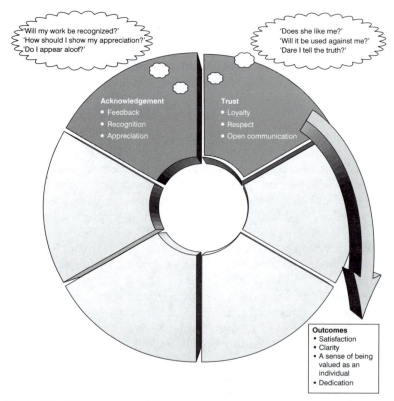

Figure 15.14   The openness model, Part 3 Openness: expression of true thoughts and feelings with respect for self and others

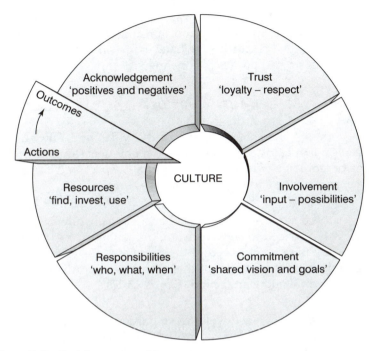

Figure 15.15   The full openness model

Figure 15.16   The negative cycle

control, openness cycle is a useful framework for helping teams to pass successfully through the forming and storming stages of team development. As teams are disbanded for whatever reason, the process reverses and the first thing that goes is the openness.

The five 'A' stages for teamwork

The awareness provided by the use of the MBTI and FIRO-B instruments helps people to appreciate their own uniqueness and the uniqueness of others – the foundation of mutual respect and for building positive, productive and high performance teams.

For any of these models or theories to benefit a team, however, the individuals within it need to become **aware** of the theory, e.g. the MBTI or FIRO-B. They then need to **accept** the principles as valid, **adopt** them for themselves in order to **adapt** their behavior accordingly. This will lead to individual and team **action** (Figure 15.17).

In the early stages of team development particularly, the assistance of a skilled facilitator to aid progress through these stages is necessary. This is often neglected, causing failure in so many team initiatives. In such cases the net output turns out to be lots of nice warm feelings about 'how good that team workshop was a year ago', but the nagging reality that no action came out and nothing has really changed.

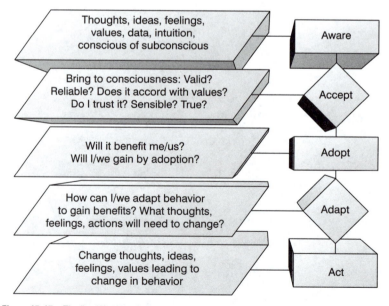

■ **Figure 15.17**   The five 'A' stages for teamwork

■ References

1. See references under *Culture change through people and teamwork* heading in Bibliography.
2. Kormonski, C. and Mozenter, A., 'A new model of team building: a technology for today and tomorrow', *The 1987 Annual: Developing Human Resources*, University Associates, San Diego, CA, USA, 1985.

Chapter highlights

The need for teamwork

- The only efficient way to tackle process improvement or complex problems is through teamwork. The team approach allows individuals and organizations to grow.
- Employees will not engage in continual improvement without commitment from the top, a quality 'climate', and an effective mechanism for capturing individual contributions.
- Teamwork for quality improvement is driven by a strategy, needs a structure, and must be implemented thoughtfully and effectively.

Running process management and quality improvement teams

- Process management and quality improvement teams are groups brought together by management to improve a process or tackle a particular problem on a project basis.
- The running of these teams involves several factors: selection and leadership, objectives, meetings, assignments, dynamics, results and reviews.

Teamwork and action-centered leadership

- Early work in the field of human relations by McGregor, Maslow, and Hertzberg was useful to John Adair in the development of his model for teamwork and action-centered leadership.
- Adair's model addresses the needs of the task, the team, and the individuals in the team, in the form of three overlapping circles. There are specific task, team and individual functions for the leader, but (s)he must concentrate on the small, central, overlap area of the three circles.
- The team process has inputs and outputs. Good teams have three main attributes: high task fulfillment, high teams maintenance, and low self-orientation.
- In dealing with the task, the team and its individuals, a situational style of leadership must be adopted. This may follow the

Tannenbaum and Schmidt, and Blanchard models through directing, coaching, and supporting to delegating.

Stages of team development

■ When teams are put together, they pass through Tuckman's forming (awareness), storming (conflict), norming (co-operation), and performing (productivity) stages of development.

■ Teams that go through these stages successfully become effective and display clear objectives and agreed goals, openness and confrontation, support and trust, co-operation and conflict, good decision making, appropriate leadership, review of the team processes, sound relationships, and individual development opportunities.

Personality types and the MBTI

■ A powerful aid to team development is provided by the Myers-Briggs Type Indicator (MBTI).

■ The MBTI is based on individuals' preferences on four scales for giving and receiving 'energy' (extroversion – E or introversion – I), gathering information (sensing – S or intuition – N), making decisions (thinking – T or feeling – F) and handling the outer world (judging – J or perceiving – P).

■ An individual's type is the combination and interaction of the four scales and can be assessed initially by completion of a simple questionnaire. There are 16 types in all, which may be displayed for a team on a type table.

Interpersonal relations – FIRO-B and the Elements

■ The FIRO-B (Fundamental Interpersonal Relations Orientation – Behavior) instrument gives insights into the needs individuals bring to their relationships with other people.

■ The FIRO-B questionnaire assesses needs for inclusion, control and openness, in terms of expressed and wanted behavior.

■ Typical manager FIRO-B profiles conflict with some of the demands of TQM and can, therefore, indicate where particular attention is needed to achieve successful implementation.

■ The inclusion, control, and openness dimensions form an 'openness' cycle which can help groups to understand how to develop their individual and joint behaviors as the team is formed. An alternative negative cycle may develop if the understanding of some of these behaviors is absent.

■ The five As: for any of the teamwork models and theories, the individuals must become aware, need to accept, adopt and adapt, in order to act. A skilled facilitator is always necessary.

■
■
■ Communications, innovation
■ and learning
■

■ Communicating the quality strategy

People's attitudes and behavior clearly can be influenced by communications; one has to look only at the media or advertising to understand this. The essence of changing attitudes is to gain acceptance for the need to change, and for this to happen it is essential to provide relevant information, convey good practices, and generate interest, ideas and awareness through excellent communication processes. This is possibly the most neglected part of many organizations' operations, yet failure to communicate effectively creates unnecessary problems, resulting in confusion, loss of interest and eventually in declining quality through apparent lack of guidance and stimulus.

Good quality management will significantly change the way many organizations operate and 'do business'. This change will require direct and clear communication from the top management to all staff and employees, to explain the need to focus on processes. Everyone will need to know their roles in understanding processes and improving their performance.

Whether a strategy is developed by top management for the direction of the business/organization as a whole, or specifically for the introduction of TQM, that is only half the battle. An early implementation step must be the clear widespread communication of the strategy.

An excellent way to accomplish this first step is to issue a message that clearly states top management's commitment to quality and outlines the role everyone must play. This can be in the form of a quality policy (see Chapters 3 and 4) or a specific statement about the organization's intention to integrate quality into the business operations. Such a statement might read:

> The board of directors (or appropriate title) believe that the successful implementation of (total) quality management is critical to achieving and maintaining our business goals of leadership in quality, delivery and price competitiveness.
>
> We wish to convey to everyone our enthusiasm and personal commitment to the (total) quality approach, and how much we need your support in our mission of business improvement. We hope that you will become as convinced as we are that business and process improvement is critical for our survival and continued success.
>
> We can become a (total) quality organization only with your commitment and dedication to improving the processes in which you work. We will help you by putting in place the necessary education, training, and teamwork development, based on business and process improvement, to ensure that we move forward together to achieve our business goals.

The quality director or manager should then assist the senior management team to prepare a quality directive. This needs to be signed by all business unit, division, or core process leaders, and distributed to everyone in the organization. The directive should include the following:

- Need for improvement.
- Concept for (total) quality.
- Importance of understanding business processes.
- Approach that will be taken and people's roles.
- Individual and process group responsibilities.
- Principles of process measurement.

The systems for disseminating the message should include all the conventional communication methods of seminars, departmental meetings, posters, newsletters, intranet, etc. First line supervision will need to review the directive with all the staff, and a set of questions and answers may be suitably prepared in support.

Once people understand the strategy, the management must establish the infrastructure (see Chapter 14). The required level of individual commitment is likely to be achieved, however, only if everyone understands the aims and benefits of quality management, the role they must play, and how they can implement process improvements. For this under-standing a constant flow of information is necessary, including:

1 When and how individuals will be involved.
2 What the process requires.
3 The successes and benefits achieved.

The most effective means of developing the personnel commitment required is to ensure people know what is going on. Otherwise they will feel left out and begin to believe that quality management has nothing to do with them, which will lead to resentment and under-mining of the whole process. The first line of supervision again has an important part to play in ensuring key messages are communicated and in building teams by demonstrating everyone's participation and commitment.

In the Larkins' excellent book *Communicating Change*, the authors refer to three 'facts' regarding the best ways to communicate change to employees.

1 Communicate directly to supervisors (first line).
2 Use face-to-face communication.
3 Communicate relative performance of the local work area.

The language used at the 'coal face' will need attention in many organi-zations. Reducing the complexity and jargon in written and spoken communications will facilitate comprehension. When written business communications cannot be read or understood easily, they receive only cursory glances, rather than the detailed study they require. *Simplify and shorten* must be the guiding principles. The communication model illustrated in Figure 16.1 indicates the potential for problems through

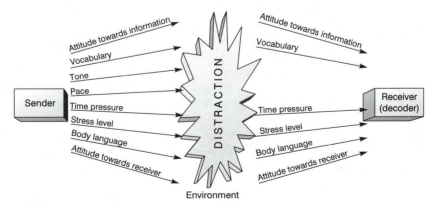

Figure 16.1   Communication model

environmental distractions, mismatches between sender and receiver (or more correctly, decoder) in terms of attitudes – towards the information and each other – vocabulary, time pressures, etc.

All levels of management should introduce and stress 'open' methods of communication, by maintaining open offices, being accessible to staff/employees and taking part in day-to-day interactions and the detailed processes. This will lay the foundation for improved interactions *between* staff and employees, which is essential for information flow and process improvement. Opening these lines of communication may lead to confrontation with many barriers and much resistance. Training and the behavior of supervisors/managements should be geared to helping people accept responsibility for their own behavior, which often creates the barriers, and for breaking down the barriers by concentrating on the process rather than 'departmental' needs.

Resistance to change will always occur and is to be expected. Again first line management should be trained to help people deal with it. This requires an understanding of the dynamics of change and the support necessary – not an obsession with forcing people to change. Opening up lines of communication through a previously closed system, and publicizing people's efforts to change and their results, will aid the process. Change can be – even should be – exciting if employees start to share their development, growth, suggestions, and questions. Management needs to encourage and participate in this by creating the most appropriate communication systems.

Case study ▪▪▪▪▪▪▪▪▪▪▪▪▪▪▪▪▪▪▪▪▪▪▪▪▪▪

The emphasis on quality is now on how BT Retail people work, not on unthinking compliance to standards. People are encouraged to participate in the Revitalizing Quality program and there is no company-wide 'sheep dip' training. Quality is integrated with other established approaches, e.g. key strategic programs, recognition, management training, and is deliberately not seen as a separate initiative. 'Management by fact' is in evidence at all levels and decisions are based on the relevant data rather than on instinct or 'gut feel'.

The BT Retail Quality Charter, explaining key aspects of quality, has been issued to the entire workforce and supported by ongoing communication. The basic quality improvement tools and techniques used by BT since the original TQM campaign launch have been reviewed and updated to provide an increased focus on root cause analysis. Computer-based training for these basic tools plus statistical process control and an advanced quality tool set are available to all employees via BT's intranet.

To demonstrate their continued commitment to TQM, the BT Retail leadership team has drawn up a set of guiding principles to place the BT Values in context for BT Retail employees.

Known as the 'BT Retail Way' these six simple principles are intended to capture the spirit of BT Retail and help guide the company in the twenty-first century. The BT Retail Way:

- Customers are at our heart – we must listen to the customer.
- We aim high – we want to be the best, not simply 'good enough'.
- Our goals are clear – based on facts not anecdotes.
- We deliver – accountability not excuses.
- We are one team – engage not tell.
- We tackle issues – honestly but sensitively.

Communicating the quality message

The people in most organizations fall into one of four 'audience' groups, each with particular general attitudes towards quality management:

- *Senior managers*, who should see quality management as an opportunity, both for the organization and themselves.
- *Middle managers*, who may see quality management as another burden without any benefits, and may perceive a vested interest in the status quo.
- *Supervisors* (first line or junior managers), who may see quality management as another 'flavor of the period' or campaign, and who may respond by trying to keep heads down so that it will pass over.
- *Other employees*, who may not care, so long as they still have jobs and get paid, though these people must be the custodians of the delivery of quality to the customer and own that responsibility.

Senior management needs to ensure that each group sees quality management as being beneficial to them. Quality training material and support (whether internal from a quality director and team or from external consultants) will be of real value only if the employees are motivated to respond positively to them. The implementation strategy must then be based on two mutually supporting aspects:

1 'Marketing' any quality management based initiatives.
2 A positive, logical process of communication designed to motivate.

There are of course a wide variety of approaches to, and methods of, quality management and business improvement. Any individual organization's quality strategy must be designed to meet the needs of its own structure and business, and the state of commitment to continuous improvement activities. These days very few organizations are starting from a green-field site. The key is that groups of people must feel able to 'join' the quality process at the most appropriate point for them. For middle managers to be convinced that they must participate, quality management must be presented as the key to help them turn the people who work for them into 'total quality employees'.

The noisy, showy, hype-type activity is not appropriate to any aspect of quality management. Quality 'events' should of course be fun, because this is often the best way to persuade and motivate, but the value of any event should be judged by its ability to contribute to understanding and change. Key words in successful exercises include discovery, affirmation, participation, and team-based learning. In the difficult area of dealing with middle and junior managers, who can and will prevent change with ease and invisibility, the recognition that progress must change from being a threat to a promise will help. In any workshops designed for them managers and supervisors should be made to feel recognized, not victimized, and the programs should be delivered by specially trained people. The environment and conduct of the workshops must also demonstrate the organization's concern for quality.

The key medium for motivating the employees and gaining their commitment to quality is face-to-face communication and *visible* management commitment. Much is written and spoken about leadership, but it is mainly about communication. If people are good leaders, they are invariably good communicators. Leadership is a human interaction depending on the communications between the leaders and the followers. It calls for many skills that can be *learned* from education and training, but must be *acquired* through practice.

■ Communication, learning, education and ■ training for quality

It may be useful to consider why people learn. They do so for several reasons, some of which include:

- Self-betterment.
- Self-preservation.
- Need for responsibility.
- Saving time or effort.
- Sense of achievement.
- Pride of work.
- Curiosity.

So communication and training can be a powerful stimulus to personal development at the workplace, as well as achieving improvements for the organization. This may be useful in the selection of the appropriate method(s) of communication, the principal ones being:

- *Verbal communication* either between individuals or groups, using direct or indirect methods, such as public address and other broadcasting systems and recordings.
- *Written communication* in the form of notices, bulletins, information sheets, reports, e-mail and recommendations.
- *Visual communication* such as posters, films, video, internet/intranet, exhibitions, demonstrations, displays and other promotional features. Some of these also call for verbal and written communication.
- *Example*, through the way people conduct themselves and adhere to established working codes and procedures, through their effectiveness as communicators and ability to 'sell' good practices.

The characteristics of each of these methods should be carefully examined before they are used in helping people to learn.

It is the author's belief that education and training is the single most important factor in actually improving quality and business performance, once there has been commitment to do so. For education and training to be effective, however, it must be planned in a systematic and objective manner to provide the right sort of learning experience. Education and training must be continuous to meet not only changes in technology but also changes in the environment in which an organization operates, its structure, and perhaps most important of all the people who work there.

Quality education and training cycle of improvement

Quality education and training activities can be considered in the form of a cycle of improvement (Figure 16.2), the elements of which are the following.

Ensure quality education and training is part of the quality policy

Every organization should define its quality policy in relation to education and training. The policy should contain principles and goals to provide a framework within which learning experiences may be planned and operated. This policy should be communicated to all levels.

Establish objectives and responsibilities for quality education and training

When attempting to set quality education and training objectives three essential requirements must be met:

1 Senior management must ensure that learning outcomes are clarified and priorities set.

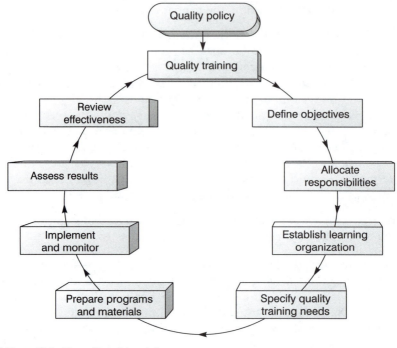

Figure 16.2   The quality training circle

2 The defined education and training objectives must be realizable and attainable.
3 The main objectives should be 'translated' for all functional areas in the organization.

Large organizations may find it necessary to promote a phased plan to identify these.

The following questions are useful first steps when identifying quality education and training objectives:

- How are the customer requirements transmitted through the organization?
- Which areas need improved performance?
- What changes are planned for the future?
- What are the implications for the process framework?

Quality education and training must be the responsibility of line management, but there are also important roles for the individuals concerned.

Establish the platform for a learning organization

The overall responsibility for seeing that education and training is properly organized must be assumed by one or more designated senior executives. All managers have a responsibility for ensuring that personnel reporting to them are properly trained and competent in their jobs. This responsibility should be written into every manager's job description. The question of whether line management requires specialized help should be answered when objectives have been identified. It is often necessary to use specialists, who may be internal or external to the organization.

Specify education and training needs

The next step in the cycle is to assess and clarify specific quality education and training needs. The following questions need to be answered:

- Who needs to be educated/trained?
- What competencies are required?
- How long will the education/training take?
- What are the expected benefits?
- Is the training need urgent?
- How many people are to be educated/trained?
- Who will undertake the actual education/training?
- What resources are needed, e.g. money, people, equipment, accommodation, outside resources?

Prepare education/training programs and materials

Senior management should participate in the creation of overall programs, although line managers should retain the final responsibility for what is implemented, and they will often need to create the training programs themselves.

Quality training programs should include:

- The training objectives expressed in terms of the desired behavior.
- The actual training content.
- The methods to be adopted.
- Who is responsible for the various sections of the program?

Implement and monitor quality education and training

The effective implementation of quality education and training programs demands considerable commitment and adjustment by the trainers and trainees alike. Quality training is a progressive process, which must take into account any learning problems of the trainees.

Assess the results

In order to determine whether further quality education or training is required, line management should themselves review performance when training is completed. However good the training may be, if it is not valued and built upon by managers and supervisors, its effect can be severely reduced.

Review effectiveness of quality education and training

Senior management will require a system whereby decisions are taken at regular fixed intervals on:

- The quality policy.
- The education and training objectives.
- The education/training organization.
- The progress towards a learning organization.

Even if the policy remains constant, there is a continuing need to ensure that new quality education and training objectives are set either to promote work changes or to raise the standards already achieved.

The purpose of quality management system audits and reviews is to assess the effectiveness of the management effort. Clearly, adequate and refresher training in these methods is essential if such checks are to be realistic and effective. Quality audits and reviews can provide useful information for the identification of changing quality training needs.

The quality education/training organization should similarly be reviewed in the light of the new objectives, and here again it is essential to aim at continuous improvement. Quality training must never be allowed to become static, and the effectiveness of the organization's education and training programs and methods must be assessed systematically.

■ A systematic approach to education and
■ training for quality

Education and training for quality should have, as its first objective, an appreciation of the personal responsibility for meeting the 'customer' requirements by everyone from the most senior executive to the newest and most junior employee. Responsibility for the training of employees in quality rests with management at all levels and, in particular, the person nominated for the co-ordination of the organization's quality effort. Education and training will not be fully effective, however, unless responsibility for the deployment of the policy rests clearly with the chief executive. One objective of this policy should be to develop a *climate* in which everyone is quality conscious and acts with the needs of the customer in mind at all times.

The main elements of effective and systematic quality training may be considered under four broad headings:

- Error/defect/problem prevention.
- Error/defect/problem reporting and analysis.
- Error/defect/problem investigation.
- Review.

The emphasis should obviously be on error, defect, or problem prevention, and hopefully what is said under the other headings maintains this objective.

Error/defect/problem prevention

The following contribute to effective and systematic training for prevention of problems in the organization:

1 An issued quality policy.
2 A written quality management system.
3 Job specifications that include quality requirements.
4 Effective steering committees, including representatives of both management and employees.
5 Efficient housekeeping standards.
6 Preparation and display of maps, flow diagrams and charts for all processes.

Error/defect/problem reporting and analysis

It will be necessary for management to arrange the necessary reporting procedures, and ensure that those concerned are adequately trained in these procedures. All errors, rejects, defects, defectives, problems, waste, etc. should be recorded and analyzed in a way that is meaningful for each organization, bearing in mind the corrective action programs that should be initiated at appropriate times.

Error/defect/problem investigation

The investigation of errors, defects, and problems can provide valuable information that can be used in their prevention. Participating in investigations offers an opportunity for training in quality. The following information is useful for the investigation:

- Nature of the problem(s).
- Date, time and place.
- Product/service with the problem(s).
- Description of the problem(s).
- Causes and reasons behind the causes.
- Action advised.
- Action taken to prevent recurrence.

Review of quality training

Review of the effectiveness of quality training programs should be a continuous process. However, the measurement of effectiveness is a complex problem. One way of reviewing the content and assimilation of a quality training course or program is to monitor behavior during quality audits. This review can be taken a stage further by comparing employees' behavior with the objectives of the quality training program. Other measures of the training processes should be found to establish the benefits derived.

Education and training records

All organizations should establish and maintain procedures for the identification of education and training needs and the provision of the actual training itself. These procedures should be designed (and documented) to include all personnel. In many situations it is necessary to employ professionally qualified people to carry out specific tasks, e.g. accountants, lawyers, engineers, chemists, etc., but it must be recognized that all other employees, including managers, must have or receive from the company the appropriate education, training and/or experience to perform their jobs. This leads to the establishment of education and training records.

Once an organization has identified the special skills required for each task, and developed suitable education and training programs to provide competence for the tasks to be undertaken, it should prescribe how the competence is to be demonstrated. This can be by some form of examination, test or certification, which may be carried out in-house or by a recognized external body. In every case, records of personnel qualifications, education, training, and experience should be developed and maintained. National vocational qualifications may have an important role to play here.

At the simplest level this may be a record of tasks and a date placed against each employee's name as he/she acquires the appropriate skill through education and training. Details of attendance on external short courses, in-house induction or training schemes complete such records. What must be clear and easily retrievable is the status of training and development of any single individual, related to the tasks that he/she is likely to encounter. For example, in a factory producing hand-machined parts that has developed a series of well-defined tasks for each stage of the manufacturing process, it would be possible, by turning up the appropriate records, to decide whether a certain operator is competent to carry out a lathe-turning process. Clearly, as the complexity of jobs increases and managerial activity replaces direct manual skill, it becomes more difficult to make decisions on the basis of such records alone. Nevertheless, they should document the basic competency requirements and assist the selection procedure.

Starting where and for whom?

Quality education and training needs occur at four levels of an organization:

- *Very senior management* (strategic decision makers).
- *Middle management* (tactical decision makers or implementers of policy).
- *First level supervision and quality team leaders* (on-the-spot decision makers).
- *All other employees* (the doers).

Neglect of education/training in any of these areas will, at best, delay the implementation of quality management and the improvements in performance. The provision of training for each group will be considered in turn, but it is important to realize that an integrated program is required, one that includes follow-up learning-based activities and encourages exchange of ideas and experience.

Very senior management

The chief executive and his team of strategic policy makers are of primary importance, and the role of education and training here is to provide awareness and instil commitment to quality. The importance of developing real commitment must be established and often this can only be done by a free and frank exchange of views between trainers and trainees. This has implications for the choice of the trainers themselves, and the fresh-faced graduate, sent by the 'package consultancy' operator into the lion's den of a boardroom, will not make much impression with the theoretical approach that he or she is obliged to bring to bear. The author recalls thumping many a boardroom table, and using all his experience and whatever presentation skills he could muster, to convince senior managers that without effective quality management they would fail. It is a sobering fact that the pressure from competition and customers has a much greater record of success than enlightenment, although dragging a team of senior managers down to the shop floor to show them the results of poor quality management was successful on more than one occasion.

Executives responsible for marketing, sales, finance, design, operations, purchasing, personnel, distribution, etc. all need to understand quality. They must be shown how to define the quality policy and objectives, how to establish the appropriate organization, how to clarify authority, and generally how to create the atmosphere in which good quality management will thrive. This is the only group of people in the organization that can ensure that adequate resources are provided and directed at:

1 Meeting customer requirements – internally and externally.
2 Setting standards to be achieved – zero failure.
3 Monitoring of quality performance – quality costs.
4 Introducing a good quality management system – prevention.
5 Implementing process control methods – SPC.
6 Spreading the idea of quality throughout the whole workforce – quality management.

Middle management

The basic objectives of management quality training should be to make managers conscious and anxious to secure the benefits of the total quality effort. One particular 'staff' manager will require special training – the quality manager, who will carry responsibility for implementation of the quality management system, including its design, operation, and review.

The middle managers should be provided with the technical skills required to design, implement, review, and change the parts of the quality management system that will be under their direct operational control. It will be useful throughout the training programs to ensure that the responsibilities for the various activities in each of the functional areas are clarified. The presence of a highly qualified and experienced quality manager should not allow abdication of these responsibilities, for the internal 'consultant' can easily create not-invented-here feelings by writing out 'procedures' without adequate consultation of those charged with their implementation.

Middle management should receive comprehensive training on the philosophy and concepts of teamwork, and the techniques and applications of statistical process control (SPC). Without the teams and tools, the quality management system will lie dormant and lifeless. It will relapse into a paper generating system, fulfilling the needs of only those who thrive on bureaucracy. They need to learn how to put this lot together in a planning – process – people – performance value chain that is sustainable for the future.

First-level supervision

There is a layer of personnel in many organizations which plays a vital role in their inadequate performance – foremen and supervisors – the forgotten men and women of industry and commerce. Frequently promoted from the 'shop floor' (or recruited as graduates in a flush of conscience and wealth!), these people occupy one of the most crucial managerial roles, often with no idea of what they are supposed to be doing, without an identity, and without training. If this behavior pattern is familiar and is continued, then quality is doomed.

The first level of supervision is where quality is actually 'managed'. Supervisors' training should include an explanation of the principles of quality, a convincing exposition on the commitment to quality of the senior management, and an explanation of what the quality policy means for them. The remainder of their training needs to be devoted to explaining their role in the operation of the quality management system, teamwork, SPC, etc., and to gaining *their* commitment to the concepts and techniques of quality management.

It is often desirable to involve the middle managers in the training of first line supervision in order to:

- Ensure that the message they wish to convey through their tactical maneuvers is not distorted.
- Indicate to the first line supervision that the organization's whole management structure is serious about quality, and intends that everyone is suitably trained and concerned about it too. One display of arrogance towards the training of supervisors and the workforce can destroy such careful planning, and will certainly undermine the educational effort.

All other employees_____

Awareness and commitment at the point of production or service delivery is just as vital as at the very senior level. If it is absent from the latter, quality will not begin to be managed; if it is absent from the shop floor, quality management will not be implemented. The training here should include the basics of quality and particular care should be given to using easy reference points for the explanation of the terms and concepts. Most people can relate to quality and how it should be managed, if they can think about the applications in their own lives and at home. Quality is really such common sense that, with sensitivity and regard to various levels of intellect and experience, little resistance should be experienced.

All employees should receive detailed training on the processes and procedures relevant to their own work. Obviously they must have appropriate technical or 'job' training, but they must also understand the requirements of their customers. This is frequently a difficult concept to introduce, particularly in the non-manufacturing areas, and time and follow-up assistance needs to be given. It is always bad management to ask people to follow instructions without understanding why and where they fit into their own scheme of things.

■ Turning education and training into learning

For successful learning quality training must be followed up. This can take many forms, but the managers need to provide the lead through the design of improvement projects and 'surgery' workshops.

In introducing statistical methods of process control, for example, the most satisfactory strategy is to start small and build up a bank of knowledge and experience. Sometimes it is necessary to introduce SPC techniques alongside existing methods of control (if they exist), thus allowing comparisons to be made between the new and old methods. When confidence has been established from these comparisons, the SPC methods will almost take over the control of the processes themselves. Improvements in one or two areas of the organization's operations, by means of this approach will quickly establish the techniques as reliable methods of controlling quality, and people will learn how to use them effectively.

The author and his colleagues have found that a successful formula is the in-company training course plus follow-up workshops. Usually a workshop or seminar is followed within a few weeks by 'surgery' workshop at which participants on the initial training course present the results of their efforts to improve processes, and use the various methods. The presentations and specific implementation problems are discussed. A series of such workshops will add continually to the follow-up, and can be used to initiate process or quality improvement teams. Wider organizational presence and activities are then encouraged by the follow-up activities.

Information and knowledge

Information and knowledge are two words used very frequently in organizations, often together in the context of 'knowledge management' and 'information technology', but how well are they managed and what is their role in supporting quality management?

Recent researchers and writers on knowledge management (e.g. Dawson 1999[1]) have drawn attention to the distinction between explicit knowledge – that which we can express to others – and tacit knowledge – the rest of our knowledge, which we cannot easily communicate in words or symbols.

If much of our knowledge is tacit, perhaps we do not fully know what we know and it can be very difficult to explain or communicate what we know. Explicit knowledge can be put into a form that we can communicate to others – the words, figures and models in this book are an example of that. In many organizations, however, especially the service sector, much of people's valuable and useful knowledge is tacit rather than explicit.

The creation and expression of knowledge takes place through social interaction between tacit and explicit knowledge and the matrix in Figure 16.3 shows this as four modes of knowledge conversion.

Socialization allows the conversion of tacit knowledge in one individual into tacit knowledge in other people, primarily through sharing experiences. The conversion of tacit knowledge to explicit knowledge is *externalization*, which is the process of making it readily communicable. *Internalization* converts explicit knowledge to tacit knowledge by translating it into personal knowledge – this could be called learning. The conversion of forms of explicit knowledge, such as creating frameworks, is *combination*.

Explicit knowledge as information

When knowledge is made explicit by putting it into words, diagrams, or other representations, it can then be typed, copied, stored, and

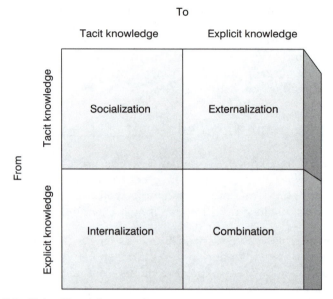

■ **Figure 16.3**   Modes of knowledge conversion
From *The Knowledge-Creating Company: How Japanese Companies Create the Dynamics of Innovation*, by Ikujiro Nonaka and Hirotaka Takeuchi © 1995, Oxford University Press, Inc. (Source: R. Dawson, reference 1)

communicated electronically – it becomes *information*. Perhaps then a useful definition for information is something that is or can be made explicit. Information, which represents captured *knowledge* has value as an input to human decision making and capabilities. Tacit knowledge remains intrinsic to individuals and only they have the capacity to act effectively in its use.

These ideas about information and knowledge enable us to substitute information for explicit knowledge, and simply knowledge – in the business sense of capacity to act effectively – as tacit knowledge. This clarification of the distinction between information and knowledge makes the knowledge conversion framework more directly applicable to inter-organizational interaction.

In the same way, externalization is capturing people's knowledge – their capacity to act in their business roles – by making it explicit and turning it into information, as in the form of written documentation or structured business processes. This remains information until other people internalize it to become part of their own knowledge – or capacity to act effectively. Having a document on a server or bookshelf does not make individuals knowledgeable, nor does reading it. Knowledge comes from understanding the document by integrating the ideas into existing experience and knowledge, and thus providing the capacity to act usefully in new ways. In the case of written documents, language and diagrams are the media by which the knowledge is transferred. The information presented must be actively interpreted and internalized, however, before it becomes new knowledge to the reader.

The process of internalization is essentially that of knowledge acquisition, which is central to the whole idea of learning, knowledge management and knowledge transfer. Understanding the nature of this process is extremely valuable in implementing effective business improvements and in adding greater value to customers.

Socialization refers to the transfer of one person's knowledge to another person, without an intermediary of captured information in documents. It is a most powerful form of knowledge transfer. As we know from childhood people learn from other people far more effectively than they learn from books and documents, in both obvious and subtle ways. Despite technological advances that allow people to telecommute and work in different locations, organizations function effectively mainly because people who work closely together have the opportunity for rich interaction and learning on an ongoing and often informal basis. This presents challenges, of course, in today's 'virtual organization'.

The learning – knowledge management cycle

One way of thinking about learning and knowledge management is as a dynamic cycle from tacit knowledge to explicit knowledge and back to tacit knowledge. In other words, people's knowledge is externalized into information, which to be useful must then be internalized by others (learned) to become part of their knowledge, as illustrated in Figure 16.4. This flow from knowledge to information and back to knowledge constitutes the heart of organizational learning and knowledge management. Direct sharing of knowledge through socialization is also vital. In large organizations, however, capturing whatever is possible in the form of documents and other digitized representations means that information can be stored, duplicated, shared, and made available to people on whatever scale desired.

The fields of learning and knowledge management encompass all the human issues of effective externalization, internalization, and socialization of knowledge. As subsets of those fields, information management and document management address the middle part of the cycle, in which information is stored, disseminated, and made easily available on demand, for example, in a quality management system. It is a misnomer to refer to information sharing technology, however advanced,

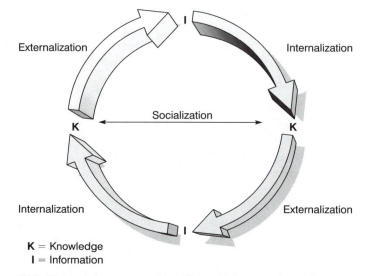

Figure 16.4   The knowledge management cycle (Source: R. Dawson, reference 1)

as knowledge management. Effective implementation of those systems must address how people interact with technology in an organizational context, which only then is beginning to address the real issues of knowledge.[1]

■ The practicalities of sharing knowledge
■ and learning

In world class organizations there is clear evidence that knowledge is shared to maximize performance, with learning, innovation and improvement encouraged. In such establishments information (explicit knowledge) is collected, structured and managed in alignment with and support of the organization's policies and strategies. These days this is often achieved in large organizations through 'intranet' and/or common network mounted file servers. These can provide common access to on-line reports, training material and performance figures versus targets.

The key is to provide appropriate access for both internal and external users to relevant information, with assurance required of its validity, integrity and security.

Cultivating, developing and protecting intellectual property can be the key in many sectors to maximizing customer value, for example in professional services, such as legal and consultancy. Firms in this sector can survive only if they are constantly seeking to acquire, increase, use and transfer knowledge effectively. This in turn has strong links with learning and innovation. The proper use and management of relevant information and knowledge resources leads to the generation of creative thinking and innovative solutions. The aim must be to make information available as widely as possible on the most appropriate basis to improve general understanding and increase efficiency. If this is to be linked to questions in employee surveys, internal valuation that knowledge is being captured and used effectively can take place.

In the EFQM Excellence Model there are now clear feedback loops of 'innovation and learning', both from the results – on people, customers, society and key performance – to the enablers and within the enablers themselves. For example, key performance outcomes such as profit market share, cash flow, can and should be used to identify problems, areas for improvements or even strengths in aspects of the organization's leadership, its policies and strategies or its processes. Equally the way people are managed and partnerships established can be improved, innovatively based on perception measures from customers. The results from the society in which we operate can lead to innovations in resource management.

In some companies and industries, for example the retail or mail order businesses, information on each customer is vital to identify specific needs and to aid communication, planning and monitoring progress. The whole area of customer's relationship management (CRM) relies on some sort of records for each customer being kept. This presents modern methods of knowledge management and information technology with the opportunity to demonstrate their powers.

There may also be learning and innovation loops and opportunities within the enabler's criteria. For example, key performance indicators related to the management of processes, such as cycle time or amount of businesses won, can help to generate innovations in strategies, particularly if there is a good alignment between the two.

Information on staff, on customers and other sources of information can be readily kept on computer databases, and their effective and efficient use can mean the difference between success and failure in many industries and sectors.

Reference

1. Dawson, R., *Developing Knowledge-based Client Relationships – The Future of Professional Services*, Butterworth-Heinemann, Oxford, 2000.

(The author has drawn heavily on the material in this excellent book for the section of this chapter on turning education and training into learning.)

Chapter highlights

■ ■ ■

Communicating the quality strategy

- ■ People's attitudes and behavior can be influenced by communication, and the essence of changing attitudes is to gain acceptance through excellent communication processes.
- ■ The strategy and changes to be brought about through quality management should be clearly and directly communicated from top management to all staff/employees. The first step is to issue a 'total quality message'. This should be followed by a signed quality directive.
- ■ People must know when and how they will be brought into the quality management process, what the process is, and the successes and benefits achieved. First line supervision has an important role in communicating the key messages and overcoming resistance to change.
- ■ The complexity and jargon in the language used between functional groups needs to be reduced in many organizations. Simplify and shorten are the guiding principles.
- ■ 'Open' methods of communication and participation should be used at all levels. Barriers may need to be broken down by concentrating on process rather than 'departmental' issues.

Communicating the quality message

- ■ There are four audience groups in most organizations – senior managers, middle managers, supervisors, and employees – each with different general attitudes towards quality management. The senior management must ensure that each group sees quality management as being beneficial.
- ■ Good leadership is mostly about good communications, the skills of which can be learned through training but must be acquired through practice.

Communication, learning, education and training for quality

- ■ There are four principal types of communication: verbal (direct and indirect), written, visual, and by example. Each has its own requirements, strengths, and weaknesses.
- ■ Education and training is the single most important factor in improving quality and performance, once commitment is

present. This must be objectively, systematically, and continuously performed.

- All quality education and training should occur in an improvement cycle of ensuring it is part of policy, establishing objectives and responsibilities, establishing a platform for a learning organization, specifying needs, preparing programs and materials, implementing and monitoring, assessing results, and reviewing effectiveness.

A systematic approach to education and training for quality

- Responsibility for education and training of employees in quality rests with management at all levels. The main elements should include error/defect/problem prevention, reporting and analysis, investigation, and review.
- Education and training procedures and records should be established to show how job competence is demonstrated.

Starting where and for whom?

- Quality education and training needs occur at four levels of the organization: very senior management, middle management, first-level supervision and quality team leaders, and all other employees.

Turning education and training into learning

- For successful learning all quality training should be followed up with improvement projects and 'surgery' workshops.
- It is useful to draw the distinction between explicit knowledge (that which we can express to others) and tacit knowledge (the rest of our knowledge which cannot be communicated in words or symbols).
- The creation and expression of knowledge takes place through social interaction between tacit and explicit knowledge, which takes the form of socialization, externalization, internalization and combination.
- When knowledge is made explicit it becomes 'information', which in turn has value as an input to human decision making and capability. Tacit knowledge (simply 'knowledge') remains intrinsic to individuals who have the capacity to act effectively in its use.
- One way of thinking about learning and knowledge management is as a dynamic cycle from tacit knowledge to explicit knowledge (information) and back to tacit knowledge.

The practicalities of sharing knowledge and learning

■ In world class organizations there is clear evidence that knowledge is shared to maximize performance, with learning, innovation and improvement encouraged. This is often achieved through an 'intranet' or common network mounted file servers, providing common on-line access to information.

■ Managing intellectual property is key to success in many sectors and this has strong links with learning and innovation. Where information must be made available as widely as possible, internal performance of this aspect can be valuable.

■ The clear feedback loops of 'innovation and learning' in the EFQM Excellence Model drive increased understanding of the linkages between the results and the so-called enablers, and between the enabler criteria themselves.

Part 6

Implementation

Joy's soul lies
in the doing.

William Shakespeare, 1564–1616,
from 'Troilus and Cressida'

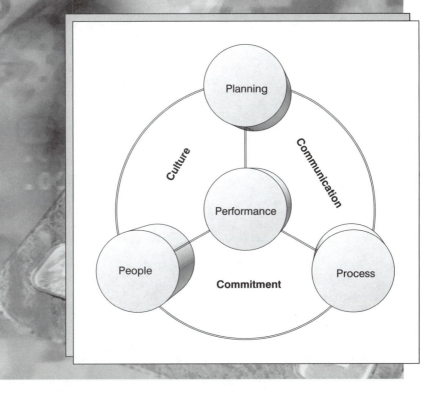

Chapter 17

Implementing quality management

■ Quality and the management of change

The author recalls the managing director of a large support services group who decided that a major change was required in the way the company operated if serious competitive challenges were to be met. The board of directors went away for a weekend and developed a new vision for the company and its 'culture'. A human resources director was recruited and given the task of managing the change in the people and their 'attitudes'. After several 'programs' aimed at achieving the required change, including a new structure for the organization, a staff appraisal system linked to pay, and seminars to change attitudes, very little change in actual organizational behavior had occurred.

Clearly something had gone wrong somewhere. But what, who, where? Everything was wrong, including what needed changing, who should lead the changes and, in particular, how the change should be brought about. This type of problem is very common in organizations which desire to change the way they operate to deal with increased competition, a changing market place, new technology, different business rules. In this situation many organizations recognize the need to move away from an autocratic management style, with formal rules and hierarchical procedures, and narrow work demarcations. Some have tried to create teams, to delegate (perhaps for the first time), and to improve communications.

Some of the senior managers in such organizations recognize the need for change to deal with the new realities of competitiveness, but they lack an understanding of how the change should be implemented. They often believe that changing the formal organizational structure, having 'culture change' programs and new payment systems will, by themselves, make the transformations. In much research work carried out by the European Centre for Business Excellence, the research and education division of Oakland Consulting plc, it has been shown that there is almost an inverse relationship between successful change and having formal organization-wide 'change programs'. This is particularly true if one functional group, such as HR, 'owns' the program.

In several large organizations in which quality management has been used successfully to effect change, the senior management did not focus on formal structures and systems, but set up *process-management* teams to solve real business or organization problems. The key to success in this area is to align the employees of the business, their roles and responsibilities with the organization and its *processes*. When an organization focuses on its key processes, that is the activities and tasks themselves, rather than on abstract issues such as 'culture' and 'participation', then the change process can begin in earnest.

An approach to change based on process alignment, and starting with the vision and mission statements, analyzing the critical success factors, *and* moving on to the core processes, is the most effective way to engage the staff in an enduring change process (see Chapter 4). Many change programs do not work because they begin trying to change the knowledge, attitudes and beliefs of individuals. The theory is that changes in these areas will lead to changes in behavior throughout the organization. It relies on a form of religion spreading through the people in the business.

What is often required, however, is virtually the opposite process, based on the recognition that people's behavior is determined largely by the roles they have to take up. If we create for them new responsibilities, team roles, and a process-driven environment, a new situation will develop, one that will force their attention and work on the processes. This will change the culture. *Teamwork* is an especially important part of the quality management model in terms of bringing about change. If changes are to be made in quality, costs, market, product or service development, close co-ordination among the marketing, design, production/operations and distribution groups is essential. This can be brought about effectively by multifunctional teams working on the processes and understanding their interrelationships. *Commitment* is a key element of support for the high levels of co-operation, initiative, and effort that will be required to understand and work on the labyrinth of processes existing in most organizations. In addition to the knowledge of the business as a whole, which will be brought about by an understanding of the mission, CSF, process breakdown links, certain *tools, techniques,* and *interpersonal skills* will be required for good *communication* around the processes. These are essential if people are to identify and solve problems as teams.

If any of these elements are missing the total quality underpinned change process will collapse. The difficulties experienced by many organizations' formal change processes are that they tackle only one or two of these necessities. Many organizations trying to create a new philosophy based on teamwork fail to recognize that the employees do not know which teams need to be formed round their process – which they begin to understand together, perhaps for the first time – and further recognition that they then need to be helped as individuals through the forming–storming–norming–performing sequence, will generate the interpersonal skills and attitude changes necessary to make the new 'structure' work.

Organizations will avoid the problems of 'change programs' then by concentrating on 'process alignment' – recognizing that people's roles and responsibilities must be related to the processes in which they work. Senior managers may begin the task of process alignment by developing a self-reinforcing cycle of *commitment*, communication, and *culture* change. In the introduction of total quality for managing change, timing can be critical and an appropriate starting point can be a broad review of the organization and the changes required by the top management team. By gaining this shared diagnosis of what changes are required, what the 'business' problems are, and/or what must be improved, the most senior executive mobilizes the initial commitment that is vital to begin the change process. An important element here is to get the top team working as a team, and techniques such as MBTI and/or FIRO-B will play an important part (see Chapter 15).

■ Planning the implementation of
■ quality management

The task of implementing quality management can be daunting and the chief executive faced with this may draw little comfort from the 'quality gurus'. The first decision is where to begin and this can be so difficult that many organizations never get started. This has been called TQP – total quality paralysis!

The chapters of this book have been arranged in an order which should help senior management manage quality effectively. The preliminary stages of understanding and commitment are vital first steps which also form the foundation of the whole quality management structure.

Too many organizations skip these phases, believing that they have the right attitude and awareness, when in fact there are some fundamental gaps in their 'quality credibility'. These will soon lead to insurmountable difficulties and collapse of the edifice.

While an intellectual understanding of quality provides a good basis, it is clearly only the planting of the seed. The understanding must be translated into commitment, policies, plans and actions for full germination. Making this happen requires not only commitment, but a competence in leadership and in making changes. Without a strategy to address quality through process management, capability, and control, the expended effort will lead to frustration. Poor quality management can become like poor gardening – a few weed leaves are pulled off only for others to appear in their place days later, plus additional weeds elsewhere. Problem solving is very much like weeding; tackling the root causes, often by digging deep, is essential for better control.

Individuals working on their own, even with a plan, will never generate optimum results. The individual effort is required in improvement but it must be co-ordinated and become involved with the efforts of others to be truly effective. The implementation begins with the drawing up of a quality policy statement, and the establishment of the appropriate organizational structure, both for managing and encouraging involvement in quality through teamwork. Collecting information on how the organization operates, including the costs of quality, helps to identify the prime areas in which improvements will have the largest impact on performance. Planning improvement involves all managers but a crucial early stage involves putting quality management systems in place to drive the improvement process and make sure that problems remain solved forever, using structured corrective action procedures.

Once the plans and systems have been put into place, the need for continued education, training, and communication becomes paramount. Organizations that try to change the culture, operate systems, procedures, or control methods without effective, honest two-way communication will experience the frustration of being a 'cloned' type of organization which can function but inspires no confidence in being able to survive the changing environment in which it lives.

An organization may, of course, have already taken several steps on the road to 'total quality'. If good understanding of quality and how it should be managed already exists, there is top management commitment, a written quality policy, and a satisfactory organizational structure, then the planning stage may begin straight away. When implementation is contemplated, priorities among the various projects must be identified. For example, a quality system which conforms to the requirements of ISO 9000 may already exist and the systems step will not be a major task, but introducing a quality costing system may well be. It is important to remember, however, that a review of the current performance in all the areas, even when well established, should be part of normal operations to ensure continuous improvement.

These major steps may be used as an overall planning aid for the introduction of quality management, and they should appear on a planning or Gantt chart. Major projects should be time-phased to suit individual organizations' requirements, but this may be influenced by outside factors, such as pressure from a customer to introduce statistical process control (SPC) or to operate a quality management system which meets the requirements of a standard. The main projects may need to be split into smaller subprojects, and this is certainly true of management system work, the introduction of SPC, and improvement teams.

The education and training part will be continuous and draw together the requirements of all the steps into a cohesive program of introduction. The timing of the training inputs, follow-up sessions, and advisory work should be co-ordinated and reviewed, in terms of their effectiveness, on a regular basis. It may be useful at various stages of the implementation to develop checks to establish the true progress. For example, before moving from understanding to trying to obtain top management commitment, objective evidence should be obtained to show that the next stage is justified.

Following commitment being demonstrated by the publication of a signed quality policy, there may be the formation of a council, and/or steering committee(s). Delay here will prevent real progress being made through teamwork activities.

The launch of process improvement requires a balanced approach and the three major components must be 'fired' in the right order to lift the campaign off the ground. If teams are started before the establishment of a good system of management, there will be nothing to which they can adhere. Equally, if SPC is introduced without a good system of data recording and standard operating procedures, the techniques will simply measure how bad things are. A quality management system on its own will give only a weak thrust which must have the boost of improvement teams and SPC to make it come 'alive'.

An effective co-ordination of these three components will result in quality improvement through increased capability. This should in turn lead to consistently satisfied customers and, where appropriate, increase or preservation of market share.

Quality management may be integrated into the strategy of any organization through an understanding of the core business processes and involvement of the people. This leads through process analysis, self-assessment and benchmarking, to identifying the improvement opportunities for the organization, including people development.

The identified processes should be prioritized into those that require continuous improvement, those that require re-engineering or redesign, and those that lead to a complete rethink or visioning of the business.

Performance-based measurement of all processes and people development activities is necessary to determine progress so that the vision, goals, mission, and critical success factors may be examined and reconstituted if necessary to meet new requirements for the organization and its customers, internal and external. This forms the basis of a practical implementation framework for quality management (Figure 17.1).

This all starts with the vision, goals, strategies and mission which should be fully thought through, agreed, and shared in the business. What follows determines whether these are achieved. The factors which are critical to success, the CSFs – the building blocks of the mission – are then identified. The key performance indicators (KPIs), the measures associated with the CSFs, tell us whether we are moving towards or away from the mission or just standing still.

Having identified the CSFs and KPIs, the organization should know what are its *core processes*. This is an area of potential bottleneck for many organizations because, if the core processes are not understood, the

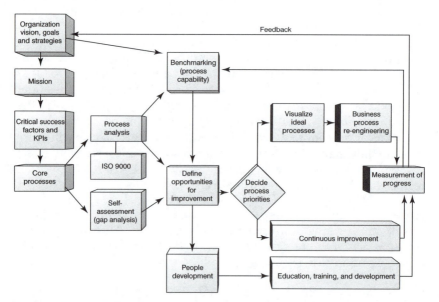

■ **Figure 17.1**   The framework for implementation of effective quality management

rest of the framework is difficult to implement. If the processes are known, we can carry out process analysis, sometimes called mapping and flow-charting, to fully understand our business and identify opportunities for improvement.

By the way, ISO 9000 standard-based systems should be easily established at this stage, rather than needing a separate and huge effort and expense.

Self-assessment to the European (EFQM) Excellence Model or Baldrige quality model, and benchmarking, will identify further improvement opportunities. This will create a very long list of things to attend to, many of which require people development, training and education. What is clearly needed next is prioritization to identify those processes which are run pretty well – they may be advertising/promoting the business or recruitment/selection processes – and subject them to a continuous improvement regime. For those processes which we identify as being poorly carried out, perhaps forecasting, training, or even financial man-agement, we may subject them to a complete revisioning and redesign activity. That is where BPR comes in. What must happen to all processes, of course, is performance measurement, the results of which fed back to the benchmarking and strategic planning activities.

World class organizations, of which there need to be more in most countries, are doing *all* of these things. They have implemented their version of the framework and are achieving world class performance and results. What this requires first, of course, is world class leadership and commitment.

In many successful companies, quality management is not the very nar-row set of tools and techniques often associated with failed 'programs' in organizations in various parts of the world. It is part of a broad-based approach used by world class companies, such as Hewlett-Packard, Milliken, TNT, and Yell, to achieve organizational excellence, based on customer results, the highest weighted category of all the quality and excellence awards. Quality management embraces *all* of these areas. If used properly, and fully integrated into the business, it will help any organization deliver its goals, targets and strategy, including those in the public sector. This is because it is about people and their identify-ing, understanding, managing and improving processes – the things any organization has to do particularly well to achieve its objectives.

Case study ▪▪▪▬▬▬▬▬▬▬▬▬▬▬▬▬

The initial efforts by the corporate management team in ST Microelectronics would have been in vain if the necessary resources had not been provided to support the implementation of TQM. A corporate TQM support group was established, budgets were allocated and the executive management, including the CEO, allocated significant time to TQM implementation. In the initial phase most of the time and effort went into training and communications with regular bulletins, emails, and brochures.

The policy deployment process allowed the corporate goals to be cascaded into local goals which were both realistic and challenging. The training programs, targeted at 50 hours per employee per year, ensured that people had the skills to accept the goals and translate them into local action plans. The management were encouraged to recognize achievements at local, national and international level. Finally strong efforts were made to break down the walls between the various parts of the organization and create an atmosphere in which cross-fertilization was not only accepted but actively encouraged, until it became a way of life.

These changes were not easily or readily accepted in all parts of the corporation. While the benefits could be seen on an intellectual plane at a cultural level some groups found it easier to move faster than others. The corporate TQM Vice President described the process as 'pulling down the walls and using the bricks to build bridges'. The difficulty of achieving success cannot be underestimated. STM started with the advantage that many of its European staff had a fundamentally Latin culture and many of the managers had been exposed to American culture, either as a result of working in American companies or interfacing with American customers. Also the semiconductor industry had its own culture which was and still is very strong. Nonetheless cultural barriers still existed and STM had to find ways of working with many different cultures while trying to overlay a common STM culture, ways of working and vision of the future.

■ Sustained improvement

Never-ending or continuous improvement is probably the most powerful concept to guide management. It is a term not well understood in many organizations, although that must begin to change if those organizations are to survive. To maintain a wave of sustained interest, it is necessary to develop generations of managers who not only understand but are dedicated to the pursuit of never-ending improvement in meeting external and internal customer needs.

The concept requires a systematic approach to quality management that has the following components:

- *Planning* the processes and their inputs.
- *Providing* the inputs.
- *Operating* the processes.
- *Evaluating* the outputs.
- *Examining* the performance of the processes.
- *Modifying* the processes and their inputs.

This system must be firmly tied to a continuous assessment of customer needs, and depends on a flow of ideas on how to make improvements, reduce variation, and generate greater customer satisfaction. It also requires a high level of commitment, and a sense of personal responsibility in those operating the processes.

The never-ending improvement cycle ensures that the organization learns from results, standardizes what it does well in a documented quality management system, and improves operations and outputs from what it learns. But the emphasis must be that this is done in a planned, systematic, and conscientious way to create a climate – a way of life – that permeates the whole organization.

There are three basic principles of sustained improvement:

- Focusing on the *customer.*
- Understanding the *process.*
- All *employees* committed to quality.

Focusing on the customer

An organization must recognize, throughout its ranks, that the purpose of all work and all efforts to make improvements is to serve the customers better. This means that it must always know how well its outputs are performing, in the eyes of the customer, through measurement and feedback. The most important customers are the external ones, but the quality chains can break down at any point in the flows of work. Internal customers therefore must also be well served if the external ones are to be satisfied.

Understanding the process

In the successful operation of any process it is essential to understand what determines its performance and outputs. This means intense focus on the design and control of the inputs, working closely with suppliers, and understanding process flows to eliminate bottlenecks and reduce waste. If there is one difference between management/supervision in the Far East and the West, it is that in the former management is closer to, and more involved in, the processes. It is not possible to stand aside and manage in never-ending improvement. Effective quality management in an organization means that everyone has the determination to use their detailed knowledge of the processes and make improvements, and use appropriate statistical methods to analyze and create action plans.

All employees committed to quality _____

Everyone in the organization, from top to bottom, from offices to technical service, from headquarters to local sites, must play their part. People are the source of ideas and innovation, and their expertise, experience, knowledge, and co-operation have to be harnessed to get those ideas implemented.

When people are treated like machines, work becomes uninteresting and unsatisfying. Under such conditions it is not possible to expect quality services and reliable products. The rates of absenteeism and of staff turnover are measures that can be used in determining the strengths and weaknesses, or management style and people's morale, in any company.

The first step is to convince everyone of their own role in total quality. Employers and managers must of course take the lead, and the most senior executive has a personal responsibility. The degree of management's enthusiasm and drive will determine the ease with which the whole workforce is motivated.

Most of the work in any organization is done away from the immediate view of management and supervision, and often with individual discretion. If the co-operation of some or all of the people is absent, there is no way that managers will be able to cope with the chaos that will result. This principle is extremely important at the points where the processes 'touch' the outside customer. Every phase of these operations must be subject to continuous improvement, and for that everyone's co-operation is required.

Never-ending improvement is the process by which greater customer satisfaction is achieved. Its adoption recognizes that quality is a moving target, but its operation actually results in quality.

Case study ■■■ ▬▬▬▬▬▬▬▬▬▬▬▬▬▬

■ SQF and business improvement

Once the Shell SQF had been developed, proven in pilots and supported by a Business Improvement System, the task then became one of deploying the SQF such that priority areas were addressed, improvements could be sustained and the whole approach could begin to permeate the organization. There is no doubt the SQF and supporting tools can be used wherever business problems exist and several areas were targeted. Perhaps one of the most powerful was in helping the whole organization come together around key management thrusts for the year – FIRST: Focus on customers, Improve billing, Reduce cost, Service excellence, Talent development. Work was carried out with the executive team to arrive at these top five thrusts by using the SQF to distil the really important strategies from a much longer 'wish list', again forcing clarity around real purpose, business impact and performance measures. Using the SQF to help with choices in this way it was possible to reinforce important aspects of organization and culture, so critical when aligning and mobilizing support from across the whole organization.

As a summary, progress in the area of business improvement is highlighted in Table 17.1.

It is worth mentioning here another area of SQF deployment in a little more detail: discretionary expenditure – The 'Business Improvement Program' in Table 17.1. Every organization spends time and money on undertaking projects, hopefully to improve key processes, market standing, profitability, reputation, etc. Faced with a wide array of some 35 IT-related projects, all seemingly justified in their own right and amounting to some $80 million, the question was – 'how do we ensure we are working on those projects that will provide best value for money and clear alignment with our business objectives?' Furthermore, 'how do we exercise some degree of control over such a disparate set of projects ranging from billing improvement, through knowledge management to hardware renewal?' And finally, 'how do we ensure a first class strategy ends up in a first class implementation?' The answers lay in using the SQF as a 'filter' to

Table 17.1

Business focus	Applied to	Purpose	Examples to date
Alignment	Plans, projects, options	To ensure alignment with business strategy and priorities	Communication strategy, leadership team priorities, strategy renewal, HSE strategy
Choice	Initiatives	To help prioritize by considering purpose, impact, resources, capacity, etc.	Business Improvement Program, HR agenda, stress management strategy, cost reduction process FIRST
Improvement	Process	To improve or redesign core processes, with a focus on best practice	Contract to billing process, account management process, property sales, joint venture accounting, service delivery
Assessment	Organization	To self-assess for continuous improvement	Customer service teams, customer satisfaction with HR
Recognition	Business	To achieve accreditation/ recognition against best in class standards	

address what may appear to be 20 simple questions, but ones that were to prove to be worth their weight in gold (Figure 17.2).

Taking just one question as an example from Figure 17.2 – the second one under People – 'what is our capacity to implement the degree of change required for successful implementation of this project?' This was a powerful question often overlooked in the rush to deliver improvements across the organization. Every organization has a limited capacity for major changes at any one time. Is it feasible to ask people to implement a new billing system while a global helpdesk is being implemented *and* a major office move is under way?

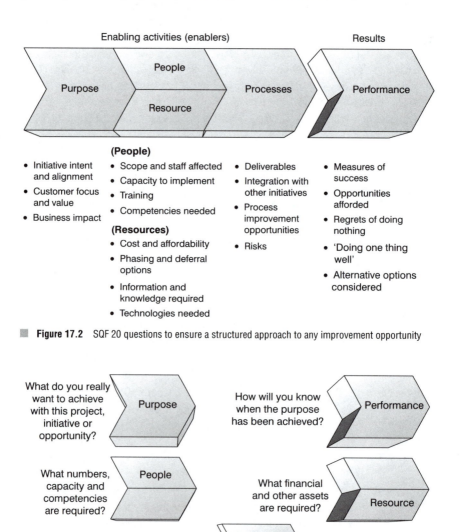

Figure 17.2   SQF 20 questions to ensure a structured approach to any improvement opportunity

Figure 17.3   Five easy questions!

Through this process of using the SQF to ensure discretionary expenditure was being properly considered, the leadership team found they were able to make better decisions, generate 100 percent support for priority projects and provide an environment for the best possible chance of successful implementation. Even more than this, some parts of the organization started to ask the question at every management team meeting – 'has this idea been SQF'd?'

On an even more simple scale, value was derived by asking just five key 'acid test' questions about any initiative under way in the organization. Sometimes these proved quite difficult to answer for even the most well-understood projects! These are shown in Figure 17.3.

A model for quality management

Quality management is basically very simple. Each part of an organization has customers, whether within or without, and the need to identify what the customer requirements are, and then set about meeting them, forms the core of a total quality approach. Good *performance* requires the three hard management necessities: *planning*, including the right policies and strategies, *processes* and supporting management systems and improvement tools, such as statistical process control (SPC), and *people* with the right knowledge, skills and training (Figure 17.4). These are complementary in many ways, and they share the same requirement for an uncompromising top level commitment, the right culture and good communications. This must start with the most senior management and

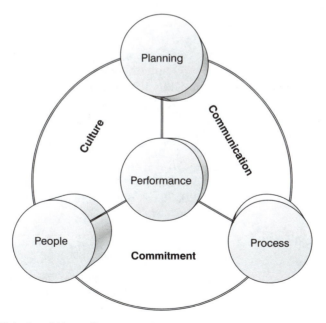

Figure 17.4   A model for quality management

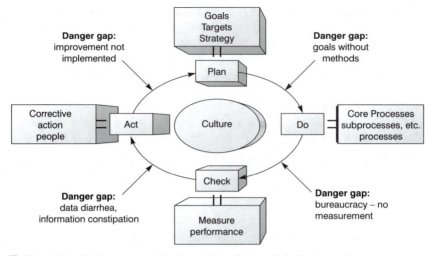

■ **Figure 17.5**   Quality management implementation – all done with the Deming continuous improvement cycle

flow down through the organization. Having said that, teamwork, SPC, or a quality management system, may be used as a spearhead to drive quality through an organization. The attention to many aspects of a company's operations – from purchasing through to distribution, from data recording to control chart plotting – which are required for the successful introduction of a good quality management system, or the implementation of SPC, will have a 'Hawthorne effect', concentrating everyone's attention on the customer/supplier interface, both inside and outside the organization.

Quality management calls for consideration of processes in all the major areas: marketing, design, procurement, operations, distribution, etc. Clearly, these each require considerable expansion and thought, but if attention is given to all areas, using the concepts of quality management, then very little will be left to chance. Much of industry, commerce and the public sector would benefit from the continuous improvement cycle approach represented in Figure 17.5, which also shows the 'danger gaps' to be avoided. This approach will ensure the implementation of the management commitment represented in the quality policy, and provide the environment and information base on which teamwork thrives.

Chapter highlights

■ ■ ■

Quality and the management of change

■ Senior managers in some organizations recognize the need for change to deal with increasing competitiveness, but lack an understanding of how to implement the changes.

- Successful change is effected not by focusing on formal structures and systems, but by aligning process management teams. This starts with writing the mission statement, analysis of the critical success factors (CSFs) and understanding the critical or key processes.
- Senior managers may begin the task of process alignment through a self-reinforcing cycle of commitment, communication and culture change.

Planning the implementation of quality management

- Making quality happen requires not only commitment but competence in the mechanics. Crucial early stages will comprise establishment of the appropriate organization structure; collecting information, including quality costs; teamwork; quality systems; and training.
- The launch of quality improvement requires a balanced approach, through systems, teams and tools.
- An implementation framework allows the integration of quality management into the strategy of an organization through an understanding of the core business processes and involvement of people. This leads through process analysis, self-assessment and benchmarking to identifying opportunities for improvement, including people development.
- The process opportunities should be prioritized into continuous improvement and re-engineering/redesign. Performance-based measurement determines progress, and feeds back to the strategic framework.

Sustained improvement

- Managers need to understand and pursue never-ending improvement. This should cover planning and operating processes, providing inputs, evaluating outputs, examining performance, and modifying processes and their inputs.
- There are three basic principles of continuous improvement: focusing on the customer, understanding the process, and seeing that all employees are committed to quality.
- In the model for quality management, performance forms the core, which is surrounded by the hard management necessities of planning, processes and people. These are complementary and share the same needs – for top level commitment, the right culture and good communications.

Bibliography

Understanding quality

Bank, J., *The Essence of Total Quality Management*, Prentice-Hall, Hemel Hempstead, 1992

Beckford, J., *Quality* (2nd edn), Routledge, London, 2002

Crosby, P.B., *Quality is Free*, McGraw-Hill, New York, 1979

Crosby, P.B., *Quality Without Tears*, McGraw-Hill, New York, 1984

Dale, B.G. (ed.), *Managing Quality* (3rd edn), Philip Alan, Hemel Hempstead, 2000

Deming, W.E., *Out of the Crisis*, MIT, Cambridge, Mass., 1993

Deming, W.E., *The New Economies*, MIT, Cambridge, Mass., 1993

Feigenbaum, A.V., *Total Quality Control* (3rd edn, revised), McGraw-Hill, New York, 1991

Garvin, D.A., *Managing Quality: The Strategic Competitive Edge*, The Free Press (Macmillan), New York, 1988

Ishikawa, K. (translated by D.J. Lu), *What is Total Quality Control? The Japanese Way*, Prentice-Hall, Englewood Cliffs, NJ, 1985

Juran, J.M. (ed.), *Quality Control Handbook*, McGraw-Hill, New York, 1988

Kehoe, D.F., *The Fundamentals of Quality Management*, Chapman & Hall, London, 1996

Macdonald, J. and Piggot, J., *Global Quality: The New Management Culture*, Mercury Books, London, 1990

Murphy, J.A., *Quality in Practice* (3rd edn), Gill and MacMillan, Dublin, 2000

Price, F., *Right Every Time*, Gower, Aldershot, 1990

Sarv Singh Soin, *Total Quality Control Essentials – key elements, methodologies and managing for success*, McGraw-Hill, New York, 1992

Stahl, M.J. (ed.), *Perspectives in Total Quality*, Quality Press, Milwaukee, 1999

Wille, E., *Quality: Achieving Excellence*, Century Business, London, 1992

■ Models and frameworks for quality management

BQF (British Quality Foundation), *The Model in Practice* and *The Model in Practice 2*, London, 2000 and 2002

Brown, M.G., *Baldrige Award Winning Quality: how to interpret the Malcolm Baldrige Award criteria* (11th edn), ASQ, Milwaukee, 2002

EFQM (European Foundation for Quality Management), *The EFQM Excellence Model*, Brussels, 2003

Hart, W.L. and Bogan, C.E., *The Baldrige: what it is, how it's won, how to use it to improve quality in your company*, McGraw-Hill, New York, 1992

Mills Steeples, M., *The Corporate Guide to the Malcolm Baldrige National Quality Award*, ASQ, Milwaukee, 1992

National Institute of Standard and Technology, *USA Malcolm Baldrige National Quality Award, Criteria for Performance Excellence*, NIST, Gaithesburg, 2003

■ Leadership and commitment

Adair, J., *Not Bosses but Leaders: How to Lead the Successful Way*, Talbot Adair Press, Guildford, 1987

Adair, J., *The Action-Centred Leader*, Industrial Society, London, 1988

Adair, J., *Effective Leadership* (2nd edn), Pan Books, London, 1988

Davidson, H., *The Committed Enterprise*, Butterworth-Heinemann, Oxford, 2002

Juran, J.M., *Juran on Leadership for Quality: An Executive Handbook*, The Free Press (Macmillan), New York, 1989

Labovitz, G., Chang, Y.S. and Rosansky, V., *Making Quality Work – A Leadership Guide for the Results-Driven Manager*, Harper Business, London, 1993

Levicki, C., *The Leadership Gene*, F.T. Management, London, 1998

Townsend, P.L. and Gebhardt, J.E., *Quality in Action – 93 Lessons in Leadership, Participation and Measurement*, J Wiley Press, New York, 1992

■ Policy, strategy and goal deployment

Collins, J.C., *Good to Great*, Random House, New York, 2001

Collins, J.C. and Porras, J.I., *Built to Last*, Random House, New York, 1998

Hardaker, M. and Ward, B.K., 'Getting things done – how to make a team work', *Harvard Business Review*, pp. 112–119, Nov/Dec., 1987

Hussey, D., *Strategic Management* (4th edn), Butterworth-Heinemann, Oxford, 1998

Johnson, G. and Scholes, K., *Exploring Corporate Strategy, Text & Cases* (6th edn), Prentice-Hall, London, 2002

Juran, J., *Juran on Planning for Quality*, Free Press, New York, 1988

Kay, J., *The Foundations of Corporate Success*, Oxford University Press, Oxford, 1995

■ Partnerships and resources

Ansari, A. and Modarress, B., *Just-in-time Purchasing*, The Free Press (Macmillan), New York, 1990

Bineno, J., *Implementing JIT*, IFS, Bedford, 1991

Harrison, A., *Just-in-Time Manufacturing in Perspective*, Prentice-Hall, Englewood Cliffs, NJ, 1992

Lysons, K., *Purchasing and Supply Chain Management*, Prentice-Hall, New York, 2000

Muhlemann, A.P., Oakland, J.S. and Lockyer, K.G., *Production and Operations Management* (6th edn), Pitman, London, 1992

Voss, C.A. (ed.), *Just-in-Time Manufacture*, IFS Publications, Bedford, 1989

■ Design for quality

Fox, J., *Quality Through Design*, MGLR, 1993

Juran, J.J., *Juran on Quality by Design*, Free Press, New York, 1992

Marsh, S., Moran, J., Nakui, S. and Hoffherr, G.D., *Facilitating and Training in QFD*, ASQ, Milwaukee, 1991

King Taylor, L., *Quality: total customer service* (a case study book), Century Business, London, 1992

Mastenbrock, W. (ed.), *Managing for Quality in the Service Sector*, Basil Blackwell, Oxford, 1991

Zeithaml, V.A., Parasuraman, A. and Berry, L.L., *Delivering Quality Service: balancing customer perceptions and expectations*, The Free Press (Macmillan), New York, 1990

■ Performance measurement frameworks

British Standard BS6143-1: 1992, *Guide to the economics of quality, Process cost model*, BSI, London, 1992

British Standard BS6143-2: 1990, *Guide to the economics of quality, Prevention, appraisal and failure model*, BSI, London, 1992

Dale, B.G. and Plunkett, J.J., *Quality Costing*, Chapman and Hall, London, 1991

Dixon, J.R., Nanni, A. and Vollmann, T.E., *The New Performance Challenge – Measuring Operations for World Class Competition*, Business One Irwin, Homewood, 1990

Hall, R.W., Johnson, J.Y. and Turney, P.B.B., *Measuring Up – Charting Pathways to Manufacturing Excellence*, Business One Irwin, Homewood, 1991

Kaplan, R.W. (ed.), *Measures for Manufacturing Excellence*, Harvard Business School Press, Boston, MA, 1990

Kaplan, R.S. and Norton, P., *The Balanced Scorecard*, Harvard Business School Press, Boston, MA, 1996

Neely, A., *Measuring Business Performance* (2nd edn), Economist Books, London, 2002

Porter, L.J. and Rayner, P., 'Quality costing for TQM', *International Journal of Production Economics*, 27, pp. 69–81, 1992

Talley, D.J., *Total Quality Management: performance and cost measures*, ASQ, Milwaukee, 1991

Zairi, M., *TQM-Based Performance Measurement*, TQM Practitioner Series, Technical Communication (Publishing), Letchworth, 1992

Zairi, M., *Measuring Performance for Business Results*, Chapman and Hall, London, 1994

Self-assessment, audits and reviews

Arter, D.R., *Quality Audits for Improved Performance* (2nd edn), Quality Press, Milwaukee, 2000

BQF, *The X-Factor, winning performance through business excellence*, European Centre for Business Excellence/British Quality Foundation, London, 1999

BQF, *The Model in Practice* and *The Model in Practice 2*, British Quality Foundation, London, 2000 and 2002 (Prepared by the European Centre for Business Excellence)

EFQM, *Assessing for Excellence – A Practical Guide for Self-Assessment*, EFQM, Brussels, 2003 International Standards Organisation: ISO 19011: *Guidelines on quality and/or environmental management systems auditing*, ISO, 2003

JUSE (Union of Japanese Scientists and Engineers), *Deming Prize Criteria*, JUSE, Tokyo, 2003

Keeney, K.A., *The ISO 9001:2000 Auditor's Companion*, Quality Press, Milwaukee, 2002

NIST (National Institute of Standards and Technology), *The Malcolm Baldrige National Quality Award Criteria*, NIST, Gaithersburg, 2003

Porter, L.J. and Tanner, S.J., *Assessing Business Excellence* (2nd edn), Butterworth-Heinemann, Oxford, 2003

Porter, L.J., Oakland, J.S. and Gadd, K.W., *Evaluating the European Quality Award Model for Self-Assessment*, CIMA, London, 1998

Pronovost, D., *Internal Quality Auditing*, Quality Press, Milwaukee, 2000

Tricker, R., *ISO 9001:2000 Audit Procedures*, Butterworth-Heinemann, Oxford, 2001

Benchmarking

Bendell, T., *Benchmarking for Competitive Advantage,* Longman, London, 1993

Camp, R.C., *Business Process Benchmarking: Finding and Implementing Best Practice,* ASQ Quality Press, Milwaukee, 1995

Camp, R.C., *Global Cases in Benchmarking: Best Practices from Organisations Around the World,* Quality Press, Milwaukee, 1998

Macdonald, J. and Tanner, S.J., *Understanding Benchmarking in a Week,* Institute of Management, London, 1996

Spendolini, M.J., *The Benchmarking Book,* ASQ, Milwaukee, 1992

Zairi, M., *Benchmarking for Best Practice,* Butterworth-Heinemann, Oxford, 1996

Zairi, M., *Effective Management of Benchmarking Projects,* Butterworth-Heinemann, Oxford, 1998

Process management

Besterfield, D., *Quality Control* (6th edn), Prentice-Hall, Englewood Cliffs, NJ, 2000

Dimaxcescu, D., *The Seamless Enterprise – Making Cross-functional Management Work,* Harper Business, New York, 1992

Francis, D., *Unblocking the Organizational Communication,* Gower, Aldershot, 1990

Harrington, H.J., *Total Improvement Management,* McGraw-Hill, New York, 1995

Rummler, G.A. and Brache, A.P., *Improving Performance: How to Manage the White Space on the Organization Chart* (2nd edn), Jossey-Bass Publishing, San Francisco, CA, 1998

Senge, P.M., *The Fifth Discipline,* Century Business, London, 1990

Senge, P.M., Roberts, C., Ross, R.B., Smith, B.J. and Kleiner, A., *The Fifth Discipline Fieldbook – Strategies and Tools for Building a Learning Organization,* Nicholas Brearley, London, 1994

Warboys, B.C., Kawalek, P., Robertson, I. and Greenwood, R.M., *Business Information Systems: A Process Approach,* McGraw-Hill, New York, 1999

Process redesign/engineering

Braganza, A. and Myers, A., *Business Process Redesign – A View from the Inside,* International Thomson Business Press, London, 1997

Hammer, M. and Champy, J., *Re-engineering the Corporation*, Nicholas Brearley, London, 1993

Hammer, M. and Stanton, S.A., *The Re-engineering Revolution – The Handbook*, BCA, Glasgow, 1995

Jacobson, I., *The Objective Advantage – Business Process Re-engineering with Object Technology*, John Wiley, Chichester, 1993

Johansson, H.J., McHugh, P., Pendlebury, A.J. and Wheeler, W.A., *Business Process Re-engineering*, John Wiley, London, 1993

Quality management systems

Born, G., *Process Management to Quality Improvement*, John Wiley, Chichester, 1994

British Standards Institute (BSI), BS EN ISO 9001:2000, *Quality Management Systems*, BSI, London, 2000

Cianfrani, C.A., Tsiakals, J.J. and West, J.E., *ISO 9001:2000 explained* (2nd edn), Quality Press, Milwaukee, 2001

Cianfrani, C.A., Tsiakals, J.J. and West, J.E., *The ASQ ISO 9000:2000 Handbook*, Quality Press, Milwaukee, 2001

Federal Information Processing Standards (FIPS), *Publications 183 and 194*, National Institute of Standards and Technology (NIST), Gaithesburg, 1993

Hall, T.J., *The Quality Manual – the Application of BS5750 ISO 9001 EN29001*, John Wiley, Chichester, 1992

Hill, N. (ed.), *Customer Satisfaction Measurement for ISO 9000:2000*, Butterworth-Heinemann, Oxford, 2002

Hoyle, D., *Integrated Management Systems*, Butterworth-Heinemann, Oxford, 2001

Hoyle, D., *ISO 9000 Quality Systems Handbook* (4th edn), Butterworth-Heinemann, Oxford, 2001

International Standards Organisation:

ISO 9000-1:1994, *Quality management and quality assurance standards – Part 1: Guidelines for selection and use*

ISO 9001:2000, *Quality management systems – Requirements*

ISO 9004:2000, *Quality management systems – Guidelines for performance improvements*

ISO 10006:1997, *Quality management – Guidelines to quality in project management*

ISO 10012:2003, *Quality assurance requirements for measuring equipment*

ISO 10013:1995, *Guidelines for developing quality manuals*

ISO/TR 10017, *Guidance on statistical techniques for ISO 9001:1994*

ISO 10241, *International terminology standards – Preparation and layout*

ISO/TR 13425, *Guide for the selection of statistical methods in standardization and specification*

ISO 14001:1996, *Environmental management systems – Specification with guidance for use*

■ Continuous improvement

Bauer, J.E., Duffy, G.Z. and Westcott, R.T. (eds), *The Quality Improvement Handbook*, Quality Press, Milwaukee, 2002

Bendell, T., Wilson, G. and Millar, R.M.G., *Taguchi Methodology with Total Quality*, IFS, Bedford, 1990

Bhote, K.R., *World Class Quality – Using Design of Experiments to Make It Happen* (2nd edn), AMACOM, New York, 1999

Caulcutt, R., *Statistics in Research and Development* (2nd edn), Chapman and Hall, London, 1991

Carlzon, J., *Moments of Truth*, Ballinger, Cambridge, Mass., 1987

Clark, T.J., *Success Through Quality: Support Guide for the Journey to Continuous Improvement*, Quality Press, Milwaukee, 1999

Harry, M. and Schroeder, R., *Six-Sigma – The Breakthrough Management Strategy Revolutionising the World's Top Corporations*, Doubleday, New York, 2000

Joiner, B., *Fourth Generation Management*, McGraw-Hill, New York, 1994

Kinlaw, D.C., *Continuous Improvement and Measurement for Total Quality – A Team-Based Approach*, Pfieffer & Business One, 1992

Logothetis, N., *Managing for Total Quality*, Prentice-Hall Intl., London, 1992

Neave, H., *The Deming Dimension*, SPC Press, Knoxville, 1990

Oakland, J.S., *Statistical Process Control: A Practical Guide* (5th edn), Butterworth-Heinemann, Oxford, 2003

Pande, P.S., Neumann, R.P. and Cavanagh, R.R., *The Six-Sigma Way – How GE, Motorola and Other Top Companies are Honing their Performance*, McGraw-Hill, New York, 2000

Price, F., *Right First Time*, Gower, London, 1985

Ranjit, Roy, *A Primer on the Taguchi Method*, Van Nostrand Reinhold, New York, 1990

Ryuka Fukuda, *CEDAC – A Tool for Continuous Systematic Improvement*, Productivity Press, Cambridge, Mass., 1990

Scherkenbach, W.W., *Deming's Road to Continual Improvement*, SPC Press, Knoxville, 1991

Shingo, S., *Zero Quality Control: Source Inspection and the Poka-yoke System*, Productivity Press, Stamford, Conn., 1986

Wheeler, D.J., *Understanding Variation*, SPC Press, Knoxville, 1993

Wheeler, D.J., *Understanding Statistical Process Control* (2nd edn), SPC Press, Knoxville, 1992

White, A., *Continuous Quality Improvement*, Piatkus, London, 1996

Tennor, A.R. and De Toro, I.J., *Total Quality Manager – Three Steps to Continuous Improvement*, Addison-Wesley, Reading, MA, 1992

Human resource management

Blanchard, K. and Herrsey, P., *Management of Organizational Behaviour: Utilizing Human Resources* (4th edn), Prentice-Hall, Englewood Cliffs, NJ, 1982

Boyatsis, R., *The Competent Manager: A Model for Effective Performance*, Wiley, New York, 1982

Choppin, J., *Quality Through People: a blueprint for proactive total quality management*, Rushmere Wynne, Bedford, 1997

Dale, B.G., Cooper, C. and Wilkinson, A., *Total Quality and Human Resources – A Guide to Continuous Improvement*, Blackwell, Oxford, 1992

Imai, M., *Gemba Kaizen: A Common Sense, Low Cost Approach to Management*, Quality Press, Milwaukee, 1997

Imai, M., *Kaizen*, McGraw-Hill, New York, 1986

Kotter, J.P. and Heskett, J.L., *Corporate Culture and Performance*, The Free Press, New York, 1992

Maguire, S., 'Learning to change', *European Quality*, Vol. 2, No. 6, p. 8, 1995

Marchington, M. and Wilkinson, A., *Core Personnel and Development*, Institute of Personnel and Development, London, 1997

Oakland, S. and Oakland, J.S., 'Current people management activities in world-class organisations', *Total Quality Management*, Vol. 12, No. 6, pp. 25–31, 2001

Ouchi, W., *Theory Z: How American Business Can Meet the Japanese Challenge*, Addison Wesley, New York, 1985

Stone, R.J., *Human Resource Management* (4th edn), Wiley, London, 2002

Culture change through people and teamwork

Adair, J., *Effective Teambuilding* (2nd edn), Pan Books, London, 1987

Atkinson, P., *Creating culture change: The key to successful Total Quality Management*, IFS, Bedford, 1990

Katzenbach, J.R. and Smith, D.K., *The Wisdom of Teams – Creating the High Performance Organisation*, McGraw-Hill/Harvard Business School, Boston, MA, 1994

Kormanski, C., 'A situational leadership approach to groups using the Tuckman Model of Group Development', *The 1985 Annual: Developing Human Resources*, University Associates, San Diego, CA, 1985

Kormanski, C. and Mozenter, A., 'A new model of team building: a technology for today and tomorrow', *The 1987 Annual: Developing Human Resources*, University Associates, San Diego, CA, 1987

Krebs Hirsh, S., *MBTI Team Building Program, Team Member's Guide*, Consulting Psychologists Press, Palo Alto, CA, 1992

Krebs Hirsh, S. and Kummerow, J.M., *Introduction to Type in Organizational Settings*, Consulting Psychologists Press, Palo Alto, CA, 1987

McCaulley, M.H., 'How individual differences affect health care teams', *Health Team News*, 1(8), pp. 1–4, 1975

Myers-Briggs, I., *Introduction to Type: A Description of the Theory and Applications of the Myers-Briggs Type Indicator*, Consulting Psychologists Press, Palo Alto, 1987

Scholtes, P.R., *The Team Handbook, Joiner Associates*, Madison, NY, 1990

Schutz, W., *FIRO: A Three-Dimensional Theory of Interpersonal Behaviour*, Mill Valley WSA, CA, 1958

Schutz, W., *FIRO: Awareness Scales Manual*, Consulting Psychologists Press, Palo Alto, CA, 1978

Schutz, W., *The Human Element – Productivity, Self-esteem and the Bottom Line*, Jossey-Bass, San Francisco, CA, 1994

Tannenbaum, R. and Schmidt, W.H., 'How to choose a leadership pattern', *Harvard Business Review*, May-June, 1973

Tuckman, B.W. and Jensen, M.A., 'States of small group development revisited', *Group and Organizational Studies*, 2(4), pp. 419–427, New York, 1977

Webb, J. and Cleary, D., *Organisational change and the management of expertise*, Routledge, London, 1994

Wellins, R.S., Byham, W.C. and Wilson, J.M., *Empowered Teams*, Jossey Bass, Oxford, 1991

Wilkinson, A. and Willmott, H. (eds), *Making Quality Critical – New Perspectives on Organisational Change*, Routledge, London, 1995

Communications, innovation and learning

Adair, J., *The Challenge of Innovation*, Talbot Adair Press, Guildford, 1990

Dawson, R., *Developing Knowledge-based Client Relationships*, Butterworth-Heinemann, Oxford, 2000

Ettlie, J.E., *Managing Technological Innovation*, Wiley, London, 2000

Francis, D., *Unblocking the Organisational Communication*, Gower, Aldershot, 1990

Larkin, T.J. and Larkin, S., *Communicating Change – Winning Employee Support for New Business Goals*, McGraw-Hill, New York, 1994

Purdie, M., *Communicating for Total Quality*, British Gas, London, 1994

Tidd, J., Bessant, J. and Pavitt, K., *Managing Innovation*, Wiley, London, 2001

Zairi, M., *Best Practice Process Innovation Management*, Butterworth-Heinemann, Oxford, 1999

■ Implementing quality management

Albin, J.M., *Quality Improvement in Employment and other Human Services – Managing for Quality Through Change*, Paul Brookes Pub., 1992

Antony, J. and Preece, D., *Understanding, Managing and Implementing Quality*, Routledge, London, 2002

Ciampa, D., *Total Quality – A User's Guide for Implementation*, Addison-Wesley, Reading, Mass., 1992

Cook, S., *Customer Care – Implementing Total Quality in Today's Service Driven Organisation*, Kogan Page, London, 1992

Economist Intelligence Unit, *Making Quality Work: Lessons from Europe's Leading Companies*, EIU, London, 1992

Fox, R., *Six Steps to Total Quality Management*, McGraw-Hill, NSW, 1994

Hiam, A., *Closing the Quality Gap – Lessons from America's Leading Companies*, Prentice-Hall, Englewood Cliffs, NJ, 1992

Kanji, G.P. and Asher, M., *100 Methods for Total Quality Management*, Sage, London, 1996

Morgan, C. and Murgatroyd, S., *Total Quality Management in the Public Sector*, Open University Press, Milton Keynes, 1994

Munro-Faure, L. and Munro-Faure, M., *Implementing Total Quality Management*, Pitman, London, 1992

Oakland, J.S., *Total Organizational Excellence*, Butterworth-Heinemann, Oxford, 2001

Senge, P.M., Kleiner, A., Roberts, C., Ross, R., Roth, G. and Smith, B., *A Fifth Discipline Resource – The Dance of Change*, Nicholas Brearley, London, 2000

Strickland, F., *The Dynamics of Change*, Routledge, London, 1998

Taguchi, G., *Systems of Experimental Design*, Vols 1 and 2, Unipub/Kraus Int., New York, 1987

Tunks, R., *Fast Track to Quality*, McGraw-Hill, New York, 1992

Whitford, B. and Bird, R., *The Pursuit of Quality*, Beaumont, Herts., 1996

Index